Future Strategies for Tissue and Organ Replacement

D0760592

WITHDRAWN
UTSA LIBRARIES

Future Strategies

for Tissue and Organ

Replacement

Editors

Julia M Polak
Imperial College School of Medicine Hammersmith Hospital, UK

Larry L Hench
Imperial College, UK

P Kemp
Intercytex, Etherley Dene House, UK

Imperial College Press

Published by

Imperial College Press
57 Shelton Street
Covent Garden
London WC2H 9HE

Distributed by

World Scientific Publishing Co. Pte. Ltd.
P O Box 128, Farrer Road, Singapore 912805
USA office: Suite 1B, 1060 Main Street, River Edge, NJ 07661
UK office: 57 Shelton Street, Covent Garden, London WC2H 9HE

British Library Cataloguing-in-Publication Data
A catalogue record for this book is available from the British Library.

FUTURE STRATEGIES FOR TISSUE AND ORGAN REPLACEMENT

Copyright © 2002 by Imperial College Press

All rights reserved. This book, or parts thereof, may not be reproduced in any form or by any means, electronic or mechanical, including photocopying, recording or any information storage and retrieval system now known or to be invented, without written permission from the Publisher.

For photocopying of material in this volume, please pay a copying fee through the Copyright Clearance Center, Inc., 222 Rosewood Drive, Danvers, MA 01923, USA. In this case permission to photocopy is not required from the publisher.

ISBN 1-86094-310-1
ISBN 1-86094-311-X (pbk)

Printed in Singapore by World Scientific Printers (S) Pte Ltd

Library
University of Texas
at San Antonio

Preface

It is clear that the capacity of the human body for self-repair reduces with age and, thus, it has been the dream of mankind for centuries to rejuvenate or replace worn out or diseased body parts. The dream started to come true in 1890 when the first total hip replacement was carried out and an ivory prosthesis was implanted and glued into place. Following this development, considerable progress was made using a variety of different materials ranging from Pyrex to Bakelite. The forerunners of the modern hip replacement however did not emerge until the 1960s when McKee & Watson-Farrar* (1966) produced the first metal prosthesis and Charnley[†] (1967) developed a metal-polyethylene construct. Organ transplantation, of course, has also revolutionized medicine. Following Alexis Carrel's Nobel prize-winning work on developing the basic surgical skills that are still used in renal transplantation today, the first human-to-human kidney transplant was performed by Voronoy in 1933.

The proportion of the world's population formed by the elderly has risen steeply over the past few decades. In addition, the introduction of new immunosuppressant regimes (VS Gorantla *et al.*, 2000), plus improvements in post-surgical care, have established transplantation as a life-saving procedure for many patients with end-stage organ failure. However, the lack of donor organs remains

---

*McKee GK, Watson-Farrar J. Replacement of arthritic hips by the McKee–Farrar prosthesis. *J Bone Joint Surgery* 1966; **48B**:245–259.
[†]Charnley J. Total prosthetic replacement of the hip. *Physiotherapy* 1967; **53**:407–409.
Gorantla VS, Barker JH, Jones JW Jr, Prabhune K, Maldonado C, Granger DK. Immunosuppressive agents in transplantation: Mechanisms of action and current anti-rejection strategies. *Microsurgery* 2000; **20**(8):420–429.

a major obstacle; it has been estimated that a new name is added every eighteen minutes to the list of patients needing a transplant (United Network for Organ Sharing). These developments have led to an even greater demand for replacement body parts.

Several books have been published on particular aspects of organ replacement and/or transplantation but none covers the major aspects of this relatively new branch of medicine as comprehensively as this volume. The Editors have done a good job both in recruiting leading figures in the field to contribute to the book and in covering the key aspects of the problems that exist currently in organ replacement, including xenotransplantation, regulatory issues and single organ engineering.

Professor Charles A. Vacanti

*Department of Anesthesiology,
University of Massachusetts Medical School,
55 Lake Ave. North, Room S2-751,
Worcester, MA 01655, USA*

Foreword

The inability of the human body to repair damaged organs and tissues had resulted in one of the largest medical problems of the twentieth century. Each year, tens of millions of people are treated for a variety of conditions ranging from tissue loss to end-stage organ failure. So far, the best treatment is to transplant healthy material either from another location or another individual. However, there is a chronic shortage of material for such treatments that shows no sign of improving. Currently, several exciting strategies are being pursued around the world to provide surgeons with new methods to address this need and these have attracted a large amount of public and media interest.

Several textbooks have been published that discuss one or other of the individual areas related to Tissue and Organ replacement. However, the Editors feel that it is appropriate to bring together all these different approaches together for the first time in one comprehensive text.

In order to be successful in any of these areas of organ replacement therapy, a multi-disciplinary approach is needed. The Editors feel that this text will appeal to both students, scientists and clinicians from a wide spectrum of disciplines who are considering a future in this area as well as those who have only so far heard of the developments in organ transplantation or organ replacements in the lay media. These fields are moving very fast indeed. Transplant surgeons continue to redefine what is possible and new products that were just laboratory curiosities a few years ago are now beginning to enter clinics around the world and improve the quality of life for many thousands of people. The promise of the various technologies described in this text, if realised, will make a profound and lasting impact on both the way that the healthcare industry operates and the way that we view the human body.

The Biotechnology revolution has so far made an enormous impact on society, providing new treatments for previously devastating diseases, altering the way virtually all pharmaceutical companies treat drug discovery and development and spawning an industry that employs tens of thousands of people and is worth hundreds of billions of dollars world-wide. However, along the way, a great deal of controversy and public distrust has been generated which has had an adverse impact in some areas. New developments in tissue and organ replacements has the potential to be even more controversial.

It is often difficult for people to have an opinion on the relative merits and ethics of research into materials at the scale of individual molecules as is the case for traditional pharmaceutical agents. However, this is not the case for tissue or organ replacements. These materials are visible structures and are also often formed from living human or animal cells. People, both professionals and lay audience alike, can see such products with the naked eye and the results are therefore at once both more dramatic and more emotive. Moreover the source of these cells could either/or be from embryonic, fetal tissue which also raises ethical issues that need to be openly addressed.

The media attention directed towards all these areas has, in the past, proven more often than not to be misdirected, oversimplified and sensationalist. It is one of the most fervent aims of the Editors that by bringing together the various advances in one text, we will be able to begin to show the skill and imagination that is being brought to bear in order to develop revolutionary new materials to treat patients who are often desperately ill and have no alternative.

The authors from this book are all established experts in their respected field and this is the first time that this distinguished spectrum of contributors has been brought together to address the following:

1) The clinical need for organ replacements
2) The future advances in organ transplantation
3) The revolutionary new therapies that are being developed
4) Regulatory issue of such product

Professor Julia M. Polak

Imperial College,
Tissue Engineering Centre,
Faculty of Medicine,
Chelsea & Westminister Hospital,
369 Fulham Road, London SW10 9NH

Professor Larry L. Hench

Imperial College,
Department of Materials,
Prince Consort Road,
London NW7 2BP

Dr P. Kemp

Intercytex Ltd, Etherley Dene House,
16 Chadkirk Road,
Romiley, Stockport, Cheshire, SK6 3JY, UK

List of Contributors

Professor P Aebischer

Division de Recherche Chirurgicale et
Centre de Therapie Genique
Pavillon 4
CHUV
1011 Lausanne
Switzerland
Tel. 00 41 21 314 2261
Fax. 00 41 21 314 2468
Email:patrick.aebischer@chuv.hospvd.ch

Dr A Atala

Director, Laboratory for Tissue Engineering and Cell Therapy
Children's Hospital and Harvard Medical School
300 Longwood Avenue
Boston, MA 02115
USA
Tel. 001-617-355-6169
Fax. 001-617-355-6587
Email:a.atala@tch.harvard.edu

Dr R A Brown

University College London
RFUCMS
Centre for Plastic & Reconstructive Surgery
Tissue Repair Unit
67-73 Riding House Street
London W1W 7EJ
Tel. 020-7679-9071

Fax. 020-7813-2829
Email:tissue.eng@ucl.ac.uk

Dr L D K Buttery
Imperial College
Tissue Engineering Centre
Faculty of Medicine
Chelsea & Westminster Hospital
369 Fulham Road
London SW10 9NH
Tel. 020-8237-2569
Fax. 020-8237-2569
Email:l.buttery@ic.ac.uk

Dr D K C Cooper
Transplantation Biology Research Center
Massachusetts General Hospital
MGH-East, Building 149, 13th Street
Boston, MA 02129
USA
Tel. 001-617-724-8313
Fax. 001-617-726-4067
Email:David.Cooper@tbrc.mgh.harvard.edu

Dr A J T George
Imperial College
Department of Immunology
Faculty of Medicine
Hammersmith Campus
Du Cane Road
London W12 0NN
Tel. 020 8383 1475
Fax. 020 8383 2786
Email:a.george@ic.ac.uk

Professor M Hammerman
Renal Division
School of Medicine
Washington University in St Louis
660 Euclid Avenue
Campus Box 8126
St. Louis, MO 63110-1093
USA
Tel. 001-314-362-8233
Fax. 001-314-362-8237
Email:mhammerm@imgate.wust1.edu

Professor L L Hench
Imperial College
Department of Materials
Prince Consort Road
London NW7 2BP
Tel. 0171 594 6745
Fax. 0171 594 6809
Email:l.hench@ic.ac.uk

Dr W Otto
Histopathology Unit
Cancer Research UK
44, Lincoln's Inn Fields
London WC2A 3PX
Tel. 0207 269 3085
Fax. 0207 269 3491
Email:bill.otto@cancer.org.uk

Dr J L Platt
Transplantation Biology
Departments of Surgery
Immunology and Pediatrics

Mayo Clinic
200 First Street SW
Rochester, Minnesota 55905
USA
Tel. 001-507-538-0313
Fax. 001-507-283-4957
Email:platt.jeffrey@mayo.edu

Dr B Rubinsky
Departments of Mechanical Engineering and Biomedical Engineering
University of California at Berkeley
Berkeley, CA 94720
USA
Tel. 001 510 642 8220
Fax. 001 510 642 6163
Email:rubinsky@me.berkeley.edu

Dr C Selden
Centre for Hepatology
Department of Medicine
Royal Free and University College Medical School
Royal Free Campus
Upper 3rd Floor
Rowland Hill Street
London NW3 2PF
Tel. 020-7433-2854
Fax. 020-7433-2852
Email:c.selden@rfc.ucl.ac.uk

Professor M Sittinger
German Rheumatism Research Center
Berlin and Department of Rheumatology
and Clinical Immunology
Charite University Hospital

Experimental Rheumatology and Tissue Engineering
Tucholskystrasse 2, 10117 Berlin
Germany
Tel. 00 49 30 2802 6553
Fax. 00 49 30 2802 6664
Email:michael.sittinger@charite.de

Professor J M W Slack

Department of Biology and Biochemistry
University of Bath
Plaverton Downs
Bath BA2 7AY
Tel. 01225 386597
Fax. 01225 386779
Email:j.m.w.slack@bath.ac.uk

Dr R Warwick

National Blood Service
Colindale Avenue
London NW9 5BG
Tel. 020-8258-2705
Fax. 020-8205-5017
Email:ruth.warwick@nbs.nhs.uk

Contents

Section 1

New Developments on Materials

Chapter 1

Bioactive Materials for Tissue Engineering Scaffolds

Larry L Hench, Julian R Jones
& Pilar Sepulveda

INTRODUCTION

Millions of orthopaedic prostheses made of bioinert materials have been implanted with excellent fifteen-year survivability of 75–85%. Improved metal alloys, special polymers and medical grade ceramics are the basis for this success, which has enhanced the quality of life for millions of patients.[1-8] However, an increasing percentage of our ageing population requires greater than thirty-year survivability of devices.[3,9] It is proposed that to satisfy this growing need for very long term orthopaedic repair that a paradigm shift is needed; a shift from emphasis on *replacement* of tissues to *regeneration* of tissues.[8] Such a shift from a materials-and-mechanics approach to biological based tissue repair requires an increase in understanding and utilisation of molecular biology. A new biological-orientated alternative based upon use of bioactive materials for tissue engineering is described.

3

The concept of the use of Class A bioactive materials to stimulate the regeneration of bone provides the scientific foundation for creating bioactive resorbable scaffolds. Class A bioactive materials exhibit 11 reaction stages that lead to enhanced proliferation and differentiation of osteoblasts and recreation of trabecular bone architecture *in situ*. Recent results presented below show that the effects of microchemical gradients on the genetic activation of bone cells are related to the molecular design of hierarchical bioactive resorbable scaffolds and can be used for tissue engineering of bone constructs.

THE NEED

During the last century, orthopaedics underwent a revolution, a shift in emphasis from palliative treatment of infectious diseases of bone to interventional treatment of chronic age-related ailments.[5,6] The evolution of stable metallic fixation devices, and the systematic development of reliable total joint prostheses were critical to this revolution in health care. The history and the current practice of these revolutionary steps are well documented[1–6] and will not be repeated here.

Devices and prostheses made from orthopaedic biomaterials ideally should survive without failure for the lifetime of the patient. The challenge is that the lifetime of patients has progressively increased during the last century[9] and will continue to do so for many years to come. Average life expectancy is currently at 80+ years, an increase of more than 15 years since the 1960s, when Professor Sir John Charnley pioneered the use of low friction total hip replacement. There is a compound effect of increased patient lifetime on the survivability of orthopaedic prostheses: i) Many more patients need prostheses; ii) The quality of bone of the patients progressively deteriorates with age, especially for women after the menopause. The two effects are multiplicative and contribute to the continuing decline in implant survivability with patient age described in Ref. 3.

There are two options to satisfy increasing needs for orthopaedic repair in the future: i) improve implant survivability by 10–20 years; or ii) develop alternative means of orthopaedic treatment that do not require implants, or at least delay the need for prostheses by 10–20 years. The objective of this chapter is to discuss use of tissue engineering to meet these needs.

THE BIOACTIVE ALTERNATIVE

During the last decade considerable attention has been directed towards the use of implants with bioactive fixation, where bioactive fixation is defined as interfacial bonding of an implant to tissue by means of formation of a biologically active hydroxyapatite layer on the implant surface.[1,4,7]

An important advantage of bioactive fixation is that a bioactive bond forms at the implant-bone interface with a strength equal to or greater than bone.

Materials for clinical use can be classified into three categories: resorbable, bioactive and nearly inert materials. A bioactive material is defined as a material that elicits a specific biological response at the interface of the material, which results in a formation of a bond between the tissue and that material.[8] The level of bioactivity of a specific material can be related to the time taken for more than 50% of the interface to bond to bone ($t_{0.5bb}$):

$$\text{Bioactivity index, } I_B = 100/t_{0.5bb} \qquad (1)$$

Materials exhibiting an I_B value greater than 8 (class A), e.g. 45S5 Bioglass®, will bond to both soft and hard tissue. Materials with an I_B value less than 8 (class B), but greater than 0, e.g. synthetic hydroxyapatite, will bond only to hard tissue.[9] A bioactive glass is one that undergoes surface dissolution in a physiological environment in order to form a hydroxycarbonate apatite (HCA) layer.[10]

The larger the solubility of a bioactive glass, the more pronounced is the effect on bone tissue growth.[11]

MECHANISM OF BIOACTIVITY

When a glass reacts with an aqueous solution, both chemical and structural changes occur as a function of time within the glass surface.[12] Accumulation of dissolution products causes both the chemical composition and pH of solution to change. The formation of HCA on bioactive glasses and the release of soluble silica and calcium ions to the surrounding tissue are key factors in the rapid bonding of these glasses to tissue, stimulation of tissue growth and use as tissue engineering scaffolds.[13–15]

There are 11 stages in the process of complete bonding of bioactive glass to bone. Stages 1–5 are chemical; stages 6–11 are the biological response:

i) Rapid exchange of Na^+ and Ca^{2+} with H^+ or H_3O^+ from solution (diffusion controlled with a $t^{1/2}$ dependence, causing hydrolysis of the silica groups, which creates silanols);

$$Si\text{-}O\text{-}Na^+ + H^+ + OH^- \rightarrow Si\text{-}OH^+ + Na^+_{(aq)} + OH^-$$

The pH of the solution increases as a result of H^+ ions in the solution being replaced by cations.

ii) The cation exchange increases the hydroxyl concentration of the solution, which leads to attack of the silica glass network. Soluble silica is lost in the form of $Si(OH)_4$ to the solution, resulting from the breaking of Si-O-Si bonds and the continued formation of Si-OH (silanols) at the glass solution interface:

$$Si\text{-}O\text{-}Si + H_2O \rightarrow Si\text{-}OH + OH\text{-}Si$$

This stage is an interface-controlled reaction with a $t^{1.0}$ dependance.

iii) Condensation and repolymerisation of a SiO_2-rich layer on the surface, depleted in alkalis and alkali-earth cations.

iv) Migration of Ca^{2+} and PO_4^{3-} groups to the surface through the SiO_2-rich layer, forming a CaO-P_2O_5-rich film on top of the SiO_2-rich layer, followed by growth of the amorphous CaO-P_2O_5-rich film by incorporation of soluble calcium and phosphates from solution.

v) Crystallisation of the amorphous CaO-P_2O_5 film by incorporation of OH^- and CO_3^{2-} anions from solution to form a mixed hydroxyl carbonate apatite (HCA) layer.

vi) Adsorption and desorption of biological growth factors, in the HCA layer (continues throughout the process), to activate differentiation of stem cells.

vii) Action of macrophages to remove debris from the site allowing cells to occupy the space.

viii) Attachment of stem cells on the bioactive surface.

ix) Differentiation of stem cells to form bone growing cells, osteoblasts.

x) Generation of extracellular matrix by the osteoblasts to form bone.

xi) Crystallisation of inorganic calcium phosphate matrix to enclose bone cells in a living composite structure.

For bone, interfacial bonding occurs because of the biological equivalence of the inorganic portion of bone and the growing HCA layer on the bioactive implant.[19] For soft tissues, the collagen fibrils are chemisorbed on the porous SiO_2-rich layer via electrostatic, ionic and/or hydrogen bonding, and HCA is precipitated and crystallised on the collagen fibre and glass surfaces.[20]

Reaction stages one and two are responsible for the dissolution of a bioactive glass, and therefore greatly influence the rate of HCA formation. Many studies have shown that the leaching of silicon and sodium to solution is initially rapid, following a parabolic relationship with time for the first six hours of reaction, then stabilises, following a linear dependence on time.[9,13,14,16,21–24]

For bioactive scaffolds, it is necessary to be able to control the solubility (dissolution rate) of the material. A low solubility material is necessary if the scaffold is designed to have a long life. A controlled solubility rate is required if it is designed to aid bone formation, such as 45S5 Bioglass® powders for bone graft augmentation or rapid formation of bone *in vitro*. Therefore, a fundamental understanding of factors influencing solubility and bioreactivity is required to develop new materials for *in situ* tissue regeneration and tissue engineering.

FACTORS AFFECTING DISSOLUTION OF BIOACTIVE GLASS TO BE USED AS TISSUE ENGINEERING SCAFFOLDS

The solution parameters of initial pH, ionic concentration and temperature have a large effect on the rate of scaffold dissolution and even type of calcium phosphate precipitated.[22] Ionic concentration, and therefore pH, will obviously change with time as dissolution progresses and this will in turn affect the dissolution rate. If pH rises above a critical value, cytotoxicity will occur.[26]

Three types of media have been used during bioactive glass dissolution experiments:

i) Tris-buffer is a simple organic buffer solution.
ii) Simulated body fluid (SBF) is a tris buffer containing similar ion concentrations to that of human blood plasma.
iii) α-MEM and D-MEM are culture media that contain both the inorganic and biological organic components of blood plasma.[16]

Zhong and Greenspan[16] found that Bioglass® underwent a faster surface reaction and exhibited a larger HCA crystal size in Tris solution than it did in SBF or culture medium. However, Pereira *et al.*[27] found that HCA nucleated faster in SBF than in Tris, which they put down to the far higher initial P concentration (40ppm) in SBF, thus aiding HCA nucleation. A similar result was found by Tsuru *et al.*,[28] but they suggested that the dissolution of Ca^{2+} ions from materials would increase the degree of supersaturation in the solution (SBF), which would already be supersaturated regarding HCA precipitation and hence make it much easier for HCA to be precipitated.

Pereira *et al.*[22,26] found that increasing the pH or Ca^{2+} concentration of SBF containing porous silica gel–glass reduced the induction time for heterogeneous HCA formation, agreeing with the early work by Wirth and Gieskes[29] on anhydrous silica powders, which found the rate of silicon release increased by two orders of magnitude with each unit increase of pH. Pereira and Hench also found evidence of homogeneous HCA precipitation in SBF, containing porous silica gel–glass, when the Ca^{2+} content increased above a critical value of 300ppm.[26]

Reaction kinetics of surface layer formation have been found to alter in culture media containing serum, due to the adsorption of serum proteins onto the surface of the bioactive glass forming a barrier to nucleation of the HCA layer.[30–32]

GEOMETRIC EFFECTS

A change in geometry of a tissue engineering scaffold will generally mean a change in the surface area to solution volume ratio (SA/V), which will affect the dissolution rate, as the amount of surface exposed to solution for ion exchange will also change.

Elsberg *et al.*[33] reacted fibres and cylinders of melt-derived 45S5 Bioglass® in tris-buffer *in vitro*, using a SA/V ratio of $0.1cm^{-1}$. Dissolution rate was found to be inversely proportional to the radius of the sample, whereas the nucleation and growth of HCA occurred earlier on surfaces with a larger radius of curvature. Release of the network modifiers, sodium and calcium, was again diffusion dependent ($t^{1/2}$). Silicon dissolution, as a function of log time, exhibited two linear regimes, rather than the parabolic/linear regimes observed in other studies. The slope of this relationship was lower during stages 1–3, formation of the silica gel layer, than for stages

4 and 5, the nucleation of the HCA layer, which implies network break-up was accelerated while the HCA layer formed.

The effect of SA/V on the dissolution of bioactive glass powders was investigated by Greenspan et al.,[13] using melt-derived 45S5 Bioglass® powders of different particle size ranges, immersed in tris-buffer solution for various periods of time. In general, the higher SA/V ratios had a faster rate of pH increase and a higher final pH when compared to the lower SA/V ratios. At high SA/V ratio, the calcium phosphate layer formed rapidly but remained thin over time. At lower SA/V ratios, the calcium phosphate layers grew more slowly, but the final thicknesses after 20 hours were greater than at high SA/V ratio. The thin layer at high SA/V ratio may be due to the layer inhibiting ion exchange at the silica-gel layer surface. Therefore, a higher SA/V favours initial dissolution, yielding a faster initial calcium phosphate layer formation. When the SA/V was held constant, the rate of formation of the HCA layer was much slower for smaller particles, which was attributed to physical differences such as radius of curvature, mass to volume ratio and test solution. There appears to be a trade-off between rapid dissolution at small particle sizes (and therefore better resorbability) and thickness of HCA formation at larger particle sizes. These results are important for processing tissue engineering scaffolds incorporating bioactive powders, fibres, meshes or foams.

Wilson and Noletti[34] found that particles of 100μm in diameter were either resorbed or phagocytosed by macrophages *in vivo*, while larger particles stimulated bone growth. This could be a problem for using fine particles, meshes or foams in scaffolds in order to attain rapid dissolution and high bioactivity *in vivo* and requires further investigation. Schepers et al.[35] also found that glass particles of less than 300μm in diameter are fully resorbed *in vivo*. Thus it may be necessary to use a range of particle sizes, fibre diameters or foam textures to produce the necessary gradients of dissolution products to enhance bone formation *in situ* or in tissue engineering. This is presently being done, as described in recent publications.[36,37]

COMPOSITION EFFECTS OF BIOACTIVE GLASS SCAFFOLDS

Until the late 1980s, bioactive glasses were generally melt-derived, with the majority of research aimed at the 45S5 Bioglass® composition (46.1% SiO_2, 24.4% NaO, 26.9% CaO and 2.6% P_2O_5, in mol%) and apatite-wollastanite (A/W) glass-ceramics.[38] The rapid rate of HCA formation exhibited by melt-derived bioactive glasses was attributed to the presence of Na_2O, or other alkali cations in the glass composition, which increased the solution pH at the implant-tissue interface as dissolution progressed.[17] Adding multivalent cations, such as alumina, stabilise the glass structure by eliminating non-bridging oxygen[25] and also retard HCA formation, but slight deviations in composition can radically alter the dissolution kinetics or even basic mechanisms of bonding.[10,14,39]

Hill found that increasing the SiO_2 content of the glass linearly decreased the glass transition temperature, the peak crystallisation temperature, the oxygen density in the glass, the glass density and increased the thermal expansion coefficient. He therefore concluded that Na_2O was a network disruptor and actually decreased the bioactivity of a glass.[40]

It is widely accepted that increasing silica content of melt-derived glass decreases dissolution rates by reducing the availability of modifier ions such as Ca^{2+} and HPO^{4-} ions to the solution and the inhibiting development of a silica-gel layer on the surface.[25,27] The result is the reduction and eventual elimination of the bioactivity of the melt-derived bioactive glasses as the silica content approaches 60%. No melt-derived glasses with more than 60mol% silica are bioactive. In order to obtain bioactivity for silica levels higher than 60mol%, the sol–gel process must be employed, which is a novel processing technique for the synthesis of tertiary bioactive glasses.

SOL–GEL-DERIVED BIOACTIVE GLASSES

The recognition that the silica gel layer plays a role in HCA nucleation and crystallisation led to the development of the bioactive three component $CaO-P_2O_5-SiO_2$ sol–gel-derived glasses by Li, Clark and Hench,[27] thus dispelling the theory that Na_2O was the active component of the bioactive glass.

A sol is a dispersion of colloidal particles in a liquid. Colloids are solid particles with diameters 1–100nm. A gel is an interconnected, rigid network with pores of submicrometer dimensions and polymeric chains whose average length is greater than 1μm. Details of sol–gel processing are given in Refs. 38–43.

There are several advantages of a sol–gel-derived glass over a melt-derived glass which are important for making tissue engineering scaffolds. Sol–gel-derived glasses have:

i) Lower processing temperatures (600–700°C).
ii) The potential of improved purity, required for optimal bioactivity due to low processing temperatures and high silica and low alkali content.
iii) Improved homogeneity.
iv) Wider compositions can be used (up to 90mol% SiO_2) while maintaining bioactivity.
v) Better control of bioactivity by changing composition or microstructure.
vi) Structural variation can be produced without compositional changes by control of hydrolysis and polycondensation reactions during synthesis.
vii) A greater ease of powder production.[17,38]
viii) A higher bioactivity.[14,15,26,27,43,44]
ix) Interconnected nanometer scale porosity that can be varied to control dissolution kinetics or be impregnated with biologically active phases such as growth factors.
x) Can be foamed to provide interconnected pores of 10–200μm, mimicking the architecture of trabecular bone.

The mechanism for dissolution and HCA formation on bioactive gel–glasses follows most of the same 11 stages as those for melt-derived glasses as listed above. The following sections compare dissolution characteristics of sol–gel-derived glasses with the melt-derived glasses described above.

Li *et al.*[45] found that, for melt-derived glasses, as the SiO_2 composition increased in CaO-P_2O_5-SiO_2 gel-glasses, the rate of HCA formation decreased. However, bioactivity of the ternary system continued up to 90mol% silica. This is due to the high concentration of nucleating sites in gel–glasses. The surface area of gel–glasses has been found to increase with silica content,[27,46] which also enhances bioactivity.

In vitro studies by Greenspan *et al.*[14,15] found more rapid dissolution and faster HCA layer formation for sol–gel derived 58S gel–glasses (60% SiO_2, 36% CaO and 4% P_2O_5, in mol%) (one hour) compared with 45S5 Bioglass® (six hours). This result was supported by *in vivo* experiments, in which after 12 weeks nearly all the silica was depleted from the 58S gel–glass and a calcium phosphate residue was observed. The 45S5 implant exhibited a loss of sodium ions, but had a lower rate of silica dissolution and a thicker HCA layer. Silica was lost at a similar rapid rate for both materials for the first six hours. After that time, the silica dissolution rate from the 58S gel–glass was more rapid than from the 45S5 glass. It took four days longer for the silica release rate from the 58S to stabilise, compared to that for the 45S5 melt-glass. Similar results were found by Hench *et al.*[43] *in vitro* for 58S and 45S5 particles (300–710μm), reacted in SBF but the *in vivo* results by Greenspan *et al.*[14,15] showed no difference in bone formation between the two glasses after 12 weeks. In contrast, Chou *et al.*[47] have shown enhanced bone formation after four weeks for *in vivo* sites treated with sol–gel derived materials compared to those treated with melt-derived glasses. The greater surface area provided by the mesoporous 58S gel–glasses allowed prolonged ion exchange and more rapid dissolution as bone formed.

Therefore, the increased rate of HCA formation and higher index of bioactivity for the sol–gel-derived glasses is attributed to more release of soluble silica that nucleates HCA crystals in the nanometer-sized pores of the gel glass.[14]

TEXTURAL EFFECTS

In addition to composition, the nanometer scale texture of sol–gel-derived scaffolds is an important class of variables in the behaviour of these materials. The unique interconnected mesoporous structure (pores with diameter between 2 and 50nm) of sol–gel-derived glasses, which increases the SA/V ratio compared to that of melt-derived glasses, has been found to be the critical factor in enhancing the dissolution rate and rate of HCA formation. *In vitro* studies[14,26] showed that the dissolution rate of 58S gel–glass increased as porosity and pore volume increased. Pore sizes greater than 2nm were required to achieve rapid kinetics. An increased SA/V ratio increases the surface exposed to the solution, improving ion exchange (stage 1) and a greater release of soluble silica (stage 2) that is required to form a porous silica rich layer. The nanometer-sized pores of the gel glass act as initiation sites for HCA crystal nucleation.[26,43] The superposition of surface potentials[44] inside the pores increases the ionic concentration and degree of supersaturation of the Ca and P ions. Thus precipitation of HCA is more likely to occur first inside the pores. The rate of nucleation is then controlled by the diffusion of ions into the pores.

The porous structure extends the silica composition range of bioactive glasses from 60 to 90mol%. Although the gel-glass network break up (stage 2) is more difficult as silica content increases, the increase in SA/V ratio means that ion exchange (stage 1) is enhanced. Thus, a high concentration of Ca^{2+} and HPO_4^- ions is released to the solution as pore volume increases,[14,26] so that a silica-gel layer can develop very rapidly on the surface of the glass.[27,45]

Sol–gel derived glasses also exhibit significant bioresorbability when their pores reach a certain size.[33] Bioresorption is defined as the resorption of a material *in vivo*, due to the action of osteoclasts, which in this case is attributed to the interconnected pore network, high surface areas and low particle density.[43] Although it is difficult to control resorbability by changing composition, controlling the pore texture significantly influences degradability.[14,22] Biodegradation is mainly governed by the crystal structure, grain size, microporosity, neck geometry and crystallinity of the material.[48]

Initially, it was thought that Ca^{2+} ion dissolution and resulting silanol formation on the surface were mandatory for a material to form an HCA layer.[28] However, Pereira and Hench[26] conclude that, as hydroxyl coverage is independent of textural characteristics, the concentration of silanol groups on the silica surface does not control the rate of HCA formation. There is no evidence to show that silanols are a requisite for HCA formation, but their involvement cannot be discounted.

Both a negative surface charge and a porous substrate are required for HCA formation.[16,26] Silica presents a negative surface charge at physiological pH, which leads to the formation of an electrical double layer with an increased number of cations at the interface.[44] This provides evidence that a porous silica layer is required for an HCA layer to form. Hence, an HCA layer can be formed on a porous pure silica gel in a solution containing Ca^{2+} and HPO_4^{-} ions.[22]

CELLULAR FEATURES OF CLASS A BIOACTIVE MATERIALS

An important feature of Class A bioactive particulates is that they are osteoproductive as well as osteoconductive. In contrast, Class B bioactive materials exhibit only *osteoconductivity*, defined as the characteristic of bone growth and bonding along a surface. Dense synthetic hydroxyapatite (HA) ceramic implants exhibit Class B bioactivity. *Osteoproduction* occurs when bone proliferates on the particulate surfaces of a mass due to enhanced osteoblast activity. Enhanced proliferation *and* differentiation of osteoprogenitor cells, stimulated by slow resorption of the Class A bioactive particles, are responsible for osteoproduction.

The biological response to bioactive gel-glasses made from the $CaO-P_2O_5-SiO_2$ system provides evidence that bone regeneration is feasible. Biological molecules can exchange with hydrated layers inside the pores of gel-glasses and maintain their conformation and biological activity.[7,42,53,57] Many enzymes remain active within a hydrated gel matrix, and in some cases exhibit enhanced activity. Such hierarchical structures and behaviour go far beyond historically important bioinert orthopaedic materials such as PMMA, ultrahigh molecular weight polyethylene, stainless steel, Co–Cr and Ti-alloys towards matching the ultrastructure and molecular chemistry of bone.

Evidence of the regenerative capacity of bioactive gel-glasses is based on comparison of the rates of proliferation of trabecular bone in a rabbit femoral defect model.[59] Melt-derived Class A 45S5 bioactive glass particles exhibit substantially greater rates of trabecular bone growth and a greater final quantity of bone than Class B synthetic HA ceramic or bioactive glass-ceramic particles. The restored trabecular bone has a morphological structure equivalent to the normal host bone after six weeks; however, the regenerated bone still contains some of the larger (>90 micrometers) bioactive glass particles.[59,60] Recent studies show that the use of bioactive gel–glass particles in the same animal model produces an even faster rate of trabecular bone regeneration with no residual gel–glass particles of either the 58S or 77S composition.[60] The gel–glass particles resorb more rapidly during proliferation of trabecular bone. The mechanical quality of the regenerated bone appears to be equivalent to that of the control sites.[64] Thus, the criteria of a regenerative allograft appear to have been met. Our challenge for the future is to extend these findings to studies in compromised bones, with osteopenia and osteoporosis, to apply the concept to humans with ageing bones and degenerative joint disease and to use the results to design the 3D architectures required for engineering of tissues.

GENETIC CONTROL BY BIOACTIVE MATERIALS

We have now discovered the genes involved in phenotype expression and bone and joint morphogenesis, and thus are on the way towards learning the correct combination of extracellular and intracellular chemical concentration gradients, cellular attachment complexes and other stimuli required to activate tissue regeneration *in situ* and in tissue engineering constructs. Professor Julia Polak's group at the Imperial College Tissue Engineering Centre has recently shown that six families of genes are up-regulated and down-regulated by bioactive glass extracts during proliferation and differentiation of primary human osteoblasts *in vitro*.[61–63] These findings should make it possible to design a new generation of bioactive materials for regeneration and tissue engineering of bone. The significant new finding is that low levels of dissolution of the bioactive glass particles in the physiological environment exert a genetic control over osteoblast cell cycle and rapid expression of genes that regulate osteogenesis and the production of growth factors.[61–63]

Xynos *et al.* have shown that within 48 hours, a group of genes was activated including genes encoding nuclear transcription factors and potent growth factors. These results were obtained using cultures of human osteoblasts, obtained from excised femoral heads of patients (50–70 years) undergoing total hip arthroplasy.[61]

In particular, insulin-like growth factor (IGF) II, IGF-binding proteins and proteases that cleave IGF-II from their binding proteins were identified.[63] The activation of numerous early response genes and synthesis of growth factors was shown to modulate the cell cycle response of osteoblasts to the bioactive glasses and their ionic dissolution products. These results indicate that bioactive glasses enhance osteogenesis through a direct control over genes that regulate cell cycle induction and progression. However, these molecular biology results also confirm that the osteoprogenitor cells must be in a chemical environment suitable for passing checkpoints in the cell cycle towards the synthesis and mitosis phases. Only a select number of cells from a population are capable of dividing and becoming mature osteoblasts. The others are switched into apoptosis. The number of progenitor cells capable of being stimulated by a bioactive medium decreases as a patient ages. These findings may account for the time delay in formation of new bone in augmented sites.

Clinical application of the use of a regenerative biomaterial in orthopaedics is beginning. In 1993, particulate bioactive glass, 45S5 Bioglass® was cleared in the USA for clinical use as a bone graft material for the repair of periodontal osseous defects. Since that time, numerous oral and maxillofacial clinical studies have been conducted to expand the material indication. More than 2,000,000 reconstructive surgeries in the jaw have been performed with the material. The same material has been used by several orthopaedic surgeons to fill a variety of osseous defects and for clinical use in orthopaedics, such as NovaBone®, which is now approved for clinical use in Europe.[64]

CONCLUSIONS

During the last century, a revolution in orthopaedics occurred which has led to a remarkably improved quality of life for millions of aged patients. Specially developed biomaterials were a critical component of this revolution. However, high rates of survivability of prostheses appear to be limited to approximately 20 years. Thus, it is concluded that a shift in emphasis from replacement of tissues to a new concept of regeneration and cellular engineering of tissues should be the research emphasis for orthopaedic materials in the years ahead. The emphasis should be on the use of materials to activate the body's own repair mechanisms, i.e. regenerative allografts and tissue engineered constructs. This concept will combine the understanding of osteogenesis and chondrogenesis at a molecular level with the design of a new generation of bioactive materials that stimulate genes that activate the proliferation and differentiation of osteoprogenitor cells and enhance rapid formation of extracellular matrix and growth of new bone *in situ*. The economic and personal benefits of *in situ* regenerative repair of the skeleton for younger patients will be profound.

ACKNOWLEDGEMENTS

The authors gratefully acknowledge the following sponsors: E.P.S.R.C, M.R.C, F.A.E.P.E and U.S. Biomaterials, Inc.

REFERENCES

1. Hench LL, Wilson J. *Introduction to Bioceramics.* World Scientific, Singapore. 1993, Chapters 4 and 6.
2. Ratner BD, Hoffman AS, Schoen FJ, Lemmons JE. *An Introduction to Materials in Medicine.* Biomaterials Science, 1996, 484.
3. Hench LL, Wilson J. *Clinical Perfomance of Skeletal Prostheses.* Chapman and Hall, London. 1996, all chapters.
4. Cao W, Hench LL. *Bioactive Materials.* Ceramics International, 1996; **22**:493–507.
5. LeFanu J. *The Fall and Rise of Modern Medicine.* Little Brown, 1999; 104–113.
6. Simon SR. *Orthopaedic Basic Science.* American Academy of Orthopedics Surgeons, 1994.
7. Hench LL, West JK. *Biological Applications of Bioactive Glasses.* Life Chemistry Reports, 1996; **13**:187–241.

8. Hench LL. Biomaterials: A forecast for the future. *Biomaterials* 1998; **19**:1419–1423.
9. Hench LL. Bioceramics: From concept to clinic. *J American Ceramic Society* 1991; **74**(7):1487–1510.
10. Wallace KE, Hill RG, Pembroke JT, Brown CJ, Hatton PV. Influence of sodium oxide content on bioactive glass properties. *J materials science: Materials in Medicine* 1999; **10**:697–701.
11. Ducheyne P, Qui Q. Bioactive ceramics: The effect of surface reactivity on bone formation and bone cell function. *Biomaterials* 1999; **20**:2287–2303.
12. Clark AE, Pantano CG, Hench LL. *Corrosion of glass*. Magazines for Industry, New York. 1979, 1.
13. Greenspan DC, Zhong JP, LaTorre GP. Effect of surface area to volume ratio on *in vitro* surface reactions of bioactive glass particulates. In *Bioceramics 7*, eds. OH Andersson, A Yli-Urpo, 1994, pp. 28–33.
14. Greenspan DC, Zhong JP, Chen ZF, LaTorre GP. The evaluation of degradability of melt and sol–gel derived Bioglass® *in vitro*. In *Bioceramics 10*, eds. L Sedel, C Rey, 1997, pp. 391–394.
15. Greenspan DC, Zhong JP, Wheeler DL. Bioactivity and biodegradability: Melt vs sol–gel derived Bioglass® *in vitro* and *in vivo*. In *Bioceramics 11*, eds. RZ LeGeros, JP LeGeros, 1998, pp. 345–348.
16. Zhong JP, Greenspan DC. Bioglass® surface reactivity: From *in-vitro* to *in-vivo*. In *Biomaterials 10*, eds. L Sedel, C Rey, 1997, pp. 391–394.
17. Li R, Clark AE, Hench LL. An investigation of bioactive glass powders by sol–gel processing. *J Applied Biomaterials* 1991; **2**: 231–239.
18. Doremus DH, Mehrotra Y, Lanford WA, Burman C. *J Materials Science* 1983; **18**(2):612.
19. Vidueau JJ, Dupuis V. Phosphates and biomaterials. *European Journal of Solid State Inorganic Chemistry* 1991; **28**:303–343.

20. Zhong JP, LaTorre GP, Hench LL. The kinetics of bioactive ceramics part VII: Binding of collagen to hydroxyapatite and bioactive glass. In *Bioceramics 7*, eds. OH Andersson, A Yli-Urpo, 1994, pp. 61–66.
21. Rinehart JD, Taylor TD, Tian Y. Real-time dissolution measurement of sized and unsized calcium phosphate glass fibres. *J Biomedical Materials Research* 1999; **48**(6):833–840.
22. Pereira MM, Clark AE, Hench LL. Effect of texture on the rate of hydroxyapatite formationon gel-silica surface. *J American Ceramic Society* 1995; **78**(9):2463–2468.
23. Mazer JJ, Walther JV. Dissolution kinetics of silica glass as a function of pH between 40 and 85C. *J Non-Crystalline Solids* 1994; **170**:32–45.
24. Douglas RW, El-shamy TM. Reactions of glasses with aqueous solutions. *J American Ceramic Society* 1967; **50**(1):1–8.
25. Andersson OH, Sodergard A. Solubility and film formation of phosphate and alumina containing silicate glasses. *J Non-Crystalline Solids* 1999; **246**:9–15.
26. Pereira MM, Hench LL. Mechanisms of hydroxyapatite formation on porous gel-silica substrates. *J Sol–gel Science and Technology* 1996; **7**:59–68.
27. Pereira MM, Clark AE, Hench LL. Calcium phosphate formation on sol–gel-derived bioactive glasses *in vitro*. *J Biomedical Material Research* 1994; **28**:693–698.
28. Tsuru K, Ohtsuki C, Osaka A, Iwamoto T, Mackenzie JD. Bioactivity of sol-gel derived organically modified silicates. *J Materials Science: Materials in Medicine* 1997; **8**:157–161.
29. Wirth G, Gieskes J. *J Colloid International Science* 1979; **68**:492.
30. Rohanizadeh R, Padrines M, Bouler JM, Couchoural D, Fortun Y, Daculsi G. Apatite precipitation after incubation of biphasic calcium-phosphate ceramic in various solutions: Influence of seed species and protein. *J Biomedical Materials Research* 1998; **42**(4):530–539.

31. Kuijer R, Bouwmeester SJM, Drees MMWE, Surtel DAM, Terwindt-Rouwenhorst EAW, Van Der Linden AJ, Van Blitterswijk CA, Bulstra SK. The polymer polyactive as a bone-filling substance: An experimental study in rabbits. *J Materials Science: Materials in Medicine* 1998; **9**:449–455.

32. Padrines M, Rohanizadeh R, Damiens C, Heymann D, Fortun Y. Inhibition of apatite formation by vitronectin. *Connective Tissue Research* 2000; **41**(2):101–108.

33. Elsberg LL, Lobel KD, Hench LL. Geometric effects on the reaction stages of bioactive glasses. (Unpublished).

34. Wilson J, Noletti D. Bonding of soft tissue to Bioglass. In *Bioceramics 3*, ed. S Hubbert, 1990, pp. 283–302.

35. Schepers E, Ducheyne P, Barbier L, Schepers S. Bioactive glass particles of narrow size range: A new material for the repair of bony defects. *Implant Dentistry* 1993; **2**(151):156.

36. Sepulveda P, Jones JR, Hench LL. Effect of particle size on Bioglass® Dissolution. In *Bioceramics 13*, eds. S Giannini, A Moroni, 2000, pp. 629–634.

37. Stamboulis A, Hench LL. Bioresorbable polymers: Their potential as scaffolds for Bioglass® composites. In *Bioceramics 13*, eds. S Giannini, A Moroni, 2000, pp. 729–734.

38. Hench LL, West JK. The sol–gel process. *Chemical Reviews* 1990; **90**: 33–72.

39. Warren LD, Clark AE, Hench LL. An investigation of Bioglass® powders: Quality assurance test procedure and test criteria. *J Biomedical Material Research: Applied Biomaterials* 1989; **23**(A2): 201–209.

40. Hill RG. *J Materials Science Letters* 1996; **15**:112.

41. Ishizaki, Komarneni, Nanko. *Porous Materials Process Technology and Applications*. Kluwer Academic Publishers, 1998, 67.

42. Livage J, Sanchez C. Sol–gel chemistry. *J Non-Crystalline Solids* **145**:11–19.

43. Hench LL, Wheeler DL, Greenspan DC. Molecular control of bioreactivity in sol–gel glasses. *J Sol–gel Science and Technology* 1998; **13**:245–250.

44. Li P, Zhang F. The electrochemistry of a glass-surface and its application to bioactive glass in solution. *J Non-Crystalline Solids* 1990; **119**:112–116.

45. Li R, Clark AE, Hench LL. Effect of structure and surface area on bioactive powders made by sol–gel process. In *Chemical Processing of Advanced Materials*, eds. LL Hench, JK West, Wiley, New York, 1992; pp. 627–633.

46. Perez-Pariente J, Balas F, Roman J, Salinas AJ, Vallet-Regi M. Influence of composition and surface characteristics on the *in vitro* bioreactivity of SiO_2-CaO-P_2O_5-MgO sol–gel glasses. *J Biomedical Materials Research* 1999; **47**(2):170–175.

47. Al-Bazie S, Chou S. The role of bioglass elements in osteogenesis at implant interface *in vivo*. *J Dental Research* 1998; **77**(839):1664. Special Issue B.

48. Oonishi H, Hench LL, Wilson J, Sugihara F, Tsuji E, Kushitani S, Iwaki H. Comparative bone growth behaviour in granules of bioceramic materials of various sizes. *J Biomedical Materials Research* 1999; **44**(1):31–43.

49. Bradley JG, Andrews CM, Lee K, Scott CA, Shaw D. Furlong Hydroxyapatite coated hip prosthesis versus the Charnley cemented hip prosthesis — A prospective randomised study. In *Bioceramics 13*, eds. S Giannini, A Moroni, 2000, pp. 1013–1020.

50. Bronzino JD. *The Biomedical Engineering Handbook*. CRC Press, Boca Raton, FL, 1995, 532.

51. Thompson I, Hench LL. Medical Applications of Composites. In *Encyclopedia of Composites*. Elsevier Press, Amsterdam, 2000, 727.

52. Bonfield W. Composite Biomaterials — Past, Present and Future. In *Bioceramics 11*, eds. R LeGeros, J LeGeros, 1998, pp. 37–40.

53. Livage J. Sol–gel processes. *Current Opinion Solid State Matter* 1997; **2**:132–138.

54. Hench LL. Sol–gel materials for bioceramic applications. *Current Opinion Solid State Materials Science* 1997; **2**:604–606.

55. Hench LL, Orefice R. Sol–gel technology. *Kirk-Othmer Encyclopedia of Chemical Technology*, 4th ed. Wiley, New York. 1997, 497.

56. Livage J, Roux C, Da Costa JM, Desportes I, Quinson JE. Immunoassays in sol–gel matrices. *J Sol–Gel Science Technology* 1996; **7**:45–51.

57. Aksay IA, Weiner S. Biomaterials — Is this really a field of research? *Current Opinion Solid State Materials Science* 1998, **3**(3):219–220.

58. Pereira MM, Clark AE, Hench LL. Homogeneity of bioactive sol–gel derived glasses in the system CaO-P2-O5-SiO2. *J Material Synthesis Proceedings* 1994; **2**(30):189–196.

59. Oonishi H, Kutrshitani S, Yasukawa E. Particulate Bioglass® and hydroxyapatite as a bone graft substitute. *J Clinical. Orthopaedics Related Research* 1997; **334**:316–325.

60. Wheeler DL, Stokes KE. *In vivo* evaluation of sol–gel Bioglass®. Part I: Histological findings. *Trans 23rd Annual Meeting of the Society for Biomaterials*, New Orleans, LA, 1997.

61. Xynos ID, Hukkanen MVJ, Batten JJ, Buttery LDK, Hench LL, Polak JM. Bioglass 45S5® stimulates osteoblast turnover and enhances bone formation *in vitro:* Implications and applications for bone tissue engineering. *Calcified Tissue International* 2000; pp. 67321–67329.

62. Hench LL, Xynos ID, Buttery LDK, Polak JM. Bioactive materials to control cell cycle. *Materials Research Innovations* 2000, (in Press).

63. Xynos ID, Edgar AJ, Buttery LDK, Hench LL, Polak JM. Ionic dissolution products of bioactive glass increase proliferation of human osteoblasts and induce insulin-like growth factor II mRNA expression and protein synthesis. *Biochemical and Biophysical Research Communications* 2000; **277**:604–610.

64. Details of the clinical cases are available from US Biomaterials Inc., Alachua, FL, USA, 32615.

Section 2

New Horizons in Grafting Human Organs

Chapter 2

Low Temperature Preservation of Biological Organs and Tissues

Boris Rubinsky

INTRODUCTION

Organ transplantation has become an established surgical technique for the treatment of many terminal diseases of organs, such as the kidney, liver and heart. The major technical difficulties in organ transplant are the lack of available organs and the logistic need for a close donor-recipient interaction in terms of time and the compatibility of the organ. It is obvious that prior to transplantation, transplanted organs must survive outside the body, for various periods of time. With the preservation techniques available today the time an organ can survive *ex vivo* is limited from between 48 hours for the kidney to about six hours for the heart. This restricts the availability of donor organs and as mentioned above, has become a significant limitation in the field of transplantation. In fact, at present, many available organs are wasted and not used for organ transplant because they cannot be preserved for sufficiently long periods of time. Tissue engineering is expected to overcome the lack of available organs. At present, the approximately 3000 annual heart transplants in the United Sates are hampered by difficulties with organ preservation. How much more acute will the problem become when all the 50,000 hearts needed annually in the USA will become available through tissue engineering? It is conceivable that the producer of the engineered tissue will be at a different location from the user. Furthermore, to optimize the use of the engineered tissue it may be necessary to establish and maintain an inventory of engineered cells, tissues and organs. Therefore, it can be anticipated that in future there will be a need for preserving the engineered tissues for extended periods of time, prior to their use. If the problem of tissue preservation is not resolved by the time tissue engineering begins to produce cells, tissues and organs, this entire field may become ineffective. The field of tissue engineering must include tissue preservation. The area of tissue preservation is covered by many journals and reviewed in numerous books. Therefore this article can provide only a cursory glimpse of this vast area of research. The article will discuss some of the key issues with tissue preservation at low temperatures.

When cells, tissues or organs are removed from their controlled environment in a functioning organism, the chemical reactions in these biological entities cease to occur in a controlled manner and they begin to deteriorate. The sum of biochemical reactions in an organism is the metabolism. Reducing the metabolism will prolong the survival of tissues and organs outside the organism. Since life processes are temperature dependent, one method for reducing the metabolism, and thereby preserving the biological material, is by lowering the temperature. Most enzymes of normothermic animals show a 1.5–2-fold decrease of metabolic activity for every 10°C decrease in temperature. Therefore a decrease in temperature from 37–0°C will decrease the metabolic rate by 12–13 fold.[1] Obviously, reducing the temperature to absolute zero will cause chemical reactions to cease entirely and could facilitate unlimited preservation in time. Low temperature preservation is affected by the chemical composition of biological materials, which contain mostly water. At atmospheric pressure, physiological saline freezes at about −0.56°C. Consequently, the phenomena that occur during cooling of biological materials differ from above and below the freezing temperature of the water in the material, giving rise to two different methods of low temperature preservation. Preservation at temperatures above freezing is called hypothermic preservation and preservation at temperatures below freezing is called cryogenic preservation.

HYPOTHERMIC PRESERVATION

Most types of mammalian cells and tissues can withstand low non-freezing temperatures for short periods of time. Cells are entities with a highly specific intracellular chemical content, separated from the non-specific extracellular milieu by the cell membrane. The cell membrane acts as a selective barrier between the intracellular and the extracellular milieu. The membrane selectively controls the transport of chemical species into and out of the cell. Therefore the membrane must be mostly impermeable except at particular sites where it can control the mass transfer. The lipid bilayer structure of the cell membrane makes it impermeable. The mass transfer through the cell membrane is controlled through membrane proteins that span the membrane. Mammalian cells have become optimized to function at the temperature in which the organism lives. As discussed earlier, one aspect of lowering the temperature of biological materials is reduced metabolism, which is beneficial to preservation. However, other aspects of temperature reduction are detrimental. A detrimental aspect of cooling the cell to temperatures lower than their normal physiological temperature is the lipid phase transition of the cell membrane.[2] The lipid membrane bilayer is in a fluid state during normal life temperatures. At lower temperatures and lower thermodynamic free energy, the lipids undergo phase transition into a gel phase or into other three-dimensional structures with lower free energy. During the process, membrane proteins become segregated and defects form between the proteins and the membrane bilayer. The lipid phase transition process makes the cell membrane more permeable and allows usually ions to enter the cell in an uncontrolled way. Detailed reviews of the effect of temperature on the cell membrane can be found in several publications.[3–5]

Normally the membrane proteins control the intracellular composition by selectively introducing and removing ionic species from the cell interior. The cells are bathed in an extracellular solution high in Na+ and low in K+. This ratio is maintained by the Na K ATPase pump which uses much of its energy from oxydative phosphorilation. The sodium pump effectively makes sodium an impermeant outside the cell that counteracts the colloid osmotic pressure derived from the intracellular proteins and other impermeable anions.[6] Hypothermic preservation suppresses the activity of the Na+ pump and decreases the membrane potential of the plasma membrane. Consequently, Na+ and Cl– enter the cell down concentration gradients and the cell swells because of water accumulation. The damage is cumulative, a function of time, and is particularly expressed when the cells are returned to their normal physiological temperature.

Additional mechanisms of damage relate to the cytoskeleton. The cytoskeleton structure depends on chemical bonds between membrane proteins and the cell scaffold. Lowering the temperature weakens these bonds and makes them particularly vulnerable to mechanical damage. A third mechanism of damage relates to the denaturation of proteins as a function of both temperature and change in the intracellular ionic content. Most cells and tissues can withstand brief cooling to above freezing temperatures. Major exceptions are cells that are highly sensitive to their ionic content, such as platelets. Cooling platelets to temperatures lower than their lipid phase transition temperature allows calcium influx, which appears to trigger platelet activation. This could lead to a cascade of events that would end in platelet aggregation and the eventual obstruction of blood vessels in the cooled region around the frozen lesions. Other cells whose function is strongly dependent on their ionic content are muscle cells, in particular in the heart and around arteries, which makes their preservation at hypothermic temperatures more difficult.

Efforts to try and prolong ischemic times by hypothermia focus on the development of preserving solutions that are perfused through the organs so as to protect them from the modes of damage discussed above. Several preservation solutions were developed over the years to help minimize the hypothermic and ischemic damage caused by cold storage.[1,7] Most of the more recent preserving solutions are based on intracellular compositions, rich in K+. The basic principles applied to the development of the cold storage solutions are the prevention of hypothermic cellular swelling and the expansion of the extracellular space by using solutions containing macromolecular impermeants, prevention of intracellular acidosis by introducing hydrogen ion buffers, prevention of injury by oxygen free radicals by including radical scavangers such as mannitol and gluthatione in the perfusate, and the supplement of high energy substrates for energy regeneration after reperfusion.[1]

A recent area of major interest to research on hypothermic preservation is the study of survival of cold tolerant organisms in nature and of hibernators. The study of cold tolerant organisms has led to the findings that two compounds found in nature, trehalose and antifreeze proteins could provide hypothermic protection to mammalian tissues.[8,9] It is anticipated that major advances in the field of hypothermic preservation will arise from further studies in the survival mechanisms of cold tolerant organisms and hibernators.

CRYOPRESERVATION

Cryopreservation involves a complex process in which the biological material is first cooled to freezing temperatures. Then it is frozen to the preservation temperature, stored at the preservation temperature for extended periods of time and at the end thawed and warmed to body temperature. To facilitate preservation, various cryoprotective chemicals are added and than removed from the preserved tissue. While the complexity of the cryopreservation processes is great and each step in the cryopreservation procedure can induce damage to the preserved biological entity, the damage can be ascertained only at the end of the entire freeze-thaw cycle. Because the number of various cryopreservation parameters and the possibility of their deleterious combinations is staggering, it is very difficult to establish optimal cryopreservation protocols, from end point estimates only. To facilitate the design of optimal cryopreservation protocols, it is important to understand the effects of each step of the protocol. To this end, I have separated the protocol in individual steps and I will discuss each step separately, for cells and organs.

1) Cryopreservation of Cells

a) Damage to cells during freezing

Most of the studies on cryopreservation were done on cells, frozen *in vitro*. Cells are usually cooled rapidly to freezing temperatures and in most cases, are not affected by hypothermic damage because of the short duration of this part of the cryopreservation protocol. Once freezing temperatures are reached, the cells are cooled to cryogenic temperatures, often in one to two steps. These steps involve cooling with one or several sequential cooling rate, i.e. change in temperature per unit time. In the field of cryopreservation, it is traditional to think that the thermal parameter that affects the survival of frozen cells is the cooling rate during freezing.[10] The relation between survival and cooling rates is traditionally depicted as an inverse U-shaped curve with optimal survival at a certain rate and decrease in viability at cooling rates above and below the optimal rate. The traditional explanation for the "inverse U-shaped" survival curve was proposed by Mazur,[10] and will be discussed next.

Experiments on cells are done primarily with cryomicroscopy, i.e. microscopes that allow controlled freezing of cells under visual observation. Diller has described cryomicroscopy in detail in several reviews.[11] The results used to explain the mechanisms of damage during freezing in this article come from experiments performed with the technique of directional solidification cryomicroscopy.[12] Figure 1 illustrates the schematic of a directional solidification apparatus and describes its principle. Figure 2 was obtained with the directional solidification stage and shows the sequence of events during freezing of red blood cells in suspension.[13] The temperature distribution in these figures is one-dimensional linear, from below freezing at the left to above freezing at the right. The change of phase interface is the vertical line separating the frozen region on the right from the unfrozen on the left. The special aspect of the process of freezing in biological materials is a consequence of the fact that water in biological materials is in the form of a solution. The process of freezing in aqueous solutions is affected by the fact that ice has a tight crystallographic structure and cannot contain any solutes. Therefore, when an aqueous solution freezes, the solutes are accumulated in front of the change of phase interface. While this phenomenon occurs on the molecular scale and is not visible, Fig. 2(a) illustrates a similar macroscopic situation. The figure shows that at the beginning of the freezing process, cells accumulate on the change of phase interface, which has the appearance of a vertical line. This is similar to what happens to solutes at the molecular scale. The increased solute concentration on the change of phase interface has the effect of lowering the temperature of the change of phase interface. Because thermal diffusion is much faster than mass diffusion, the increased concentration and related change in phase transformation temperature leads to a phenomenon known as "constitutional supercooling" and the so-called Mullins–Sekerka interface instability. This phenomenon is discussed in detail in many

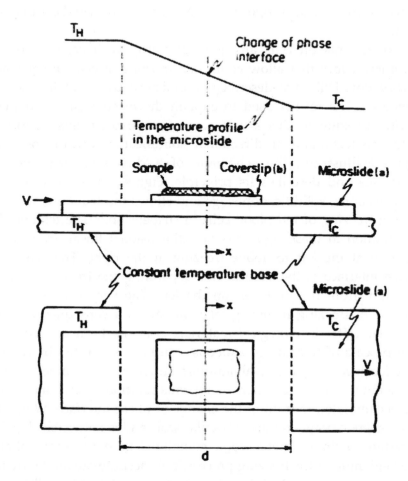

Fig. 1. Scheme of the directional solidification stage. The directional stage is attached to a microscope and consists of two bases each maintained at a different constant temperature. The one on the left is at a temperature above freezing and the one on the right at a temperature below freezing. The bases are separated by a gap and support a substrate (microslide) that has a linear temperature distribution. The change of phase interface is in the middle of the substrate under the focal point of a microscope. When a sample is set on the microslide and when the microslide is moved at a constant velocity across the bases, the sample will take the temperature of the microslide and freeze under controlled conditions under the microscope.

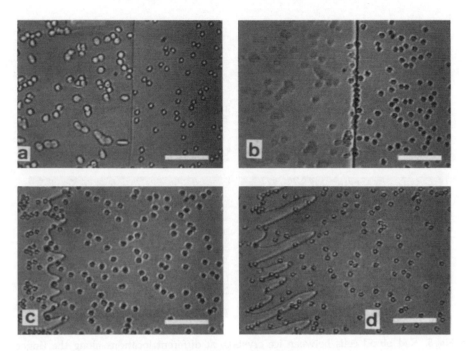

Fig. 2. A sequence of photographs showing the freezing of red blood cells (round circles) in a suspension. The freezing interface, vertical curve in the middle of the figure propagates from left to right. The frozen region is to the right of the figure and the unfrozen to the left.

material science texts.[14] It causes the planar freezing interface to become unstable and take the finger-like shapes shown in Figs. 2(b), 2(c) and 2(d). In this configuration, the concentration of the solution at the tip of the finger-like ice crystal structure is very close to the bulk solution concentration and the rejected solutes become accumulated between the finger-like ice crystal structures. Figures 2(c) and 2(d) show that the cells in the freezing solution are unfrozen and find themselves in the hypertonic, high concentration solute channels, between the ice crystals. Although referred to as freezing of tissue or cells, during many freezing processes the freezing in fact begins in the extracellular milieu and the interior of the cell is

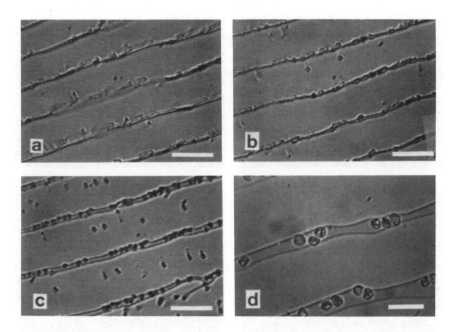

Fig. 3. Red blood cells between ice crystals, at different locations along the finger like ice crystal structures. The temperatures from which the images were taken increase from (a) to (d). It is possible to observe that at lower temperatures, and higher extracellular concentrations the cells have shrunk more.

unfrozen. The exposure of cells to hypertonic solutions, in the channels between ice crystals is a major cause of cell damage during freezing.

Figure 3 illustrates a sequence of events that cells experience in the hypertonic solutions, between ice crystals.[13] The figure shows that at lower temperatures, as the extracellular concentration increases, cells shrink. This shrinkage is caused by the fact that the unfrozen cells are supercooled relative to the extracellular solution, which is in thermodynamic equilibrium with the ice. To equilibrate the difference in chemical potential between the extracellular and intracellular solutions, water will leave the cell through the cell membrane that is readily permeable to water. This causes an increase in the intracellular solute concentration, with a decrease in temperature. It was originally proposed by Lovelock,[15] and later incorporated in his comprehensive theory by Mazur,[10] that increased hypertonic extracellular solutions damage the cells. The mechanisms are not entirely clear and they could relate to denaturation of proteins when the intracellular solution becomes hyperosmotic, to the osmolality-induced changes in the cell structure, or both.

The exposure of cells to hyperosmotic concentrations during freezing explains the first part of the inverse U shaped survival curve. Because water transfer across the cell membrane is a rate-dependent process and because chemical damage is time-dependent, it should be expected that exposing cells more rapidly to cryogenic temperatures will reduce these modes of damage. Therefore, at first, increasing the cooling rates will improve the survival of cells preserved at cryogenic temperatures, as shown by the inverse U shaped survival curve.

However, the inverse U shaped survival curve shows that as the cooling rates are further increased, above a certain optimum, the survival of frozen cells begins to decrease with an increase in cooling rates. Experiments have shown that this sudden increase in cell destruction corresponds to sudden formation of intracellular ice. Formation of intracellular ice has been proposed to be responsible for the decrease in cell survival at above optimal cooling.[10] The condition under which intracellular ice forms was investigated in several studies, starting with the work of Diller.[16] It is thought that intracellular ice forms because the water transport through the cell membrane is a rate dependent process. When cells are cooled too rapidly to equilibrate in concentration with the extracellular solution, the intracellular solution becomes thermodynamically supercooled and unstable. The probability for intracellular ice formation increases with supercooling. It is not clear if the nucleation sites for intracellular ice formation are intracellular, extracellular or on the membrane.[17] However, whatever the cause of intracellular ice may be, it appears that it is almost always lethal to the cell. It is unclear if the intracellular ice per se is lethal or if the processes that led to the formation of the intracellular ice, such as damage to the cell membrane, are lethal.

There are additional mechanism of damage in the region of temperatures and cooling rates associated with hypertonic solution damage. These modes of damage were originally observed by Nei[18] and later by Mazur[19] in his so-called "unfrozen fraction hypothesis". Experiments have shown that the percentage of death cells after freezing is larger than the percentage of death cells after exposure to a similar extracellular hypertonic solution. This suggests that mechanical interaction between ice and cells may contribute to cell death. This is a reasonable assumption, since ice rejects cells in the space between ice crystals, as shown in Figs. 2 and 3. This may generate a mechanical force on the cells, whose cellular cytoskeleton is weakened by cold, and destroy them.[20] Another possible mode of damage is the contact and interaction between ice and the lipid bilayer, which by itself may be damaging.

b) Damage to cells during storage, thawing and warming

Storage, thawing and warming has been studied much less than freezing. They are all interrelated phenomena. While at a temperature of absolute zero, processes cease. At any temperature above absolute zero, they take place. The higher the temperature the faster the rate of these processes. Therefore storage, thawing and warming can cause damage, which is a function of temperature. While at cryogenic temperatures, the storage damage is negligible in the time scale of years, during thawing and warming, as the biological material experiences elevated temperatures, the rate of chemical reactions will accelerate and damage will occur. In addition, during warming, in a frozen state, ice has a tendency to recrystalize at high subzero temperatures, to minimize the Gibbs free energy.[21] Recrystallization will cause further disruption of the extracellular space and may disrupt the macroscopic structure of the tissue.[22] Furthermore, during thawing, as ice melts, the extracellular solution can be briefly and locally hypotonic causing water to enter some cells and expand them, thereby rupturing the membrane.[23] It is usually thought that slow rates of warming are more detrimental to cryopreserved cells than higher rates.

c) Damage to cell by apoptosis

While most of the studies on the process of cell death during freezing have employed viability tests that evaluated survival of cells immediately after freezing and thawing, it appears that some cooling and freezing conditions may produce less lethal modes of damage, which eventually result in gene regulated cell death (apoptosis). Apoptosis can be triggered by a variety of conditions present during cryopreservation, such as hyperosmolality. Apoptosis will take place after the end of cryopreservation and can produce further cell death.[24]

d) Cryopreservation of cells

Luyet was the first to report survival of frozen cells in the presence of cryoprotective chemicals. In 1938, he reported cryogenic preservation of frog sperm frozen, dehydrated in sucrose and in 1941, he reported the survival of frozen vinegar eels when first immersed in an ethylene glycol solution.[25] However, the great breakthrough in the field was made in 1949 when Polge, Smith and Parkes discovered that a 10% solution of glycerol could preserve cattle sperm at $-79°C$, and that this protective effect can be applied to other cells.[26] Subsequently additional cryoprotective chemicals were found, such as dymethyl sulfoxide by Lovelock.[27] Currently many polyalcohols are being used as cryoprotectants with different effectiveness and for various cells. Most of the cryoprotectants, with a few exceptions, are chemicals that penetrate the cell membrane and replace part of the water in the cell. Their effect is primarily to

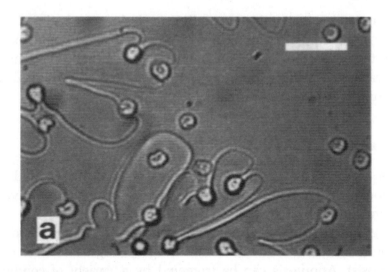

Fig. 4. Red blood cells frozen in 10% glycerol. Note the significant change in ice crystal pattern in comparison to freezing in physiological saline. The cells are also less dehydrated. Scale bar 100 micron.

lower the freezing temperature and they afford protection against hyperosmotic chemical damage during freezing with low cooling rate damage, i.e. the ascending part of the inverse U-shaped survival curve. The presence of these cryoprotectants inside the cells decreases the concentration of salts and the change in cell volume at each sub-freezing temperature and thereby reduces chemical damage. These chemicals also modify the structure of ice crystals during freezing and may inhibit the other modes of damage to cells. A typical image of red blood cells frozen in the presence of glycerol is shown in Fig. 4. It should be emphasized that while cryoprotectants protect against hyperosmotic freezing damage, they are themselves toxic. Therefore in designing cryopreservation protocols, an optimal cryoprotectants concentration must be found that maximizes the cryoprotective effect of the additive versus the chemical damage it induces.

2) Cryopreservation of Tissue

a) Damage to tissue by freezing

In tissue, cells are in a different configuration from cells in suspension. They are in an organized structure, which is important for proper function. The volume of the extracellular space is usually smaller than that around cells in a suspension. It is natural to question whether the process of freezing and the mechanism of damage experienced by cells frozen in a suspension are similar to those in tissue. While experiments with tissue are more difficult than with cells, the few experimental results that exist show that the process of freezing cells in tissue and in suspension is roughly similar. Experiments in which different types of tissue were frozen with controlled thermal conditions and then viewed by electron microscopy or freeze-substitution show that in tissue, ice also usually forms first in the extracellular space. Ice appears to form usually in the vasculature and propagates in the general direction of temperature gradients, but in and along blood vessels.[28] The pattern of freezing seems to be also tissue specific. In addition to freezing along blood vessels, it was found that in the prostate, ice forms in the ducts,[29] the breast in the connective tissue[30] and the kidney in the ducts.[31] The cells in the various tissues appear to experience cellular dehydration and intracellular ice formation in a similar manner to cells frozen *in vitro*, in cellular suspensions. Figure 5 illustrates a typical appearance of slowly frozen liver tissue as seen with electron microscopy. The figure shows dehydrated hepatocytes, surrounding expanded sinusoids.

Mathematical models were developed for studying the process of freezing in tissue.[32,33] An analysis was performed which compared the process of freezing of hepatocytes in the liver and in a cellular suspension.[32] The results demonstrate that in both cases, hepatocytes experience a similar dehydration process and a similar probability for intracellular ice formation. Therefore, cells in tissue will probably experience both qualitatively and quantitatively similar mechanisms

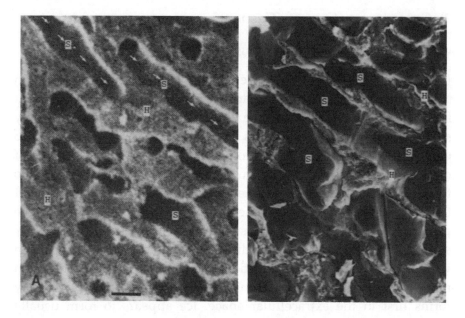

Fig. 5. A comparison of longitudinal cross-sections of sinusoids (S) between live rat liver (left) and rat liver tissue frozen with a low cooling rate of 4°C/min (right). Single ice crystals are evident in the sinusoids of the frozen liver. The sinusoids in the frozen liver are expanded relative to those in the normal liver. The hepatocytes (H) in the frozen liver have dehydrated (Scale bar 10 microns).

of hypertonic solution damage and intracellular ice formation damage like cells frozen in cellular suspensions. However, the analysis suggests that in tissue there will be additional modes of damage. The dehydration of cells and the expansion of the blood vessels will most likely result in a disruption of the vasculature and of the connective tissue. The consequence of this mode of damage to cryopreservation is important. Obviously if the vasculature is destroyed, the cryopreserved organ will not be functional even if the cells of the organ survive cryopreservation. Support for this mechanism of tissue damage comes from work on cryosurgery, freezing of undesirable tissue for destruction. Experiments show that when tissue is frozen *in vivo*, immediately after thawing there is edema on the outer margin of the previously frozen lesion.[34] Shortly thereafter, the endothelial cells in the previously frozen region appear damaged, probably by the mechanism of blood vessel expansion during freezing, discussed earlier. Within a period of several hours after thawing, the endothelial cells become detached, with increased permeability of the capillary wall, platelets aggregation and blood flow stagnation.[35] Many small blood vessels are completely occluded within a few hours after cryosurgery. This mechanism of damage occurs during cryopreservation of organs also, and will be discussed later.

b) Damage to tissue by thawing

Experiments with cells have shown that rapid heating is beneficial to cryopreserved cells. Slow warming is detrimental. However, both

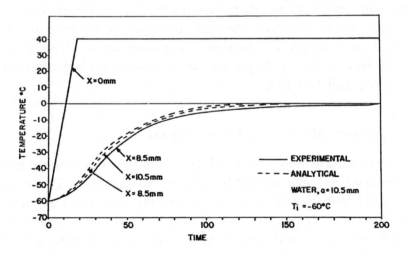

Fig. 6. Temperature as a function of time at different locations inside a frozen sample, thawed by conduction heat transfer with the temperature boundary condition at x = 0 mm. Note that inside the sample, the temperature approaches the phase transition temperature rapidly and stays at this value throughout the thawing process.

analytical and experimental studies have shown that in tissue the pattern of warming is peculiar, different from that of cells and conducive to damage.[36] Because of the different thermal conductivities of frozen and unfrozen tissue, it has been found that during thawing by conduction heating, the temperature of the frozen tissue will reach the phase transition temperature rapidly and stay at that temperature, in a frozen state, for the entire period the organ thaws. Figure 6 illustrates this. These are conditions in which maximal damage will occur. Therefore, alternative methods of heating than conduction may be needed for organs, such as microwave or ultrasound.

c) Additional mechanisms of damage to tissue

Freezing and thawing can also induce another mode of macroscopic damage, related to the thermal stresses. Temperature variations in frozen tissue can cause thermal stresses. These stresses can cause fractures. They have been shown to occur experimentally and were studied analytically.[37,38] During cryopreservation, these fracture should be avoided as they could lead to uncontrollable damage to the tissue and bleeding.

d) Cryopreservation of tissue and organs

Our studies have shown that cells frozen in tissue experience similar phenomena as cells frozen in cellular suspensions. Therefore, it should be anticipated that the parameters that affect survival of cells frozen in suspension should also similarly affect the survival of cells in tissues and organs. Obviously when designing a cryo-preservation protocol for organs and tissues, one should consider the special constraints due to their geometry and the morphology, with respect to the cooling rates during freezing, heating rates during thawing and the introduction of cryoprotectants. Because organs are much larger in volume than cells, the cooling rates that

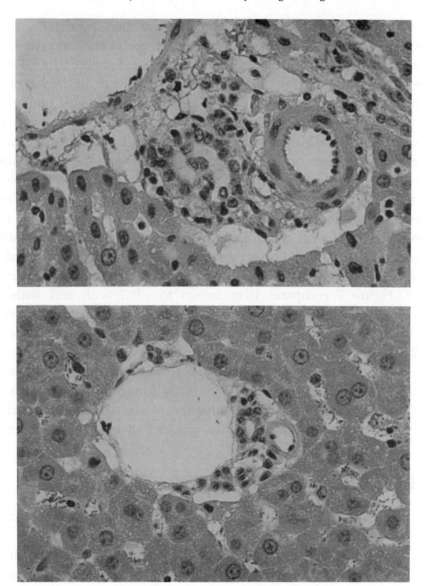

Fig. 7. Livers frozen to −3°C. Top without cryoprotectants, bottom with glycerol protection. Note the intact hepatocytes in the protected liver and the damaged hepatocytes in the top figure.

can be imposed on an organ are much lower than those that can be imposed on cells. The limitations in the range of cooling rates that can be imposed on organs have been determined analytically.[39] From geometric limitation, it is obvious that organs can be frozen only with cooling rates that are low. The geometry-constrained limitation in heating rates has been discussed earlier. The introduction of cryoprotectants is also limited by the fact that in organs and tissue, the cryoprotectants can be introduced only through the vasculature, by perfusion. This mode of cryoprotectants introduction has been analyzed and models were developed.[40] On the basis of these studies, we have developed preliminary cryopreservation protocols for the liver, preserved with glycerol and frozen to relatively mild high sub-freezing conditions.[41] The cryopreserved livers were transplanted.[42] The hepatocytes in all the cryopreserved livers survived cryopreservation. Figure 7 compares liver frozen to −3°C with and without glycerol protection. The figure illustrates our finding that hepatocytes in the liver appear to have survived cryopreservation and produced bile. However, the transplantation study showed that the liver ceased to function after transplantation due to failure of the vasculature, caused by endothelial detachment. Our future work in this field will focus on protecting the integrity of the vasculature during cryopreservation, an area that we consider of major importance.

REFERENCES

1. Belzer FO, Southard JH. Principles of solid organ preservation by cold storage. *Transplantation* 1988; **45**:673–676.
2. Drobins EZ, *et al.* Cold shock damage is due to lipid phase transition in cell membranes, a demonstration using sperm as a model. *J Experimental Zoology* 1993; **265**:432–437.
3. McGrath JJ. Membrane transport properties. In *Low Temperature Biotechnology: Emerging Applications and Engineering Contributions*, eds. JJ McGrath, KR Diller, ASME, New York, 1988, pp. 273–331.

4. Quinn PJ. A lipid phase separation model of low temperature damage to biological membranes. *Cryobiology* 1985; **22**:128–147.
5. Wolfe J, Bryant G. Drying and/or vitrification of membrane-solute-water systems. *Cryobiology* 1999; **39**:103–130.
6. MacNight ADC, Leaf A. Regulation of cellular volume. *Physiol Rev* 1977; **57**:510–562.
7. Muhlbacher F, Langer F, Mittermayer C. Preservation solutions for transplantation. *Transplantation Proc* 1999; **5**:2069–2070.
8. Singer MA, Lindquist S. Thermotolerance in Saccharomyces cerevisiae: The yin and yang of trehalose. *Trends in Biotechnology* 1998; **16**:460–468.
9. Rubinsky B, Arav A, Fletcher GL. Hypothermic protection — A fundamental property of antifreeze proteins. *Biochem, Biophys Res Comm* 1991; **180**:566–571.
10. Mazur P. Cryobiology: The freezing of biological systems. *Science* 1970; **68**:939–949.
11. Diller KR. Cryomicroscopy. In *Low Temperature Biotechnology: Emerging Applications and Engineering Contributions*, eds. JJ McGrath, KR Diller, ASME, New York, 1988, pp. 347–363.
12. Rubinsky B, Ikeda M. A Cryomicroscope Using Directional Solidification for the Controlled Freezing of Biological Material. *Cryobiology* 1985; **22**:56–68.
13. Ishiguro H, Rubinsky B. Mechanical interactions between ice crystals and red blood cells during directional solidification. *Cryobiology* 1994; **31**:483–500.
14. Kurtz W, Fisher DJ. *Fundamentals of Solidification*. Trans Tech SA, Switzerland. 1984, 242.
15. Lovelock JE. The haemolysis of human red blood cells by freezing and thawing. *Biochem Biophys Acta* 1953; **10**:414–426.
16. Diller KR, Cravalho EG. A cryomicroscope for the study of freezing and thawing process in biological cells. *Cryobiology* 1970; **7**:191–199.
17. Toner M, Cravalho EG, Karel M. Thermodynamics and kinetics of intracellular ice formation during freezing of biological cells. *J Appl Phys* 1990; **69**:1582–1593.

18. Nei T. Mechanism of hemolysis by freezing at near zero temperatures. II Investigation of factors affecting hemolysis by freezing. *Cryobiology* 1967; **4**:303–308.

19. Mazur P, Rall WF, Rogopoulos N. Relative contributions of the fraction of unfrozen water and of salt concentration to the survival of slowly frozen human erythrocytes. *Biophysical J* 1981; **36**:653–665.

20. Takamatsu H, Rubinsky B. Viability of deformed cells. *Cryobiology* 1999; **39**:243–251.

21. Whitaker D. Electron microscopy of the ice crystals formed during cryosurgery: Relationship to duration of freeze. *Cryobiology* 1978; **15**:603–607.

22. Merryman HT. *Cryobiology*. Academic Press, New York. 1966, 966.

23. Gage AA, Baust J. Mechanisms of tissue injury in cryosurgery. *Cryobiology* 1998; **37**:171–186.

24. Hollister WR, *et al.* The effects of freezing on cell viability and mechanisms of cell death in an *in vitro* human prostate cancer cell lne. *Mol Urol* 1998; **2**:13–18.

25. Prehoda RW. *Suspended Animation*. Philadelphia, New York, London. Chilton Book Company, 1969, 211.

26. Polge C, Smith AV, Parkes A. Revival of spermatozoa after vitrification and dehydration at low temperature. *Nature* 1949; **164**:666.

27. Lovelock JE, Bishop MWH. Prevention of freezing damage to living cells by dimethyl sulfoxide. *Nature* 1959; **183**:1394–1395.

28. Rubinsky B, *et al.* The process of freezing and the mechanism of damage during hepatic cryosurgery. *Cryobiology* 1990; **27**:85–97.

29. Onik G, *et al. Percutaneous Prostate Cryoablation*. Quality Medical Publishing, Inc., St. Louis, MO. 1994, 172.

30. Hong JS, Rubinsky B. Patterns of ice formation in normal and malignant breast tissue. *Cryobiology* 1994; **31**:109–120.

31. Bischof J, *et al.* Effects of cooling rate and glycerol concentration on the structure of the frozen kidney: Assessment by cryoscanning electron microscopy. *Cryobiology* 1990; **27**:301–310.

32. Rubinsky B, Pegg DE. A mathematical model for the freezing process in biological tissue. *Proc. of the Royal Society* 1988; **234**:343–358.

33. Rubinsky B. The energy equation for freezing of biological tissue. *ASME Trans J Heat Transfer* 1989; **111**:988–996.

34. Gilbert JC, *et al.* MRI-monitored cryosurgery in the rabbit brain. *Magnetic Resonance Imaging* 1993; **11**:1155–1164.

35. Whittaker DK. Mechanisms of tissue destruction following cryosurgery. *Ann Rev Cool Surg England* 1984; **66**:313–318.

36. Rubinsky B, Cravalho EG. Analysis for the temperature distribution during the thawing of a frozen biological organ. *A.I.Ch.E. Symposium Series* 1979; **75**:81–88.

37. Rubinsky B, Cravalho EG, Mikic B. Thermal Stresses in Frozen Organs. 1980; **17**:66–74.

38. Najimi S, Rubinsky B. Non-invasive detection of thermal stress fractures in frozen biological materials. *Cryo-letters* 1997; **18**: 209–216.

39. Rubinsky B, Cravalho EG. An Analytical Method to Evaluate Cooling Rates During Cryopreservation Protocols for Organs. *Cryobiology* 1984; **21**:303–320.

40. Lee CYC, Rubinsky B. A multi-dimensional model of momentum and mass transfer in the liver. *Int J Heat and Mass Transfer* 1989; **32**:2421–2434.

41. Ishine N, Rubinsky B, Lee CYC. A histological analysis of liver injury in freezing storage. *Cryobiology* 1999; **39**:271–277.

42. Ishine N, Rubinsky B, Lee CYC. Transplantation of mammalian livers following freezing: Vascular damage and functional recovery. *Cryobiology* 2000; **40**:84–89.

Chapter 3

Tissue Engineering: Clinical Applications and Mechanical Control

Robert A Brown

INTRODUCTION

Tissue Engineering offers a compelling new approach to major clinical problems. This review aims to clarify its definitions and likely practical uses by examining rational surgical design processes and the inevitable control engineering. It is proposed that the most important developments in present tissue engineering will be towards the supply of small, relatively simple connective tissue parts and for this, the provision of cell-control cues will be a major limitation. Cytomechanical cues, applied through the substrate materials, are key to normal tissue organisation and probably also the disorganisation of poorly functional scars. It is concluded that input from clinical tissue repair and reconstructive surgery is essential in the design of engineering solutions to problems of tissue organisation, dynamic function and integration after implantation.

Without this, the next engineered tissues may be a little better than implantable scar-tissues.

There is a revolution in the wings of modern medicine, the shape and size of which we are only slowly coming to understand. It is called tissue engineering. It can be defined as the application of enginee-ring principles to new biology, for the purposes of constructing body tissues. Variation in interpretation remains, for example in the importance of using the patient's own cells or donor cells, the extent to which tissue engineered (TE) constructs perform mature functions straightaway. It is sometimes suggested that TE should be seen as part of a larger field of "reconstructive and regenerative medicine", including cell therapies, cyto-monitoring and gene therapies.

In fact, this sense of rapid flux is consistent with the idea that we are presently passing through one of those rare periods in the evolution of a scientific discipline, when preconceived boundaries and definitions are suspended. This is said to persist whilst conventional science becomes readjusted to the new landscape, made possible by a significant change in thinking. This was described by Kuhn[1] as the stage of "scientific revolution" which interrupts evolutionary progression. In key areas, tissue engineering is approaching a new stage where a fresh view of its progress and direction is needed. We are now at such a point, where diverse early attempts of seeding cells into scaffolds have been extensively tested, and it is possible to design new strategies. Here, we examine this progress and the future need for cell-control CUES: in particular cytomechanical cues for connective tissue engineering.

Engineering is control. Controls are applied through understanding and application of physical and chemical processes to achieve a required endpoint. This may be the manufacture of a "device", such as an engineered tissue structure, or induction of changes in structure and function by incorporation of new control cues to achieve simple *in vivo* tissue engineering. Present examples of the latter are surgical tissue expanders or conduits to guide cell

(a)

Fig. 1. (a) An idealistic, but currently unpromising strategy for tissue engineering is to mimic as closely as possible, the natural process of tissue/organ embryonic development. Difficulties include insufficiently detailed understanding of the cell/molecular control processes and inability to replicate the "tissue bioreactor" involved. Much non-biological work will be needed to engineer this process. **(b)** Perhaps the simplest and most familiar bioreactor, incorporating spatial and behavioural signalling, is the culture of skin epithelium in culture flasks.[17] Layered sheets of keratinocytes assume a normal polarity from the inherent basal and surface cues of the dish (air–liquid interface and feeder layer substrate). This system is sufficient for epithelial type cells with strong cell–cell interactions (sheet forming) and little extracellular matrix. **(c)** Recent forms of "tissue bioreactor" have relied on fluid mixing to improve perfusion of 3D constructs. Static culture is insufficient to supply nutrients to deep cells (diffusion path > approx. 0.4 mm, depending on cell type). Mixing of culture medium can use adapted spinner flasks (left hand diagram), though problems arise from complex fluid shear forces at the surface. Perfusion with minimal fluid shear is a feature of modern, concentric rotating cylinder bioreactors, for example developed by Synthecon[TM] (right hand diagram). These systems are not designed to provide controlled mechanical cues (indeed external loading is minimised). **(d)** For application of mechanical cues to 3D tissue constructs, a range of devices have been used,[48–51,71,72,76,77] mechanical cues being primarily to stimulate cell responses. The force direction, or vector, is critical. It is shown here as a uniaxial load (double arrow) into a high aspect ratio construct, using a customised bioreactor chamber (Brown, Mudera, Blunn, in preparation). **(e)** Some forms of "bioreactor" operate largely in the body, for example, tissue engineered conduits for peripheral nerve repair to bridge gaps after injury. Because of the nature of nerve regeneration, these devices are inevitably designed to control the neural tissue regrowth *in vivo*, rather than to supply finished tissue. This diagrammatic example shows an implanted conduit using substrate guidance (with growth factor impregnation) to provide spatial cues for regeneration[59,61] (thanks to Dr. Györgyi Talas for the photograph).

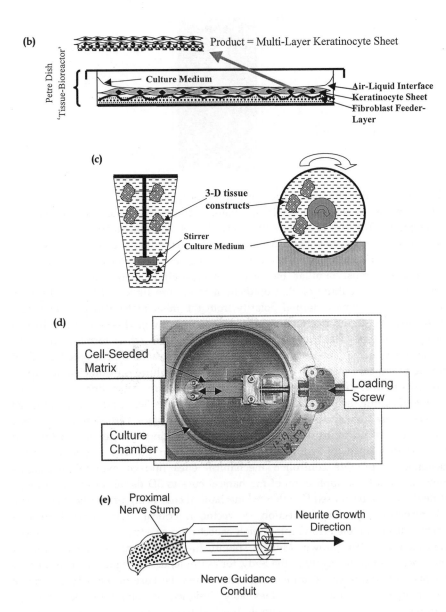

Fig. 1 (*Continued*)

growth (e.g. in nerve regeneration). Throughout, the fundamental principle of "engineering is control" remains paramount, but in order to achieve this, it is important that the normal mechanisms involved *at the cell level* can be defined with precision. In particular, these are likely to involve both biochemical and mechanical cues. Mechanics at the cell level (cytomechanics) also involve the extra cellular matrix and, even in primarily mechanical connective and contractile tissues, consideration of its role in cell cuing is surprisingly rare and poorly understood. In most cases, practical application of cellular control cues involves some form of "tissue bioreactor" (examples shown in Fig. 1) which can range from simple 2 dimensional cell culture flasks through complex engineered vessels for prolonged 3D culture to forms of *in vivo* development where tissue constructs develop and mature *in situ* prior to use. Among current trends in this technology, Fig. 1 illustrates two traits which may not incorporate well into current TE strategies; (a) reliance *primarily* on recapitulation of foetal tissue growth and (b) approaches which do not take into account of scarring and contracture; the default processes in normal post-foetal tissue repair (Fig. 2).

Two contrasting philosophies to tissue engineering have been applied.[2] One assumes a great potential for *biological regeneration* within living cells. The second is more sceptical on this point and assumes that a considerable level of external biological control and mimicry will be needed for making higher order structures. These parallel our concepts of tissue regeneration (in foetal and amphibian injuries) versus repair, leading to scarring, seen in post natal mammals. The first hopes that by putting suitable cell types into an appropriate support matrix, an organised and functional tissue will be produced through inherent cellular repertoires.[3,4] This is the simplest and where appropriate, most economic approach exemplified by seeding of a resorbable polymer with chondrocytes for implantation into cartilage defects.[5-7] The second view holds that resident cells need rate-limiting, spatial and organisational cues

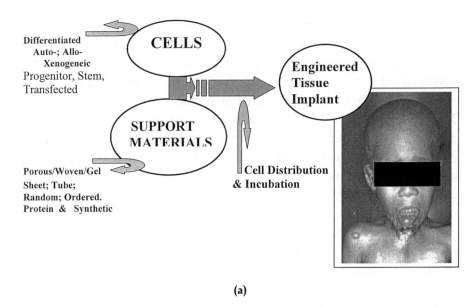

(a)

Fig. 2. Summary diagram of strategies for tissue and cell engineering. **(a)** The simplest approach is to select a suitable cell source, isolate and expand the culture to provide cells for seeding. In most cases, a suitable, resorbable material is developed as a 3-dimensional implantable support, having appropriate porosity, shape, micro-architecture (resorption of most protein-based materials is *cell-dependent* whilst synthetic polymers are largely degraded through *cell-independent* hydrolysis). Cell-seeded constructs can then be cultured to encourage cell dispersal through the material and matrix elaboration (note culture expansion (proliferation stage) is ideally achieved prior to seeding). In the case of connective tissues, this process most closely resemble adult wound repair, which is characterised by contracture and scarring (inset: Thanks to Dr. Jaysheela Mudera the photograph showing classic scar contracture deformity in a mouth-to-shoulder burn scar). **(b)** Further engineering control begins to be needed as progressively more complex, larger or biologically sophisticated implants are envisaged. These can be divided into essentially repair tissues (or implantable templates — see Fig. 1(e)) and mature tissue engineered implants, needing short- and long-term culture, respectively. In each, the control processes can be predicted and designed in detail with biological cues being incorporated to regulate long-term events at the host site in primary (repair) implants. Secondary, mature tissue implants require the development of more elaborate tissue bioreactors to control the growth and perfusion of large, dense tissues with appropriate spatial and differentiation cues.

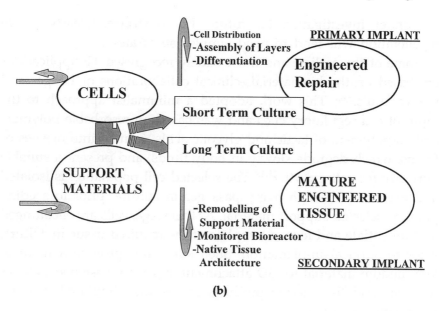

Fig. 2 (*Continued*)

in order to elaborate functional tissue architectures,[8] and inevitably need more complex, engineered solutions.

Whatever approach is adopted, there are two central biomedical questions. The first is availability of suitable, rapidly dividing cells to populate the new tissue. This has led to a search for active cells and, most recently, to the use of "stem", or at least poorly committed progenitor cells which can be directed into a suitable phenotype. Embryonic stem cells pose substantial practical and ethical problems at present and marrow or adult tissue progenitor cells may represent a more promising avenue.[9,10] To an extent, the search for more potent cell types to enhance regeneration is a progression of the "cell-centric" or regenerative idea of tissue engineering. The second key area is that of support materials. These have so far been divided into synthetic biodegradable polymers and modified natural polymers (chiefly proteins) though increasingly, smart substrates are under investigation for future TE constructs. Clearly, tissue regeneration will tend to need simpler substrates.

Some of the earliest and most widely recognised TE applications are based on the biomaterials-clinical collaborations of Langer and the Vacanti's.[3,4,7] This work adopted a minimalist approach to the input of cell regulatory cues and was based on resorbable polymers and co-polymers of lactic and glycolic acids, formed into meshes or foams, which degrade slowly in body fluids, and present a suitable pore size for cell growth.[11,12] The selected cell population is isolated and expanded through many passages in culture, prior to seeding onto the selected support material. At this stage, cells are encouraged to differentiate and form a template of the required tissue in culture, prior to surgical implantation. Such substrates provide a more or less random material for 3D attachment and were designed to leave the cells and their tissue products as the final implant.[11] Control information consists of

(i) The gross 3D shape into which the polymer is formed (i.e. as an ear, heart valve, etc.).
(ii) Its rate of disappearance.
(iii) Type/density of cells grown in the material.
(iv) Implantation site.

Example 1. Engineered articular cartilage tissue is an active area of research. In this, chondrocytes (frequently autologous) are isolated from biopsy sites and cultured to expand the cell population. These are then introduced back into a surgically prepared articular cartilage defect, either as a cell suspension (i.e. with no material support at all)[4] or after seeding into a polymer support mesh.[6,7] The clinical outcome of these techniques are currently being assessed. Major clinical issues here are the difficulties of establishing medium term clinical effectiveness without the use of invasive analysis and the difficulties in extending from young trauma patient to older osteoarthritis cases.

Where more *cellular control* is important, identification of appropriate cues (spatial, rate, etc.) is essential. Clearly, greater control adds to complexity, cost and difficulty making it important to determine minimal levels of effective control, across four major areas:

i) Cells used.
ii) Support material (surface chemistry, mechanical properties and spatial structure).
iii) Physical and (bio)chemical cues supplied and how they develop chronologically.
iv) Method of assembly of the implant/device.

Example 2. (New, clinically available forms of artificial skin; Table 1) illustrates this increasing level of control. Perhaps the most complex Apligraf™ (for burns and chronic wounds), consisting of a layer of neo-dermis (allogeneic dermal fibroblasts contracted into a native collagen gel) covered by a differentiating sheet of allogeneic keratinocyte.[13,14] These (nurse cells) are regarded as a temporary source of diffusible factors, though a small proportion can persist for weeks *in vivo*. The two structural layers also provide spatial cues and a fibrillar collagen template for host cells.[15] Dermagraft™ (also clinically available) is a single layer implant of allogeneic dermal fibroblasts in a resorbable polymer scaffold.[16] Perhaps the earliest TE construct in this series is the keratinocyte sheet, developed a quarter century ago.[17] This still applied spatial/directional and biochemical cues from the culture dish, air interface and underlying fibroblast feeder layer (Fig. 1). From these examples of control, it is reasonable to examine the surgical perspective.

THE SURGICAL/CLINICAL PERSPECTIVE

Tissue engineering is quintessentially collaborative (and indeed ceases to be a distinct subject where it is not interdisciplinary), involving biologists, engineers, material scientists and surgeons. Arguably, a radical re-assessment of the potential for "novel replacement tissue parts" is needed. In this way it is possible to imagine reconstructing larger, more complex structures gradually, *in vivo*, from their components. This concept of surgical reconstruction based on the *minimal defective tissue* unit is ideally suited to tissue engineering and distinguishes it from transplantation. Clinical issues presently focus on the cost-benefit analysis (in Example 2: high cost of the constructs versus long term nursing for wound closure) or the lack of effective alternatives, such as in burns. Interestingly, few truly novel surgical applications have so far emerged.

The most effective applications begin with a detailed analysis of the surgical/pathological problem, which it is hoped to improve.[8] For any new tissue construct, then, a small number of key surgical imperatives must first be defined. For example, early replacement of mechanical function is normally important in orthopaedics. Urological implants need to form liquid-tight layers but must not seed crystals from urine.[18,19] Bioartificial drug depots require rapid vascular perfusion[3,4] as will an artificial pancreas, in diabetes, where long-term activity of donor islet cells must also be segregated from immune surveillance.[20] For replacement of heart valves and major vessels in cardiac-vascular surgery, long-term mechanical and non-thrombogenic properties are imperatives[21] whilst in peripheral nerve repair, the rate and orientation of axon regeneration and reduction of mechanical tension are key factors.[8,15,22] Clearly, the clinical collaborator should be a key driver at this stage, outlining the principal biological and engineering problems.

Echoing the presence of two approaches to control, described at the outset, the clinical perspective identifies TE constructs from their position in the spectrum of "functional competence at implantation". At the extremes of this spectrum, constructs can be designed to be **mature tissue grafts**, able to take on much of the original tissue function soon after surgery, or **engineered repair tissues**, functioning as templates for the repair process (Fig. 2(b)).

Table 3.1. Basic evolution of complexity and control in tissue engineered constructs (illustrated through Examples 1–3 in the text).

	"Repair" Cells	Support Substrate	Growth Factors	Guidance Cues	Support Cells	Mechanical Cues
Example 1 *CARTILAGE*	Chondrocytes	Nil PLA/PGA	BMP/TGFβ	Nil	Nil	Compressive Loading
Example 2 *SKIN*	Fibroblasts Keratinocytes melanocytes	Nil PLA/PGA Collagen-Fibrin Hyaluronan Fibronectin	TGF-β	Nil Fibronectin-Fibres Collagen-Fibres	Nil	Tensile Loading
Example 3 *NERVE*	Nil	PLA/PGA Fibronectin Collagen	NGF NT-3	Nil Fibronectin-Fibres Collagen-Fibres	Schwann Cells Transfected Fibroblasts	Nil

This is an inevitable compromise between the ideal of a finished mature graft (frequently not attained at present) and an increasing reliance on the repair process (with consequent scarring). A TE heart valve, for example, would need to have some function almost immediately after surgery, whereas there is no credible TE design envisaged to implant a mature, functional peripheral nerve (Table 1).

Example 3. Peripheral nerve has to be repaired by endogenous processes with the assistance of a tissue engineered template.[8,22] Work on tissue engineering solutions to peripheral nerve repair presents a good example of obligate tissue repair devices, evolving to incorporate increasing levels of control (guidance, rate-restricting factors, duration).

Throughout this development, a crucial issue, particularly for the clinical collaborator, is cell sourcing, acquisition and processing (with its implied dependence on specialist laboratory facilities). Use of autologous cells can carry heavy penalties of delay, additional surgery and local laboratory costs. One solution to the last and most costly of these problems has been developed for cartilage cell grafting, where commercial companies accept autologous cartilage and return suitably extracted and expanded cultures of the patients own chondrocytes, reminiscent of conventional blood fractionation services. Where cost, clinical delay and commercial production are important, seeding with allogeneic cells is frequently the most attractive approach, as in commercial skin substitutes, Apligraf™ and Dermagraft™.

In the simplest designs, crude cell homogenates can be added to the implant material, just prior to, or during surgery. At the other end of the complexity spectrum tissues are harvested and cell cultures isolated and expanded, prior to seeding in the support material and maturation tissue bioreactor. Examples of this include incorporation of Schwann cells to nerve implants,[23,24] chondrocytes into cartilage constructs[25] and fibroblasts and keratinocytes into TE skin equivalents.[13,26] This spectrum of complexity is well-illustrated in TE skin implants, based on the collagen-GAG dermal substitute material, *INTEGRA™*.[27] First used as a cell-free "dressing" for burns, it has now been clinically tested with dermal fibroblasts, keratinocytes and hair follicles.

Clearly, there is a trade-off between the tight control which can be maintained during longer term culture to form more mature or complex constructs and the poorer control but rapid growth, low cost and natural integration achieved by maturation of engineered repair tissues *in vivo*. In each case, control of the repair and scarring process is the issue, balanced against manufacturing cost and complexity. Early attempts at such complex constructs, requiring tissue bioreactor technology, include 3D co-cultured liver and endothelial cells under fluid flow.[28] This TE process design scheme is summarised in diagram form in Fig. 2(b) and shows the familiar TE design scheme, now extended to incorporate the concepts of primary (repair type) and secondary (mature type) TE constructs. An inescapable consequence of longer term maturation in culture is the need to develop ever more complex forms of perfused, tissue-bioreactors. Though there are few examples of convincing secondary constructs to date, one candidate has been described by Atala *et al.*[29] in the form of a pre-formed and partially functional bladder implanted into dogs. The scene is also set for development of novel surgical approaches, particularly of hybrid techniques. An example here is implantation of an early stage, immature TE construct into a "nursery" anatomical site for later reconstruction of the injured site. In this case, the patient acts as their own bioreactor.

However constructs are used and implanted, a major surgical priority in mechanically functional tissues is the control of scarring. The twin processes of integration and gliding, forming fixed and non-fixed interfaces, are crucial to dynamic function in surgical reconstruction. For example, scarring and adhesion (integration) at tendon, spinal and abdominal implant sites is just as likely to lead to failure as poor integration (rupture) at implant sites in ligament, cartilage or blood vessels. This is only one example of the underlying research agenda in TE. Progressive ability to control (engineer) complex tissues in the lab is a potent strategy for unravelling the mechanisms of natural control, and their failure for example in tumour development. It may be that, in addition to bioartificial spare parts, TE offers a novel means to tackle head-on the failures of tissue control. By replicating entire working tissue controls we should identify more productive approaches to pathology than traditional dissection of the failed control system.

DISTINCT ROLES FOR TISSUE ENGINEERING AND TRANSPLANTATION

The case for a "generic component" or the spare-part-tissue approach to tissue engineering has been previously outlined.[8,15] This also serves as a useful niche definition, distinguishing TE and transplantation technology. This concept proposes that in the foreseeable future, TE will be better adapted to production of relatively simple component body parts rather than whole organs (e.g. a heart valve or vessel rather than the whole heart; tendons, nerves and bone implants rather than whole limbs). The supply of functionally complex, whole organs seems likely to be best served by transplantation, perhaps with developments in xenotransplantation from genetically modified pigs,[30] more effective organ retrieval or use of temporary or partial-function prostheses, such as extracorporeal hepatic devices or supplementary, artificial hearts. Heart and liver transplantations, in reality, affect a relatively small patient group yet the technical and economic challenge of engineering complete replacements in one leap, is enormous. Rather, the exciting and novel role for TE is in its application to tissue degeneration in the huge and progressively ageing patient base (presently, 17 million are over 75 years old in the US). This involves a supply of relatively simple (i.e. safe and inexpensive) tissue grafts or implants for surgical **reconstruction** for non life-threatening but life-impairing conditions.[31] This reconstructive approach leads to the concept that TE implants could be designed to be simple platform structures, rather than for a specified operation. Ideally, some of these would be basic structures (flat patches, tubes, interface surfaces), representing tools for surgical innovation.

THE BIO-ENGINEER'S PERSPECTIVE

The basic tools available to tissue engineers are the cells, resorbable materials and the control cues.

1) Cell Acquisition and Selection

Biological collaborators have the particular problem of identifying suitable cell sources, ideally from the patient, which will grow rapidly and controllably. Once obtained (either from autologous biopsy or allogeneic sources), these must then be selected, and the population expanded (to provide sufficient density for seeding) under conditions suitable either to preserve or to return cells to the required phenotype. Chondrocytes are a particularly clear example, in that they normally shift to a fibroblastic phenotype during expansion and need to be returned to a chondrogenic phenotype.[7,32] Techniques are now widely used for cell types including fibroblasts, peripheral nerve Schwann cells, endothelial cells, chondrocytes, bone and urothelial cells.[24,33–37] The slow growth of some of these cell types, particularly from older patients, represents a major downstream hurdle.

Cells are used in TE constructs either to provide sufficient numbers to contribute to the repair process or to provide a **temporary** source of helper cells as a source of growth factors for host cell recruitment or tissue deposition. The most basic level of cell use in TE is to implant either as a suspension or a sheet. Examples of these are chondrocyte cell suspension, keratinocyte and urothelial cell sheets as discussed above.[5,33,34] Seeding of peripheral nerve repair implants with supporting Schwann cells is used to provide neurotrophic growth factors (e.g. nerve growth factor, neurotrophin-3) and it supports the ingrowth and regeneration of neurites.[24] Secondarily, such cells would be expected to participate in re-myelination. TE blood vessels are often designed to be lined with endothelial cells to restore a biological non-thrombogenic surface.[37] In contrast, marrow cells have been seeded, as nurse-cells, into vascular constructs to provide migration and angiogenic factors for recruitment of local endothelial cells.[38]

The use of active but poorly differentiated cells from the marrow is an attractive possibility for construction of future TE grafts, not only because they are relatively easy to obtain but also since they have the potential to undergo rapid population expansion (not always the case with differentiated cells). Stromal cells from this source have long been known to differentiate into either osteoblasts or chondrocytes in culture. However, the current research aim is to identify molecular or physical controls able to direct differentiation of such adult progenitor cells towards a range of new phenotypes in culture. Comparable possibilities exist for the utilisation of poorly committed mesenchymal progenitor cells from non-match tissues.[39] For example, it may be that certain populations of dermal fibroblasts can be stimulated to produce extracellular matrix suitable for tendon or ligament or smooth muscle cells from bladder may prove useful in vascular constructs. Clearly, such strategies would provide a ready source of autologous cells without ethical difficulties.[9,10]

2) Cell Support Materials

The growing array of bio-resorbable materials for tissue engineering have been reviewed elsewhere.[2,8,35] Materials can be divided into synthetic and semi-natural polymers and aggregates. Of the resorbable synthetic polymers, poly-lactic and -glycolic acids have been most widely applied.[11,12] These have advantages of ease of use, cheapness and simplicity, but may produce undesirable biological responses in some situations, including poor cell attachment and ingrowth. Materials made from natural proteins collagen,[13,14,40-43] fibrin,[44] fibronectin,[45,46] or polysaccharides such as hyaluronan[47] tend to have better biocompatibility but are more complex to prepare in a controlled manner.

3) Control Processes

Control processes can be applied to both the cellular and material elements of any given TE construct. Controls can involve three main activities; (i) rate control, (ii) spatial control, (iii) cell phenotype. Rate control represents, for example, the regulation of how fast cells accumulate or deposit new matrix. Spatial control is necessary to regulate the final tissue 3D architecture and cell phenotype determines a range of outcomes, from tissue composition to integration with surrounding tissues. Unfortunately unlike many conventional engineering problems, these control elements are inextricably interlinked in tissue engineering, such that no one rate- or spatial cue can be reliably expected to control a single part of the tissue growth process. For example, aligned mechanical loading can be used to give spatial cues, but it also influences matrix deposition.[36,48-51]

Techniques for stimulating cell division (or suppression of cell death) are chiefly restricted to cell population expansion, prior to seeding into the tissue construct. Rate stimulating factors at this stage tend to be well-known mitogenic growth factors. In contrast, stimulation of matrix production rates would be after cell-seeding of the tissue construct, using factors such as Insulin-like growth factor (IGF-1) and transforming growth factor beta (TGF-β). Yet another group of rate controlling factors, responsible for promotion of tissue integration can be incorporated into the implant design, but to operate after implantation into the body site. These are the factors which promote new blood vessel or neural ingrowth (eg. VEGF and NGF). Recruitment, either of seeded cells deep into the tissue construct or from surrounding host tissues into the implant, can be a key to success. Cell migration and direction can be promoted by chemotaxis, haptotaxis and mechanotaxis, for example, PDGF and FGF are chemotactic for fibroblasts and endothelial cells, respectively.[52-54]

Regulation of the direction of motion of recruited cells inevitably gives both spatial and rate control. Spatial control of tissue architecture can use substrate adhesion or shape, on cell guidance substrates. The detailed effects of fine structure[55-57] are now used in applications to orientate tendon growth[56] and nerve regeneration.[58-61] Surface topography and adhesivity have been used in the form of shear-aligned fibronectin fibres[45,46] and magnetically-aligned collagen fibrils[62] for cell orientation, mimicking aspects of natural cell guidance.

Aligned mechanical loads can also act as guidance cues.[8] There is ample evidence that many cells are mechano-sensitive, including matrix production responses in chondrocytes and fibroblasts.[50,51,63,64] More importantly, the directional information in load alignment can lead to tissue orientation. The detailed mechanism underlying mechano-sensitivity of cells is far from clear but may be mediated by combinations of cytoskeletal or integrin-based systems[65,66] and stretch-sensitive ion-channels.[67,68] Although tension, compression and shear act in combination, different cell types are adapted to one principle type of force (e.g. fibroblasts, chondrocytes and vascular endothelial cells, respectively). Vascular endothelial cells take on a bipolar shape in response to the main shear force vector[69,70] and such responses are proving important in the tissue engineered vascular implants. Application of aligned tensional loads to muscle cells[71] and fibroblasts[49,72] in culture has been used to produce predictable alignments. Fibroblasts in simple 3D tissue constructs are aligned parallel with the maximum principal strain by low frequency uniaxial loading,[49] though completely opposing biochemical responses were elicited by complex or multi-axial cues.[49,51]

These two interdependent spatial cues, substrate guidance and force vectors, turn out, then, to be key to many of our goals in tissue engineering, as they are important in repair. Their role can be described as an overarching hypothesis (Fig. 3). This states that,

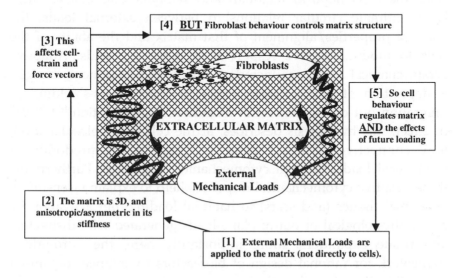

Fig. 3. Summary of the essentially circular (M25*) hypothesis of cytomechanics. In this, it is proposed that external forces regulate cell function, in particular spatial organisation (architecture) and composition of their extracellular matrix. External loading, though, is always applied to cells *through* the medium of that extracellular matrix. That matrix is almost always anisotropic, with asymmetry of stiffness. This means that the matrix inevitably alters loading of resident cells (particularly loading direction) *at the cell level*. However, those same cells are responsible for the matrix and its architecture, making the whole control process a self-regulating loop in which matrix structure and material properties act as the key intermediary. It is proposed that basic cell behaviours, in response to strain at the cell level (cytomechanical loading), tend to promote stress shielding of the cells in terms of deposition of matrix fibres, migration and shape change.[49] This has basic, predictable consequences for regulation. In short, the spatial orientation of cells, fibrous matrix and the propagation of forces through both are completely interdependent, making remodelling of *established matrices* improbable, just as scar tissue reorganisation by these same cells *in vivo* is poor. [* M25, around London, is Europe's largest circular motorway].

since most applied stress on tissues is taken by the matrix (not the cells), it is axiomatic that resident cells can only respond to those strains *which they perceive,* i.e. at the micro-scale. This is dependent on both loading and matrix, structure (e.g. fibre alignment). But, since the cells deposit, maintain and remodel that matrix, there is an intimate circular relationship between external loads, the material properties/alignment of that matrix and the ability of the cells to mount adaptive responses to perceived loads. Adaptive cytomechanical responses include — shape and alignment changes, — deposition of new collagen fibres to effect stress-shielding or -migration out of the area, with local matrix degradation.[48,49,51,64,73] In effect, this hypothesis describes, at the cell and molecular level, how connective tissues become adapted to the predominant mechanical loads, through cytomechanical responses. Furthermore, these adaptive cytomechanical responses help to explain insensitivity of normal tissues (and scars) to habitual loading, but activation by supra-physiological or out-of-plane loads generated with distractors and tissue expanders. More importantly here, they provide a framework for rational design of bioreactors to engineer organised connective tissues, rather than scar tissues.

CONCLUSIONS AND FUTURE DIRECTIONS

TE aims to develop solutions (free from adherent ethical problems) to the looming problems of healthcare in an ageing, high-expectation population without reliance on fixed-life prostheses.[84] In the light of recent reports,[31,74,75] TE will have diverse impacts on surgery and biomaterials. Future **biomaterials** will provide bio-mimetic controls including topographic, dynamic cytomechanical and biochemical cues. **Cell acquisition** is likely to be tackled using allogeneic committed tissue cells or autologous mesenchymal progenitor cells, potentially with gene modification. **Bio-engineering** of advanced tissue bioreactors is essential to produce large tissues to near-maturity stage. **The clinician,** however, should be key to perhaps the most critical question; that of control of bio-integration of the tissue constructs into recipient sites. Of the three principal tissue systems which must become integrated after implantation; vascular-, neural- and mechano/connective-, it is mechanical connectivity to surrounding tissues which is most unpredictable and challenging. This is exemplified by the tissue we least want to engineer for any site in the body; the scar.

ACKNOWLEDGEMENTS

The author is grateful to Professor Gus McGrouther (UCL; Plastic & Reconstructive Surgery) for helpful comments on the manuscript, Rebecca Porter (TRU) and for financial support from the Trusthouse Charitable Foundation and the European Union (awards, QLK3-CT-1999-00625.and –00559).

REFERENCES

1. Kuhn TS. *The Structure of Scientific Revolutions*, 2nd ed. University Chicago Press, Chicago, 1962, 210.
2. Brown RA, Smith KD, McGrouther DA. Strategies for cell engineering in tissue repair. *Wound Repair Regen* 1997; 5:212–221.
3. Langer R, Vacanti JP. Tissue engineering. *Science* 1993; 260:920–926.
4. Vacanti JP, Langer R. Tissue engineering: The design and fabrication of living replacement devices for surgical reconstruction and transplantation. *Lancet* 1999; 354:32–34.
5. Brittberg M. Autologous chondrocyte transplantation. *Clin Orthop* 1999; 367:S147–155.
6. Sittinger M, Braunling J, Kastenbauer E, Hammer C, Burmester G, Bujia J. Proliferative potential of nasal septum chondrocytes

for *in vitro* culture of cartilage transplants. *Laryngorhinootologie* 1997; **76**:96–100.

7. Vacanti CA, Upton J. Tissue-engineered morphogenesis of cartilage and bone by means of cell transplantation using synthetic biodegradable polymer matrices. *Bone Repair Regen* 1994; **21**:445–462.

8. Brown RA. Bioartificial implants: design and tissue engineering. In *Structural Biological Materials*, ed. M. Elices M, Elsevier Science, Oxford, 2000, pp. 105–160.

9. Petersen BE, Bowen WC, Patrene KD, *et al.* Bone marrow as a potential source of hepatic oval cells. *Science* 1999; **284**:1168–1170.

10. Bjornson CRR, Rietze RL, Reynolds BR, Magli MC, Vescovi AL. Turning brain into blood: A hematopoietic fate adopted by adult neural stem cells *in vivo*. *Science* 1999; **283**:534–537.

11. Mooney DJ, Mazzoni CL, Breuer C, *et al.* Stabilised polyglycolic acid fibre-based tubes for tissue engineering. *Biomaterials* 1996; **17**:115–124.

12. Putnam AJ, Mooney DJ. Tissue engineering using synthetic extracellular matrices. *Nature Medicine* 1996; **2**:824–826.

13. Eaglestein WH, Falanga V. Tissue engineering and the development of Apligraf, a human skin equivalent. *Cutis* 1998; **62** suppl:1–8.

14. Falanga V, Margolis D, Alvarez O, *et al.* Rapid healing of venous ulcers and lack of clinical rejection with an allogeneic cultured human skin equivalent. *Arch Dermatol* 1998; **134**:293–300.

15. Brown RA, Porter RA. Tissue engineering. In *Animal Cell Culture*, ed. JRW Masters, Oxford University Press, Oxford, 2000, pp. 149–173.

16. Naughton G, Mansbridge J, Gentzkow G. A metabolically active human dermal replacement for the treatment of diabetic foot ulcers. *Artif Organ* 1997; **12**:1203–1210.

17. Rheinwald JG, Green H. Serial cultivation of strains of human epidermal keratinocytes: The formation of keratinizing colonies from single cells. *Cell* 1975; **6**:331–334.

18. Scott R, Gorham SD, Aitcheson M, Bramwell P, Speakman J, Meddings RN. First clinical report of a new biodegradable membrane for use in urological surgery. *Br J Urol* 1991; **68**:421–424.
19. Sabbagh W, Masters JRW, Duffy PG, Herbage D, Brown RA. *In vitro* assessment of a collagen sponge for engineering urothelial grafts. *Br J Urol* 1998; **82**:888–894.
20. Sun AM. Methods for the immuno-isolation and transplantation of pancreatic cells. In *Tissue Engineering Methods and Protocols*, eds. JR Morgan, ML Yarmush, Humana Press, New Jersey, 1999, pp. 469–481.
21. Oyer PE, Stinson EB, Miller DC, Jamieson SW, Mitchell RS, Shumway NE. Thromboembolic risk and durability of the Hancock bioprosthetic cardiac valve. *Eur Heart J* 1984; **5**:81–85.
22. Furnish EJ, Schmidt CE. Tissue Engineering of the peripheral nervous system. In *Frontiers in Tissue Engineering*, eds. CW Patrick, AG Mikos AG, LV McIntire, Pergamon Press, Oxford, 1998, pp. 514–535.
23. Brown RE, Erdmann D, Lyons SF, Suchy H. The use of cultured Schwann cells in nerve repair in a rabbit hind-limb model. *J Reconstr Microsurg* 1996; **12**:149–152.
24. Ansellin AD, Fink T, Davies DF. Peripheral nerve regeneration through nerve guides seeded with adult Schwann cells. *Neuropath. Appl Neurobiol* 1997; **23**:387–398.
25. Vacanti CA, Langer R, Schloo B, Vacanti JP. Synthetic polymer seeded with chondrocytes provide a template for new cartilage formation. *J Plast Reconstr Surg* 1991; **88**:753–759.
26. Anreassi L, Casini L, Trabucchi E, *et al.* Human keratinocytes cultured on membranes composed of benzyl ester of hyaluronic acid. *Wounds* 1991; **3**:116–126.
27. Chamberlain LJ, Yannas IV. Preparation of collagen-glycosaminoglycans co-polymers for tissue regeneration. In *Tissue Engineering Methods and Protocols*, eds. JR Morgan, ML Yarmush, Humana Press, New Jersey, 1999, pp. 3–17.

28. Pollok JM, Kluth D, Cusick RA, *et al.* Formation of spheroidal aggregates of hepatocytes on biodegradable polymers under continuous-flow bioreactor conditions. *Eur J Pediatr Surg* 1998; 8:195–199.

29. Oberpenning F, Meng J, Yoo JJ, Alata A. De novo reconstitution of a functional mammalian urinary bladder by tissue engineering. *Nature Biotechnol* 1999; **17**:149–155.

30. Lin SS, Weidner BC, Byrne GW, *et al.* The role of antibodies in acute vascular rejection of pig-to-baboon cardiac transplants. *J Clin Invest* 1998; **101**:1745–1756.

31. Peckham M. Foresight: Making the future work for you. Healthcare in 2020 panel. (2000). UK Department of Trade & Industry, (Pub.0.3k/12/00/NP) URN/00/1187, London. (www.foresight.gov.uk).

32. Mayne R, van der Rest M, Bruckner P, Schmid TM. The collagens of cartilage and the type IX-related collagens of other tissues. In *Extracellular Matrix*, eds. MA Haralson MA, JR Hassell JR, IRL Press, Oxford, 1995, pp. 73–97.

33. Prunieras M, Regnier R, Woodley D. Methods for cultivation of keratinocytes with an air-liquid interface. *J Invest Dermatol* 1983; **81**:S28–S33.

34. Cilento BG, Freeman MR, Schneck FX, Retik AB, Atala A. Phenotypic and cytogenetic characterization of human bladder urothelial expanded *in vitro*. *J Urol* 1994; **152**:665–670.

35. Silver FH. *Biomaterials, Medical Devices and Tissue Engineering: An Integrated Approach.* Chapman Hall, London, 1994:92–119.

36. Bishop JE, Butt R, Dawes K, Laurent G. Mechanical load enhances the stimulatory effect of PDGF on pulmonary artery fibroblast procollagen synthesis. *Chest* 1998; **114**:25S.

37. L'Heureux N, Paquet S, Labbe R, Germaine L, Auger FA. A completely biological tissue-engineered human blood vessel. *FASEB J* 1998; **12**:47–56.

38. Noishiki Y, Yasuko T, Yamane Y, Matsumoto A. Autocrine vascular prosthesis with bone marrow transplantation. *Nature Medicine* 1996; **2**:90–93.

39. Young RG, Butler DL, Weber W, Caplan AI, Gordon SL, Fink DJ. Use of mesenchymal stem cells in a collagen matrix for Achilles tendon repair. *J Orthop Res* 1998; **16**:406–413.

40. Yannas IV, Lee E, Orgill DP, Skrabut EM, Murphy GF. Synthesis and characterisation of a model extracellular matrix that induces partial regeneration of adult mammalian skin. *Proc Natl Acad Sci, USA*, 1989; **86**:933–937.

41. De Vries HJC, Middlekoop E, Mekkes JR, Dutrieux RP, Wildevuur CHR, Westerhof W. Dermal regeneration in native non-cross linked collagen sponges with different extracellular matrix molecules. *Wound Rep Reg* 1994; **2**:37–47.

42. Guido S, Tranquillo RT. A methodology for the systematic and quantitative study of cell contact guidance in oriented collagen gels. Correlation of fibroblast orientation and gel birefringence. *J Cell Sci* 1993; **105**:317–331.

43. Hull BE, Finley RK, Miller SF. Coverage of full thickness burns with bilayered skin equivalents: A preliminary clinical trial. *Surg* 1990; **107**:496–502.

44. San-Galli F, Deminiere C, Guerin J, Rabaud M. Use of biodegradable elastin-fibrin material, Neuroplast, as a dural substitute. *Biomaterials* 1996; **17**:1081–1085.

· 45. Brown RA, Blunn GW, Ejim OS. Preparation of orientated fibrous mats from fibronectin: Composition and stability. *Biomaterials* 1994; **15**:457–464.

46. Ejim OS, Blunn GW, Brown RA. Production of artificial-orientated mats and strands from plasma fibronectin: A morphological study. *Biomaterials* 1993; **14**:743–748.

47. Campoccia D, Hunt JA, Dohetery PJ, Zhong SP, O'Regan M, Benedetti L, Williams DF. Quantitative assessment of the tissue response to films of hyaluronan derivatives. *Biomaterials* 1996; **17**:963–975.

48. Mudera VC, Pleass R, Eastwood M, Tarnuzzer R, Schultz G, Khaw P, McGrouther DA, Brown RA. Molecular responses of

human dermal fibroblasts to dual cues: Contact guidance and mechanical load. *Cell Motil Cytoskel* 2000; **45**:1–9.

49. Eastwood M, Mudera VC, McGrouther DA, Brown RA. Effect of precise mechanical loading on fibroblast populated collagen lattices: Morphological changes. *Cell Motility Cytoskel* 1998; **49**:13–21.

50. Prajapati RT, Chavally-Mis B, Herbage D, Eastwood M, Brown RA. Mechanical loading regulates protease production by fibroblasts in 3-dimensional collagen substrates. *Wound Repair Regeneration.* 2000; **8**:227–238.

51. Parjapati RT, Eastwood M, Brown RA. Duration and alignment of mechanical stress are critical to activation of fibroblasts. *Wound Repair Regeneration* 2000; **8**:239–247.

52. Choquet D, Felsenfeld DP, Sheetz MP. Extracellular matrix rigidity causes strengthening of integrin-cytoskeleton linkages. *Cell* 1997; **88**:39–48.

53. Sterpetti AV, Cucina A, Fragale A, Lepidi S, Cavallaro A, Santoro D'Angelo L. Shear stress influences the release of platelet derived growth factor and basic fibroblast growth factor by arterial smooth muscle cells. *Eur J Vasc Surg* 1994; **8**:138–142.

54. Stokes CL, Rupnick MA, Williams SK, Lauffenburger DA. Chemotaxis of human microvessel endothelial cells in response to acidic fibroblast growth factor. *Lab Invest* 1990; **63**:657–668.

55. Clark P, Connolly P, Curtis AS, Dow JA, Wilkinson CD. Cell guidance by ultrafine topography *in vitro*. *J Cell Sci* 1991; **99**:73–77.

56. Wojciak-Stothard B, Crossan J, Curtis ASG, Wilkinson CD. Grooved substrata facilitate *in vitro* healing of completely divided flexor tendons. *J Mater Sci Mater Med* 1995; **6**:266–271.

57. den Braber ET, de-Ruijter JE, Smits HT, Ginsel LA, von Recum AF, Jansen JA. The effect of parallel surface microgrooves and surface energy on cell growth. *J Biomed Mater Res* 1995; **29**:511–518.

58. Whitworth IH, Brown RA, Dore C, Green CJ, Terenghi G. Orientated mats of fibroncetin as a conduit material for use in peripheral nerve repair. *J Hand Surg* 1995; **20B**:429–436.

59. Sterne GD, Brown RA, Green CJ, Terenghi G. Neurotropin-3 delivered locally via fibronectin mats enhances peripheral nerve regeneration. *Eur J Neurosci* 1997; **9**:1388–1396.

60. Hobson MI, Brown RA, Green CJ, Terenghi G. Interrelationships between angiogenesis and nerve regeneration: A histochemical study. *Br J Hand Surg* 1997; **50**:125–131.

61. Sterne GD, Coulton GR, Brown RA, Green CJ, Terenghi G. An immunohistochemical study of the myosin heavy chains in gastrocnemius muscle following NT-3-enhanced nerve regeneration. *J Cell Biol* 1997; **139**:709–713.

62. Dubey N, Letourneau PC, Tranquillo RT. Guided neurite elongation and Schwann cell invasion into magnetically aligned collagen in simulated peripheral nerve regeneration. *Exp Neurol* 1999; **158**:338–350.

63. Urban JP. The chondrocyte, a cell under pressure. *Br J Rheum* 1994; **33**:901–908.

64. Parsons M, Kessler E, Laurent GJ, Brown RA, Bishop JE. Mechanical load enhances procollagen processing in dermal fibroblasts by regulating levels of procollagen C-protein ase. *Exp Cell Res* 1999; **252**:319–331.

65. Kain HL, Reuter U. Release of lysosomal protease from retinal pigment epithelium and fibroblasts during mechanical stresses. *Graefes Arch Clin Exp Ophthalmol* 1995; **233**:236–243.

66. Wang N, Butler JP, Ingber DE. Mechanotransduction across the cell surface and through the cytoskeleton. *Science* 1993; **206**:1124–1127.

67. Ohno M, Cooke JP, Dzau VJ, Gibbons GH. Fluid shear stress induces endothelial transforming growth factor beta-1 transcription and production. Modulation by potassium channel blockade. *J Clin Invest* 1995; **95**:1363–1369.

68. Markin VS, Tsong TY. Reversible mechanosensitive ion pumping as part of a mechano-electrical transduction. *Biophys J* 1991; **59**:1317–1324.

69. Levesque MJ, Nerem RM. The elongation and orientation of cultured endothelial cells in response to shear stress. *J Biomech Eng* 1985; **107**:341–347.

70. Davies PF. Flow-mediated endothelial mechanotransduction. *Physiol Rev* 1995; **75**:519–560.

71. Vandenberg HH. A computerised mechanical cell stimulator for tissue culture: Effects on skeletal muscle organisation. *In Vitro Cell Devel Biol* 1988; **24**:609–619.

72. Brown RA, Prajapati R, McGrouther DA, Eastwood M. Mechanical responses of dermal fibroblasts to mechanical loading in 3-dimensional collagen matrices: Tensional homeostasis. *J Cell Physiol* 1998; **175**:323–332.

73. Parsons M, Bishop JE, Laurent GJ, Eastwood M, Brown RA. Mechanical loading of collagen gels stimulates dermal fibroblast collagen synthesis and reorganisation. *J Invest Dermatol* (in press).

74. Bentley G, *et al.* "Tissue Engineering in Orthopaedics". Report of the London Medicine/DTI Mission to Boston, Pittsburgh & Mayo Clinic, Sept. 1998.

75. Butler BRR. A review of the UK research base in biomaterials. The Royal Academy of Engineering. RAE, 29, Great Peter St., London SW1P 3LW; 1998. [ISBN 1-871-634-58-X].

76. Huang D, Chang TR, Aggarwal AA, Lee RC, Ehrlich P. Mechanisms and dynamics of mechanical strengthening in ligament-equivalent fibroblast-populated collagen matrices. *Ann Biomed Eng* 1993; **21**:289–305.

77. Langelier E, Rancourt D, Bouchard S, Lord C, Stevens P-P, Germain L, Auger FA. Cyclic traction machine for long term culture of fibroblast-populated collagen gels. *Ann Biomed Eng* 1999; **27**:67–72.

Section 3

Tissue Engineering: The Construction of Living Organs for Replacement

Chapter 4

Recent Developments in Skin Substitutes

William R Otto

INTRODUCTION

Patients who are burned or involved in accidents, those suffering chronic ulcerations or requiring skin grafting for tumour-clearing purposes all need a supply of replacement skin. The historical gold standard solution to this problem has been, and continues to be, the split-thickness autograft. There is no rejection problem, but the patient does receive a second injury which can have its own morbidity, and the healing of either wound may leave something to be desired cosmetically. This is especially true where skin is in short supply in a severely burned patient, who may have the split-skin graft (SSG) meshed and stretched up to one in six to increase its effective area. Such outcomes have led to efforts to produce skin substitutes which overcome these limitations. The holy grail would be to replace skin with something "off-the-shelf", but which would perform at least as well as the patient's own skin. The earlier literature has been well reviewed several times.[1-4] This review sifts some of the more recent literature on how far the search for the grail has gone.

It is nearly 20 years since the first cultured keratinocyte sheet graft was used clinically to save the life of a burned patient.[5] The early trials were optimistic that the clonal keratinocyte culture method introduced by Rheinwald and Green[6] would revolutionise burns clinics. Since then, it has become clear that epidermal cells alone are not sufficient to permit a cosmetically pleasing result — a dermal component is essential.[1] Keratinocyte sheets did save lives and cover the patient's wounds, but there were very poor cosmetic outcomes, coupled with contractures and incomplete contour correction. This often necessitated further corrective surgery.

Attempts to overcome the deficiencies associated with keratinocyte sheet cultures as skin substitutes began soon after it became clear from the pioneering work of Bell and colleagues[7-9] that a collagen-based matrix could be remodelled *in vitro* by embedded dermal fibroblasts, and that the behaviour of the epidermal component was considerably enhanced when grown on top of this support. This could be further improved by raising the whole organotypic culture to the air–liquid interface such that the epidermis was fed from below.[10-14] This mimicked the situation *in vivo*, and a further improvement in epidermal differentiation characteristics ensued.

It was clear that the market for a skin substitute was potentially large and several companies began researching this area, seeking ways to modulate cellular behaviour and get around each other's patents. As a result, there are now a handful of products at various stages of clinical development. These products are summarized below (Table 1). They are classified as "devices" rather than "drugs", and are licenced for use as dressings, particularly for venous ulcers, although several are also in use for the treatment of burn wounds. The "device" definition is becoming less meaningful since these products at least partly integrate into and/or become remodelled by the patient. In addition, their properties may lead to significant biological effects.

Table 4.1. Skin substitute products and experimental models.

Product	Composition	Selected Refs.
In Clinical Use		
Alloderm	Acellular human cadaver dermis	15,16
(LifeCell Corp, The Woodlands, TX, USA)		
Apligraf (Graftskin)	Cellular bovine collagen bilayer with keratinocytes/fibroblasts	17–19
(Novartis, East Hanover, NJ, USA)		
Biobrane	Acellular silicone/nylon/collagen	20,21
Dermagraft-TC	Vicryl (Biobrane) mesh + fibroblasts, silicone overlay	22–24
(Advanced Tissue Sciences, Inc., La Jolla, CA, USA)		
Integra Artificial Skin	Acellular bovine collagen chondroitin sulfate, silicone overlay	25–28
(Integra Lifesciences Corp., Plainsboro, NJ, USA)		
TransCyte	Dermagraft-TC rebranded	29–31
(Advenced Tissue Sciences)		
Laserskin	Hyaluronic acid based membrane, optional cell covering	32–34
(Fidia Advanced Biopolymers, Abano Terme, Italy)		
Other Experimental Models		
Chimerism	Mixed allo-auto-geneic cultured cells	35
Pig	Cellular/decellularized allogeneic dermal grafts	36
"Upside-down"	Recombined human keratinocytes/pig skin (RHPS)	37,38
Polyactive	Acellular poly(ethyleneoxide)-poly(butyleneterephthalate) copolymer	39–42
Sacchachitin	Acellular β-1,3-glucan + N-acetylglucosamine copolymer (from Ganoderma tsugae)	43,44
In vitro	Crosslinked collagen	45–47

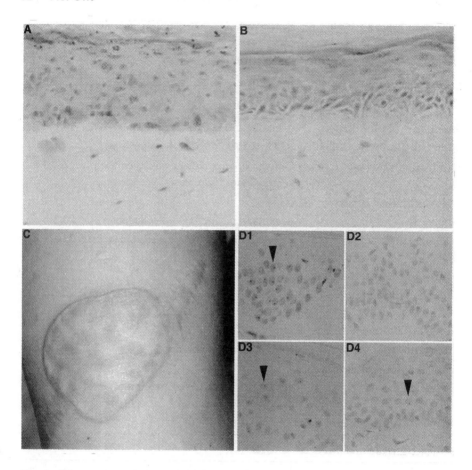

Fig. 1. Organotypic skin substitutes. A. Cultured skin substitute using rat tail collagen gel with embedded fibroblasts, overlain with keratinocytes, raised to the air–liquid interface for seven days (based on model of Bell *et al*[8]). B Similar to A, keratinocytes cultured on de-epidermalised Alloderm. C. Cultured autograft similar to A, four years after grafting.[69] D. Cultured allografts as A, showing epidermis at various times after grafting, stained for Y-specific DNA probe[69]: D1 Male on male control allograft at 18 months (80% Y-probe positive cells, not 100% due to sectioning artifact); D2 Female on female control autograft at four weeks (all cells Y-probe negative); D3 Male on female allograft at eleven weeks. Arrowheads indicate positively scored male cells (single peripheral nuclear spot); D4 Additional male on female allograft at 2.5 years. C and D reproduced by kind permission of Lippincott Williams & Wilkins, publishers of Plastic and Reconstructive Surgery.

ALLODERM AND ALLODERMIS

This product is an acellular human cadaveric dermal slice without an epidermal cover, but which retains basement membrane components over which epidermal cells can migrate. It has been used widely in the burn clinic since it has the inherent strength to withstand the hostile wound environment.[16,48–50] It is seen to "take" well and elicit little or no immunological reaction, while retaining its original surface area and contour correction. The dermal component is repopulated by the host's cells, with revascularization along original and neogenic pathways.[50] In an *in vitro* study, human allodermis was more successfully repopulated by dermal fibroblasts when these were added to the papillary surface in the presence of keratinocytes.[51] However, this route is not expected to be significant *in vivo*, where ingress would be from basal or lateral aspects of the graft. Prior reseeding with fibroblasts is not essential when grafting experimental animals if sufficient collagen IV is retained in the basement membrane region.[52] If surrounding tissues are irradiated (20 Gy) prior to engraftment with Alloderm, there is an initial (3 weeks) reduction in fibroblast and vascular ingress, which over-compensates at 14 weeks, but this normalises with time, and such grafts are similar to controls without an irradiated bed.[53] Alloderm material is also successful when covering joints, and resists contracture formation.[16] In these characteristics, it is very successful as a dermal replacement. However, Alloderm does require a second surgical intervention to cover the surface with an epidermal layer, either by split-thickness autograft[49,50] or by cultured keratinocytes.[54] If using the former, there is clearly the disadvantage of donor site morbidity, but delay till the second operation may be less than waiting for cultured keratinocyte sheets. The combination of Alloderm and ultrathin skin autografts has been successfully used to cover hand and foot burns, with good cosmetic and functional outcomes.[49,55] This will depend greatly on the areas to be covered. In a series of patients with 94% total body surface area burns, there were further problems when using cultured autologous keratinocytes to cover allodermis, since the risk of bacterial infections was high, and led to the failure of the epidermal cover.[54]

A porcine model was recently reported[36] studying the use of cryopreserved allogeneic dermal grafts, either cellular or treated to remove cells ("decellularized"). The authors wished to create a one-step operation by using split-thickness autograft to cover the allogeneic dermis. Experimental grafts were compared to autografted skin alone. Takes of either graft were similar to control sites, at over 83% for cellular and 93% for acellular grafts. The former grafts did not inhibit wound contracture (a significant porcine feature), whereas the acellular ones showed both less contraction at one week (43 versus 31%) and improved cosmetic outcomes. A similar approach using Alloderm has been translated into the human burn clinic, where the outcomes were judged to be fair to good when used to cover several articular surfaces.[16] Patients retained good joint mobility at one year post operation.

A further experimental approach to early burn wound closure has been the report of the use of a combined acellular porcine dermis overlaid with allogeneic human keratinocytes and applied "upside-down" to partial-thickness debrided burn wounds. The authors claim faster adnexal healing than in tulle gras and dry gauze treated wounds, while providing immediate cover and closure. However, full-thickness wounds did not heal well.[37,38]

APLIGRAF

Ulcer Treatment

Apligraf is a living allogeneic composite of foreskin-derived epidermal keratinocytes overlying a contracted dermal substitute based on bovine tendon collagen seeded with dermal fibroblasts.[19,56] This is the outcome of much research following the early work of

Bell and co-workers.[8,9] It has been used on venous leg ulcer wounds in a variety of studies.[56-67]

These studies have shown Apligraf to be broadly acceptable by the clinicians and the patients, and if the economics justify it,[66] also by the financial offices of hospitals. There is as yet no data on whether Apligraf persists on patients on whom it has "taken". It is claimed not to be rejected.[19] If cells are found to survive, it may have something in common with the tolerance to allogeneic cells induced by mosaicism or chimaerism, obtained by intermingling allogeneic and autogeneic tissues (reviewed by Starzl, 2000, but see also Refs. 68–71). A recent report was made of a non-commercial laboratory based organotypic skin model similar to Apligraf which was used on a small series of burns patients.[72] This was compared side-by-side with a split-thickness meshed autograft to achieve a direct comparison. Unfortunately, the experimental skin substitute was not successful in establishing a permanent cover for the full-thickness burn wound. Despite both the lack of bacterial infection and attempts to regraft the organotypic skin one week later, there was little "take", with only a few islands of surviving keratinocytes. The experimental grafts had to be removed and the patients were regrafted using meshed autografts, which healed well. This "negative" study highlights the need for a dermal component which is strong enough to withstand the exceedingly hostile burn wound environment.

It is difficult to distinguish two effects of Apligraf that may be occurring. One is the direct integrative effect of the tissue ("take" is a close approximation),[73] the other is the slightly less direct effect of factors which Apligraf may secrete into the wound environment and which may benefit that process. From the patient's point of view, such differences may be meaningless if there is clinical benefit. Nevertheless, there is a strong case to understand the biology of such interactions, and indeed whether a "pure" (removable) dressing could deliver growth and other factors so efficiently. The "dressing effect" is exemplified by keratinocyte sheet preparations secreting large amounts of TGF-α, TGF-β, IL-6, and IL-8 when compared to cryopreserved skin.[74] When such sheets have been applied to ulcers, the rate of healing increased.[75] This effect was depressed when the keratinocyte sheets were cryopreserved.[76] However, other research found no differences between cytokine (GM-CSF, IL-1α, IL-1β, IL-6) and growth factor (PDGF, bFGF) levels in the ulcer fluids collected from healing and non-healing ulcers dressed with Tegaderm.[77] Similar profiles of factors were found in the exudate collected five days after split-thickness skin grafts were applied to patients whose wounds were covered by a hydrocolloid dressing.[78] However, neither study used a keratinocyte sheet preparation as comparator. In a pig burn model, a combination of rPDGF-BB and rKGF produced faster and more stable epithelial and dermal repairs than either alone, and this was better than EGF.[79] On balance, whatever the nature of the effect(s), it is likely that a biological overlay, whether integrated into the patient or not, will tend to produce faster healing than none. Whether this is actually "fast" is still questionable, since many such studies still take several months for ulcers to heal, and may be compared to compression bandages only, rather than a competitor product.[61] In contrast, less than two weeks would be needed for a meshed skin autograft to be applied to a suitable bed (Nanchahal, personal communication). A recent report of the use of (unmeshed) Apligraf *as a dressing* over meshed autografts applied to excised burn wounds showed some beneficial effects compared to either a meshed autograft dressing or no further cover than a non-biological dressing.[80] One case report exists of the successful use of Apligraf as a graft substitute in a patient with a small area of full thickness burns to the dorsum of the foot.[81] It remains to be seen whether this product will fulfil the promise of regular engraftment within the burn context.

DERMAGRAFT

Dermagraft is a cellular composite bilayer made of a nylon polymer mesh, vicryl (Biobrane), into which collagen is cast and human neonatal fibroblasts are seeded. There is a top covering of silastic to protect the patient until a further operation is carried out to replace the epidermal layer. It was reported that Biobrane itself (nylon mesh plus collagen, overlaid with silastic) could reduce contractures in rat skin wounds studies,[82] and was used successfully for covering partial thickness burn[20,83] and toxic epidermal necrolytic wounds.[84] There were reports of some adverse effects also, such as toxic shock syndrome[85] and contact dermatitis.[86] The uses of Biobrane *per se* have been reviewed recently.[87]

Dermagraft has an extensive literature since its introduction, with much use in the ulcer[24,63,88-92] and burn[22,23,93,94] clinics.

Ulcers

Diabetic foot ulcers, treated for 8 weeks with Dermagraft as a topical dressing, healed faster and without recurrences. It was noted that there was a dose effect of the treatment — more frequent replacement resulted in faster healing, often by 14 weeks, compared to conventional 25 weeks treatment, and that the costs of purchase were offset by the speed of recovery.[88,91,92] It has been cryopreserved successfully, but many biochemical parameters decline after this treatment. Thus viability of cells dropped to 60%, protein synthesis fell by over 70%, but did recover to 45–85% after 48 hours, post-thawing. Several cytokines also displayed decreased synthesis (vascular endothelial growth factor, granulocyte colony stimulating factor, platelet-derived growth factor).

Burns

There have been several reports of using Dermagraft on full-thickness burns to replace the dermis, and over which a split-thickness autograft has been placed. In a study of 17 patients with paired meshed grafts, one over Dermagraft, it was found that the "take" was slightly better without the dermal substitute, and that there was some expulsion of the vicryl fibres in the meshed interstices, though in general these were dissolved in two to four weeks. Elastic fibre appearance was not seen in either group after one year.[22] A further study compared frozen cadaver allograft and Dermagraft (not frozen) for the temporary cover of burn wounds in patients with an average of 44% total body surface area (TBSA), of which 1% received the experimental grafts for a minimum of 5 days before removal and autografting. The Dermagraft-treated site was the same or better than the allografted site regarding "take" of the autograft.[23] Another trial compared frozen Dermagraft (without viability control), cryo-preserved Dermagraft and human cadaver allograft for temporary cover of burn wounds as above. The subsequent meshed split-thickness autografts all took equivalently, without evidence of Dermagraft rejection, a feature noted for four of the ten patients' allografts.[94] The role of bacteria was assessed in another study, which found that mice grafted with Dermagraft or a simple skin graft both retained their grafts to the same extent, and their viability was equal.[93]

These studies suggest that the performance of Dermagraft is not quite good enough to justify its general use. This may be due to the lack of any long-term study which needs to show persistence of the graft, good cosmesis and contour correction without rejection and vicryl expulsion or other defect. Its use as a temporary dressing, while adequate, has not resulted in a quantum leap in patient outcomes. Time will tell whether the potential will be realised.

TRANSCYTE (ADVANCED TISSUE SCIENCES)

This product is a rebranding of Dermagraft-TC (temporary cover). There have been three reports of its use to date on burns patients.[29-31] A study of 11 patients with partial thickness flame burns compared Transcyte with 10 patients with bacitracin ointment two to three times a day open treatment. There was a halving of time to re-epithelialise (7 versus 13 days), and the patients suffered less pain. It was not clear whether any incorporation of the graft occurred. However, the study seems to show a marked "dressing effect".[29] The second study used 14 patients with paired wound sites which either received Transcyte or silver sulfadiazine. Time to heal was faster with the dressing (11 versus 18 days), without infections and with less hypertrophic scarring at up to 12 months.[30] This study did not assess any incorporation into the patient, and can be regarded as another dressing experiment. Nevertheless, there was improvement over a non-biological dressing.

The third study used 24 patients with full-thickness burns who were treated with either autograft, allograft or Transcyte to study their metabolic energy expenditures when on parenteral nutrition. All patients suffered raised metabolism on each type of treatment regimen. No attempt was made to assess how far the grafts incorporated and were remodelled or how many cells survived, however.

INTEGRA (INTEGRA LIFESCIENCES CORP., PLAINSBORO, NJ, USA)

This dermal substitute material has been used for over a decade in the treatment of skin wounds (reviewed by Sefton[95]). It consists of a bovine collagen membrane with crosslinked chondroitin-6-sulphate to which an outer removable polymer of polysiloxane (silastic) is bonded. Its use generally consists of removal of damaged tissue, dermal replacement with Integra, waiting a period for blood vessel ingrowth, followed by surface epidermal recovering, either with a split-thickness autograft or cultured autologous keratinocyte sheets. Two early multicentre hospital trials showed good "takes" and little immunological effects in a large number of patients with full thickness burns.[25,26] By contrast a model using guinea pigs[96] showed good "take" as such at two weeks, but less satisfactory blood vessel ingrowth, in only 14 of 20 animals grafted. Four animals displayed no neovasculature.

Nevertheless, Integra has shown considerable value in treating burns victims with an otherwise poor prognosis. Boyce et al.[27] reported three patients with greater than 60% body surface area burns who had Integra followed by autografts of keratinocyte sheets 14 days later instead of split thickness autografts. No allodermis was used. Post-operative histology showed a stable epidermis and a remodelling dermis. The skin remained hypopigmented at 28 days. The grafts survived dorsal positioning with the aid of air-beds.

There remains a problem with contractures which may mean further refining of the product for its use. Hunt et al. reported using Integra on patients undergoing reconstruction in the neck region.[97] While Integra engraftment resulted in an initially good cosmetic outcome, contractures recurred in over half the patients within about four years. This was put down to insufficient immobilisation at early post-operative times.

Integra has also been used in extreme clinical settings. One report of a case of purpura fulminans, a condition characterised by haemorrhagic necrosis of the skin, often requiring amputations, was treated using Integra.[98] The patient's upper amputation stumps were successfully grafted with Integra, whereas the lower extremities were not, and needed replacement. The artificial skin acted as a predictor of which areas could be successfully grafted with the patient's own split thickness autograft.

Further development of the use of Integra continued with its use in an animal model of human skin grafts, where the dermal substitute was covered with keratinocytes prior to grafting.[99] The histological findings were good, with little contracture and a good functional outcome. This raises the possibility of one-step engraftment, which would be of benefit if translated into patient practice.

The combined use of a mineralised bone matrix (Grafton Flex or Putty) to repair the cranium, over which Integra was placed to reconstitute the scalp has been reported in a rabbit model.[100] The outcomes were promising, and auger well for more complex surgery using artificial tissues.

LASERSKIN (FIDIA ADVANCED BIOPOLYMERS, ABANO TERME, ITALY)

This material is based on hyaluronic acid membranes with a series of tiny holes made with a laser, to allow exudate to pass through. It has been used as a matrix upon which to grow skin cells, which can be grafted. There have been three reports of its use so far.

Andreassi *et al.*[34] tested it on a patient with vitiligo, using cultured autologous keratinocytes and melanocytes on a Laserskin membrane, which was applied to an unpigmented area from which the epidermis had been cryogenically removed by liquid CO_2. Repigmentation was variable, with six of eleven patients being complete by 18 months, in four there was 40–71% repigmentation, and in one the graft was lost due to infection. The authors concluded that the graft was a further option for patients with refractory vitiligo.

Lam *et al.*[33] examined the use of Laserskin on rats, and on patients with full-thickness burns. The material was seeded on either side with combinations of autologous and allogeneic cultured keratinocytes or fibroblasts. The seeding of rat and human keratinocytes was improved by the presence of allogeneic fibroblasts, still further by mitomycin C-treated mouse 3T3 cells. The rat grafts showed highest takes (80%) when allogeneic fibroblasts were seeded on both sides of the Laserskin, with co-cultured autologous keratinocytes on the outer aspect. The groups receiving outer cultures of allogeneic fibroblasts and autologous keratinocytes, or autologous keratinocytes alone had lower "takes" of 40% and 35% respectively. Two burns patients were studied using either autologous fibroblasts and keratinocytes or allogeneic fibroblasts and autologous keratinocytes, both on the outer surface of the Laserskin. Takes were 60–100%, which was considered good.

A third report of Laserskin use was by Chan et al.[32] on elective reconstruction in three patients. Scar revision was performed by use of Integra as a dermal substitute, followed by Laserskin cultured with autologous keratinocytes. "Takes" of 50–100% were reported, with reepithelialisation at up to three weeks in all patients.

OTHER MODELS

Chimaerism

Several reports exist on the use of mixtures of grafts and/or cells to induce some form of graft tolerance in the host. Intermingling of autogenic skin into allogeneic full-thickness grafts has resulted in survival of the latter, where complete rejection along classical lines would be expected.[101,102] Instead, the dermal matrix is accepted, whereas the adnexae are lost, and do not reform. Elasticity is retained on a comparable level to meshed autografts,[101] while the levels of T helper and suppressor cells are higher in the islands of autogeneic tissue rather than the allograft, as were the Langerhans cells in the epidermis over allografted areas.[102]

The concept of intermingling cells has been extended recently using different mouse strains as a source of keratinocytes which were grown in culture for seven days in various ratios to generate sheets of cells, and then grafted subcutaneously. C3H/He and Balb/C keratinocytes were mixed from 1:1 up to 1:15 ratios. Control 100% allogeneic sheets implanted into C3H/He mice were rejected, whereas all other sheets survived. The allogeneic cells were gradually replaced by host cells.[35] A similar gradual replacement was suggested when organotypic grafts generated in vitro using human allogeneic skin cells were seen to survive for up to 2.5 years on sex mismatched hosts.[69] If more widely studied, this method of mixing keratinocytes could be promising for extending the acceptability of allogeneic skin, especially when combined with the use of allodermis, so long as suitable screening for transmissible diseases is successful.[103]

Polyactive

This is a synthetic biodegradable dermal substitute consisting of a co-polymer of polyethyleneoxide and polybutyleneterephthalate (polyether and polyester), having a dense upper layer and a porous lower layer.[39,40] The material is permeable to water vapour and to molecules up to the size of IgG (150,000 Da). It becomes well vascularised at two weeks after implanting, but very little of the material persists one year after grafting. A modified graft dermal material with variable pore sizes (Biskin-M) was grafted to Yucatan mini-pigs, and studied for up to two years after seeding with auto- and allogeneic fibroblasts. Initial wound adherence was good, revascularization was extensive at two weeks, and contractures were small. Polymer fragmentation was seen at three months, with particles ingested by macrophages, a process accelerated by the absence of fibroblasts. Neo-collagen deposition was similar to host tissue.[41] This model has been refined to enable human cultured fibroblasts and keratinocytes to be grown, with differentiation qualities of the epidermal component similar to the hyperproliferative state *in vivo*, but without basement membrane formation.[42] The model has not yet been reported for use in the burn clinic.

Sacchachitin

This material is a chitinous polysaccharide extracted from the fungal fruiting bodies of the basidiomycete *Ganoderma tsugae*. It is a copolymer of β-1,3-glucan and N-acetylglucosamine (60:40 approximately). After grafting to full thickness excisional wounds in rats, little contraction occurred by 28 days, compared to control.[43] *In vitro*, the material shows some toxicity to dermal fibroblast proliferation at a concentration of over 0.1%. On subcutaneous injection into rats, neither a suspension of the material nor a woven membrane produced an immune response, but there was vascularization and granulation tissue formation. However, it did attract significant numbers of inflammatory cells.[104]

IN VITRO CROSSLINKED COLLAGEN

This material is based on a collagen gel, but with added chondroitin-6-sulphate crosslinked by various bifunctional reagents. The crosslinking was needed to prevent leaching of the chondroitin from the gel. Putrescine was found to be the most "beneficial" linker on the grounds that it promoted the growth of keratinocytes grown on the top, over that seen with diamino hexane and 1-ethyl-3-(3-dimethylaminopropyl) carbodiimide (EDAC).[47] 20% chondroitin-6-sulphate incorporation resulted in a significant rise in the *in vitro* strength of the gel, while crosslinking with EDAC, putrescine or 1,1-carbonyldiimidazole (CDI) had little additional benefit. CDI was inhibitory to the growth of fibroblasts into the gel.[46] However, the crosslinked gels are still contracted under the influence of fibroblasts. The presence of chondroitin-6-sulphate did not alter this parameter. The crosslinking polyamines putrescine and diaminohexane proved to limit gel contraction.[45]

CONCLUSIONS

There are several commercially available skin substitutes and more models in development. It remains to be seen which will be successful over the long term, particularly as far as the patient is concerned. The grafting done to date has fallen short often in not comparing the skin substitute with the current gold standard of a split-thickness autograft. There are also rather a lot of "dressing" type experiments, which do not fully address the biological consequences of the graft.

Nevertheless, there are grounds for optimism that with time the goal of a skin replacement available "on tap" will be achieved. The way the models are progressing, it is likely that future refinements will permit fully allogeneic engraftment in a one-step procedure,[105] but using a matrix that can fully withstand the hostile burn wound environment and yet not result in contractures that of necessity need revisions. This would permit early wound debridement and engraftment, which should benefit both the patient biologically, and reduce the costs of post-operative treatment if early discharge is enabled. The risks of immune and allergic reactions, while not yet eliminated, are looking smaller than in the past, and should still improve further. The use of cloned human collagens as a basis for dermal substitutes ought to be the final goal, though the market for purified animal collagens may persist for patent reasons. No doubt the Collagen Corporation is developing its answer to this.

REFERENCES

1. Gallico GG. III 3rd. Biologic skin substitutes. *Clin Plast Surg* 1990; **17**:519–526.
2. Nanchahal J, Ward CM. New grafts for old? A review of alternatives to autologous skin. *Br J Plast Surg* 1992; **45**: 354–363.
3. Cairns BA, deSerres S, Peterson HD, Meyer AA. Skin replacements. The biotechnological quest for optimal wound closure. *Arch Surg* 1993; **128**:1246–1252.
4. Phillips TJ. Biologic skin substitutes. *J Dermatol Surg Oncol* 1993; **19**:794–800.
5. Gallico GG, O'Connor NE, Compton CC, Kehinde O, Green H. Permanent coverage of large burn wounds with autologous cultured human epithelium. *N Engl J Med* 1984; **311**:448–451.
6. Rheinwald JG, Green H. Serial cultivation of strains of human epidermal keratinocytes: The formation of keratinizing colonies from single cells. *Cell* 1975; **6**:331–344 (1975).

7. Bell E, Ivarsson B, Merrill, C. Production of a tissue-like structure by contraction of collagen lattices by human fibroblasts of different proliferative potential *in vitro*. *Proc Natl Acad Sci, USA* 1979; **76**:1274–1278.

8. Bell E, Ehrlich HP, Buttle DJ, Nakatsuji T. Living tissue formed *in vitro* and accepted as skin-equivalent tissue of full thickness. *Science* 1981; **211**:1052–1054.

9. Bell E, *et al.* The reconstitution of living skin. *J Invest Dermatol* 1983; **81**:2s–10s.

10. Harriger MD, Hull BE. Cornification and basement membrane formation in a bilayered human skin equivalent maintained at an air-liquid interface. *J Burn Care Rehabil* 1992; **13**:187–193.

11. Maruguchi T, *et al.* A new skin equivalent: Keratinocytes proliferated and differentiated on collagen sponge containing fibroblasts. *Plast Reconstr Surg* 1994; **93**:537–544; discussion 545–546.

12. Nolte CJ, *et al.* Ultrastructural features of composite skin cultures grafted onto athymic mice. *J Anat* 1994; **185**:325–333.

13. Auger FA, *et al.* Skin equivalent produced with human collagen. *In Vitro Cell Dev Biol Anim* 1995; **31**:432–439.

14. Lafrance H, Yahia L, Germain L, Guillot M, Auger FA. Study of the tensile properties of living skin equivalents. *Biomed Mater Eng* 1995; **5**:195–208.

15. Terino EO. Alloderm acellular dermal graft: Applications in aesthetic soft-tissue augmentation. *Clin Plast Surg* 2001; **28**:83–99.

16. Tsai CC, Lin SD, Lai CS, Lin TM. The use of composite acellular allodermis-ultrathin autograft on joint area in major burn patients — one year follow-up. *Kaohsiung J Med Sci* 1999; **15**: 651–658.

17. Eaglstein WH, Falanga V. Tissue engineering and the development of Apligraf, a human skin equivalent. *Clin Ther* 1997; **19**:894–905.

18. Eaglstein WH, Falanga V. Tissue engineering and the development of Apligraf a human skin equivalent. *Adv Wound Care* 1998; **11**:1–8.
19. Trent JF, Kirsner RS. Tissue engineered skin: Apligraf, a bilayered living skin equivalent. *Int J Clin Pract* 1998; **52**:408–413.
20. Hansbrough JF, et al. Clinical experience with Biobrane biosynthetic dressing in the treatment of partial thickness burns. *Burns Incl Therm Inj* 1984; **10**:415–419.
21. Purdue GF, et al. Biosynthetic skin substitute versus frozen human cadaver allograft for temporary coverage of excised burn wounds. *J Trauma* 1987; **27**:155–157.
22. Hansbrough JF, Dore C, Hansbrough WB. Clinical trials of a living dermal tissue replacement placed beneath meshed, split-thickness skin grafts on excised burn wounds. *J Burn Care Rehabil* 1992; **13**:519–529.
23. Purdue GF, et al. A multicenter clinical trial of a biosynthetic skin replacement, Dermagraft-TC, compared with cryopreserved human cadaver skin for temporary coverage of excised burn wounds. *J Burn Care Rehabil* 1997; **18**:52–57.
24. Mansbridge J, Liu K, Patch R, Symons K, Pinney E. Three-dimensional fibroblast culture implant for the treatment of diabetic foot ulcers: metabolic activity and therapeutic range. *Tissue Eng* 1998; **4**:403–414.
25. Michaeli D, McPherson, M. Immunologic study of artificial skin used in the treatment of thermal injuries. *J Burn Care Rehabil* 1990; **11**:21–26.
26. Stern R, McPherson M, Longaker MT. Histologic study of artificial skin used in the treatment of full-thickness thermal injury. *J Burn Care Rehabil* 1990; **11**:7–13.
27. Boyce ST, Kagan RJ, Meyer NA, Yakuboff KP, Warden GD. The 1999 clinical research award. Cultured skin substitutes combined with Integra Artificial Skin to replace native skin autograft and allograft for the closure of excised full-thickness burns. *J Burn Care Rehabil* 1999; **20**:453–461.

28. Orgill DP, Straus FH, Lee RC. The use of collagen-GAG membranes in reconstructive surgery. *Ann N Y Acad Sci* 1999; **888**:233–248.

29. Demling RH, DeSanti L. Management of partial thickness facial burns (comparison of topical antibiotics and bio-engineered skin substitutes). *Burns* 1999; **25**:256–261.

30. Noordenbos J, Dore C, Hansbrough JF. Safety and efficacy of TransCyte for the treatment of partial-thickness burns. *J Burn Care Rehabil* 1999; **20**:275–281.

31. Noordenbos J, Hansbrough JF, Gutmacher H, Dore C, Hansbrough WB. Enteral nutritional support and wound excision and closure do not prevent postburn hypermetabolism as measured by continuous metabolic monitoring. *J Trauma* 2000; **49**:667–671; discussion 671–672.

32. Chan ES, et al. A new technique to resurface wounds with composite biocompatible epidermal graft and artificial skin. *J Trauma* 2001; **50**:358–362.

33. Lam PK, et al. Development and evaluation of a new composite Laserskin graft. *J Trauma* 1999; **47**:918–922.

34. Andreassi L, Pianigiani E, Andreassi A, Taddeucci P, Biagioli M. A new model of epidermal culture for the surgical treatment of vitiligo. *Int J Dermatol* 1998; **37**:595–598.

35. Suzuki T, Ui K, Shioya N, Ihara S. Mixed cultures comprising syngeneic and allogeneic mouse keratinocytes as a graftable skin substitute. *Transplantation* 1995; **59**:1236–1241.

36. Reagan BJ, Madden MR, Huo J, Mathwich M, Staiano-Coico L. Analysis of cellular and decellular allogeneic dermal grafts for the treatment of full-thickness wounds in a porcine model. *J Trauma* 1997; **43**:458–466.

37. Konigova R, Matouskova E, Broz L. Burn wound coverage and burn wound closure. *Acta Chir Plast* 2000; **42**:64–68.

38. Matouskova E, et al. Treatment of burns and donor sites with human allogeneic keratinocytes grown on acellular pig dermis. *Br J Dermatol* 1997; **136**:901–907.

39. Beumer GJ, van Blitterswijk CA, Bakker D, Ponec M. A new biodegradable matrix as part of a cell seeded skin substitute for the treatment of deep skin defects: A physico-chemical characterisation. *Clin Mater* 1993; **14**:21–27.

40. Beumer GJ, van Blitterswijk CA, Ponec M. Biocompatibility of a biodegradable matrix used as a skin substitute: An *in vivo* evaluation. *J Biomed Mater Res* 1994; **28**:545–552.

41. Van Dorp AG, *et al.* Dermal regeneration in full-thickness wounds in Yucatan miniature pigs using a biodegradable copolymer. *Wound Repair Regen* 1998; **6**:556–568.

42. van Dorp AG, Verhoeven MC, Koerten HK, van Blitterswijk CA, Ponec, M. Bilayered biodegradable poly(ethylene glycol)/poly(butylene terephthalate) copolymer (Polyactive) as substrate for human fibroblasts and keratinocytes. *J Biomed Mater Res* 1999; **47**:292–300.

43. Su CH, *et al.* Fungal mycelia as the source of chitin and polysaccharides and their applications as skin substitutes. *Biomaterials* 1997; **18**:1169–1174.

44. Su CH, *et al.* Development of fungal mycelia as skin substitutes: Effects on wound healing and fibroblast. *Biomaterials* 1999; **20**, 61–68.

45. Osborne CS, Reid WH, Grant MH. Investigation into the biological stability of collagen/chondroitin-6-sulphate gels and their contraction by fibroblasts and keratinocytes: The effect of crosslinking agents and diamines. *Biomaterials* 1999; **20**:283–290.

46. Osborne CS, Barbenel JC, Smith D, Savakis M, Grant MH. Investigation into the tensile properties of collagen/chondroitin-6-sulphate gels: The effect of crosslinking agents and diamines. *Med Biol Eng Comput* 1998; **36**:129–134.

47. Hanthamrongwit M, Reid WH, Grant MH. Chondroitin-6-sulphate incorporated into collagen gels for the growth of human keratinocytes: The effect of cross-linking agents and diamines. *Biomaterials* 1996; **17**:775–780.

48. Barret JP, Dziewulski P, McCauley RL, Herndon DN, Desai MH. Dural reconstruction of a class IV calvarial burn with decellularized human dermis. *Burns* 1999; **25**:459–462.

49. Lattari V, et al. The use of a permanent dermal allograft in full-thickness burns of the hand and foot: A report of three cases. *J Burn Care Rehabil* 1997; **18**:147–155.

50. Wainwright DJ. Use of an acellular allograft dermal matrix (AlloDerm) in the management of full-thickness burns. *Burns* 1995; **21**:243–248.

51. Ghosh MM, Boyce S, Layton C, Freedlander E, Mac Neil S. A comparison of methodologies for the preparation of human epidermal-dermal composites. *Ann Plast Surg* 1997; **39**:390–404.

52. Chakrabarty KH, et al. Development of autologous human dermal-epidermal composites based on sterilized human allodermis for clinical use. *Br J Dermatol* 1999; **141**:811–823.

53. Dubin MG, et al. Allograft dermal implant (AlloDerm) in a previously irradiated field. *Laryngoscope* 2000; **110**:934–937.

54. Sheridan RL, Tompkins RG. Cultured autologous epithelium in patients with burns of ninety percent or more of the body surface. *J Trauma* 1995; **38**:48–50.

55. Achauer BM, VanderKam VM, Celikoz B, Jacobson DG. Augmentation of facial soft-tissue defects with Alloderm dermal graft. *Ann Plast Surg* 1998; **41**:503–507.

56. Dolynchuk K. et al. The role of Apligraf in the treatment of venous leg ulcers. *Ostomy Wound Manage* 1999; **45**:34–43.

57. Alvarez OM, Fahey CB, Auletta MJ, Fernandez-Obregon A. A novel treatment for venous leg ulcers. *J Foot Ankle Surg* 1998; **37**:319–324.

58. Fahey C. Experience with a new human skin equivalent for healing venous leg ulcers. *J Vasc Nurs* 1998; **16**:11–15.

59. Kirsner RS. The use of Apligraf in acute wounds. *J Dermatol* 1998; **25**:805–811.

60. Falabella AF, Schachner LA, Valencia IC, Eaglstein WH. The use of tissue-engineered skin (Apligraf) to treat a newborn with epidermolysis bullosa. *Arch Dermatol* 1999; **135**:1219–1222.

61. Falanga V, Sabolinski M. A bilayered living skin construct (Apligraf) accelerates complete closure of hard-to-heal venous ulcers. *Wound Repair Regen* 1999; **7**:201–207.

62. Langemo DK. Venous ulcers: etiology and care of patients treated with human skin equivalent grafts. *J Vasc Nurs* 1999; **17**:6–11.

63. Edmonds M, Bates M, Doxford M, Gough A, Foster A. New treatments in ulcer healing and wound infection. *Diabetes Metab Res Rev* 2000; **16 Suppl 1**:S51–54.

64. Falanga VJ. Tissue engineering in wound repair. *Adv Skin Wound Care* 2000; **13**:15–19.

65. Mathias SD, Prebil LA, Boyko WL, Fastenau J. Health-related quality of life in venous leg ulcer patients successfully treated with Apligraf: A pilot study. *Adv Skin Wound Care* 2000; **13**: 76–78.

66. Schonfeld WH, Villa KF, Fastenau JM, Mazonson PD, Falanga V. An economic assessment of Apligraf (Graftskin) for the treatment of hard-to-heal venous leg ulcers. *Wound Repair Regen* 2000; **8**:251–257.

67. Streit M, Braathen LR. Apligraf — a living human skin equivalent for the treatment of chronic wounds. *Int J Artif Organs* 2000; **23**:831–833.

68. Starzl TE, Murase N, Demetris A, Trucco M, Fung J. The mystique of hepatic tolerogenicity. *Semin Liver Dis* 2000; **20**:497–510.

69. Otto WR, Nanchahal J, Lu QL, Boddy N, Dover R. Survival of allogeneic cells in cultured organotypic skin grafts. *Plast Reconstr Surg* 1995; **96**:166–176.

70. Hafemann B, Frese C, Kistler D, Hettich R. Intermingled skin grafts with *in vitro* cultured keratinocyte — experiments with rats. *Burns* 1989; **15**:233–238.

71. Kistler, D., Hafemann, B. & Hettich, R. Cytogenetic investigations of the allodermis after intermingled skin grafting. *Burns* 1989; **15**:82–84.

72. Nanchahal J, Dover R, Otto WR. Allogeneic skin equivalents applied to burns patients. *Burns* 2002 *(in press)*.

73. Sabolinski ML, Alvarez O, Auletta M, Mulder G, Parenteau NL. Cultured skin as a "smart material" for healing wounds: Experience in venous ulcers. *Biomaterials* 1996; **17**:311–320.

74. Rennekampff HO, Kiessig V, Loomis W, Hansbrough JF. Growth peptide release from biologic dressings: A comparison. *J Burn Care Rehabil* 1996; **17**:522–527.

75. Marcusson JA, Lindgren C, Berghard A, Toftgard R. Allogeneic cultured keratinocytes in the treatment of leg ulcers. A pilot study. *Acta Derm Venereol* 1992; **72**:61–64.

76. Lindgren C, Marcusson JA, Toftgard R. Treatment of venous leg ulcers with cryopreserved cultured allogeneic keratinocytes: A prospective open controlled study. *Br J Dermatol* 1998; **139**:271–275 (1998).

77. Harris IR, *et al.* Cytokine and protease levels in healing and non-healing chronic venous leg ulcers. *Exp Dermatol* 1995; **4**: 342–349.

78. Ono I, Gunji H, Zhang JZ, Maruyama K, Kaneko F. Studies on cytokines related to wound healing in donor site wound fluid. *J Dermatol Sci* 1995; **10**:241–245.

79. Danilenko DM, *et al.* Growth factors in porcine full and partial thickness burn repair. Differing targets and effects of keratinocyte growth factor, platelet-derived growth factor-BB, epidermal growth factor, and neu differentiation factor. *Am J Pathol* 1995; **147**:1261–1277.

80. Waymack P, Duff RG, Sabolinski M. The effect of a tissue engineered bilayered living skin analog, over meshed split-thickness autografts on the healing of excised burn wounds. The Apligraf Burn Study Group. *Burns* 2000; **26**:609–619.

81. Hayes DW, Jr, Webb GE, Mandracchia VJ, John KJ. Full-thickness burn of the foot: Successful treatment with Apligraf. A case report. *Clin Pediatr Med Surg* 2001; **18**:179–188.

82. Frank DH, Brahme J, Van de Berg JS. Decrease in rate of wound contraction with the temporary skin substitute biobrane. *Ann Plast Surg* 1984; **12**:519–524.

83. Phillips LG, *et al*. Uses and abuses of a biosynthetic dressing for partial skin thickness burns. *Burns* 1989; **15**:254–256.

84. Bradley T, Brown RE, Kucan JO, Smoot EC, 3rd, Hussmann J. Toxic epidermal necrolysis: A review and report of the successful use of Biobrane for early wound coverage. *Ann Plast Surg* 1995; **35**:124–132.

85. Egan WC, Clark WR. The toxic shock syndrome in a burn victim. *Burns Incl Therm Inj* 1988; **14**:135–138.

86. Reed BR, Zapata-Sirvent RL, Kinoshita JR, Hansbrough J. Contact dermatitis to Biobrane. *Plast Reconstr Surg* 1985; **76**: 124–125.

87. Smith DJ. Use of Biobrane in wound management. *J Burn Care Rehabil* 1995; **16**:317–320.

88. Gentzkow GD, *et al*. Use of dermagraft, a cultured human dermis, to treat diabetic foot ulcers. *Diabetes Care* 1996; **19**: 350–354.

89. Naughton G, Mansbridge J, Gentzkow G. A metabolically active human dermal replacement for the treatment of diabetic foot ulcers. *Artif Organs* 1997; **21**:1203–1210.

90. Sacks MS, Chuong CJ, Petroll WM, Kwan M, Halberstadt C. Collagen fiber architecture of a cultured dermal tissue. *J Biomech Eng* 1997; **119**:124–127.

91. Grey JE, Lowe G, Bale S, Harding KG. The use of cultured dermis in the treatment of diabetic foot ulcers. *J Wound Care* 1998; **7**:324–325.

92. Allenet B, *et al*. Cost-effectiveness modeling of Dermagraft for the treatment of diabetic foot ulcers in the French context. *Diabetes Metab* 2000; **26**:125–132.

93. Economou TP, Rosenquist MD, Lewis RW, 2nd, Kealey GP. An experimental study to determine the effects of Dermagraft on skin graft viability in the presence of bacterial wound contamination. *J Burn Care Rehabil* 1995; **16**:27–30.

94. Hansbrough JF, *et al*. Clinical trials of a biosynthetic temporary skin replacement, Dermagraft-Transitional Covering, compared with cryopreserved human cadaver skin for temporary coverage of excised burn wounds. *J Burn Care Rehabil* 1997; **18**:43–51.

95. Sefton MV, Woodhouse KA. Tissue engineering. *J Cutan Med Surg* 1998; **3 Suppl 1**:S1-18–23.

96. King WW, Lam PK, Liew CT, Ho WS, Li AK. Evaluation of artificial skin (Integra) in a rodent model. *Burns* 1997; **23 Suppl 1**:S30–32.

97. Hunt JA, Moisidis E, Haertsch P. Initial experience of Integra in the treatment of post-burn anterior cervical neck contracture. *Br J Plast Surg* 2000; **53**:652–658.

98. Besner GE, Klamar JE. Integra Artificial Skin as a useful adjunct in the treatment of purpura fulminans. *J Burn Care Rehabil* 1998; **19**:324–329.

99. Kremer M, Lang E, Berger AC. Evaluation of dermal-epidermal skin equivalents ("composite-skin") of human keratinocytes in a collagen-glycosaminoglycan matrix (Integra artificial skin). *Br J Plast Surg* 2000; **53**:459–465.

100. Shermak MA, Wong L, Inoue N, Nicol T. Reconstruction of complex cranial wounds with demineralized bone matrix and bilayer artificial skin. *J Craniofac Surg* 2000; **11**:224–231.

101. Kistler D, Hafemann B, Hettich R. Morphological changes of intermingled skin transplants on rats. *Burns Incl Therm Inj* 1988; **14**:115–119.

102. Kistler D, Kauhl W, Hafemann B, Hofstadter F, Hettich R. Distribution of lymphocytes in intermingled skin grafts. *Burns* 1989; **15**:85–87.

103. Clarke JA. HIV transmission and skin grafts. *Lancet* 1987; **1**:983.

104. Hung WS, *et al.* Cytotoxicity and immunogenicity of Sacchachitin and its mechanism of action on skin wound healing. *J Biomed Mater Res* 2001; **56**:93–100.
105. Nanchahal J, Otto WR, Dover R, Dhital SK. Cultured composite skin grafts: Biological skin equivalents permitting massive expansion. *Lancet* 1989; **2**:191–193.

Chapter 5

Tissue Engineering in the Musculoskeletal System

Michael Sittinger, Olaf Schultz
& Thomas Häupl

INTRODUCTION

Skeletal tissue has to fulfil essential functions for movement. These depend on static properties, given by bone, and tight but flexible articulations, given by joints between bones. Both elements build a dynamic entity. Static of bone depends on mechanical load to form an internal trabecular network and to withstand mechanical load, bending, tension and traction. Articulations between individual bones are exposed to the same mechanical forces. In addition, there are special requirements needed to minimise friction on the joint surfaces. Therefore, internal and/or external ligaments together with the form of the articulation are essential for the geometric distribution of the mechanical forces. Furthermore, the cartilage, as highly specialized tissue, has developed on the articulating surfaces and is lubricated by the joint fluid produced from the synovial membrane.

In contrast to many organs like liver or kidney, where cells are the central unit of function, all elements of the skeletal tissues manage the mechanical requirements by an appropriate extracellular matrix. This matrix therefore builds the major component of these tissues, and cells are responsible for maintenance of these tissues by remodelling and repair of the complex extracellular network of organic and inorganic molecules. There are striking differences between bone, cartilage and ligaments concerning repair. In general, bone fractures will be organised by the powerful regenerative capacity of the periost. In contrast, a comparable repair of cartilage does not exist. Cartilage defects penetrating into the subchondral bone will be unspecifically replaced by a fibrous cartilage like tissue. Non-penetrating defects will not be reorganized at all. Both the scar and the defect are unable to fulfil their physiologic function and will become a centre of progressive cartilage degeneration and destruction. Similarly to cartilage, ligaments have also poor regenerative capacity and rupture usually needs surgical repair by adaptation of both ends and fixation to allow at least a fibrous but less robust repair.

According to these specificities of the skeletal tissues, certain requirements arise for tissue engineering. Cells need to be enabled to produce an enormous amount of matrix products, which have to accumulate around the cells. Quantity and quality of the individual tissue specific matrix components may be influenced by nutritional aspects, by factors supporting differentiation of the cells and by exposure to appropriate mechanical load. Vascularisation can be neglected for joint cartilage and ligaments as long as their volume–surface ratio allows sufficient nutrient supply by diffusion. Bone, however, cannot be produced as one piece in a size needed for transplantation, except that diffusion is facilitated or a pre-formed vascular network can be connected to the arterial and venous vessel system. With respect to these specific requirements, the following chapters will summarise and discuss the prerequisites and techniques for tissue engineering of skeletal organs.

AUTOLOGOUS ADULT CELLS

Tissue Engineering protocols generally require extensive handling of isolated autologous cells. Primarily, tissue samples from adult patients have to be dissociated by enzymes such as collagenase and hyaluronidase to remove the extracellular connective tissue matrix containing mainly collagens, proteoglycans and hyaluronic acid. Tissue handling, cell isolation and subsequent proliferation has to be carried out with special care to avoid contaminations or potential infections by media supplements. So far, most approaches for tissue repair by autologous cells use biopsies from healthy sites of the corresponding tissue such as joint cartilage for articular cartilage repair and nasal septal chondrocytes for facial plastic reconstruction. It is not clear, whether also nasal or auricular chondrocytes could potentially be used for joint repair. Meanwhile, research is increasingly focused on tissue regeneration by relevant precursor or multipotent stem cells. (For details, see Chapter 13, Buttery *et al.*) Three-dimensional tissue cultures have to provide an environment that triggers (re)differentiation and appropriate formation of connective tissue matrix. Mainly factors of the TGF-β superfamily are known and currently further investigated for their potential to differentiate *in vitro* formed artificial mesenchymal tissues. For a succesful transfer into clinics, two major criteria will have to be considered:

i) a simple and minimal invasive procedure to collect cells from the patient.
ii) differentiation of the functional key properties (e.g. mechanical stability) *in vitro* or *in vivo* within a short time.

CELL AND TISSUE DIFFERENTIATION BY MORPHOGENETIC FACTORS

The quality of the tissues is certainly influenced by the kind of nutrient supply, but it is strictly dependent on factors, which signal the differentiation towards a specific phenotype. Using expanded primary cells from the tissue to be constructed, the inherited determination supports this differentiation process. However, as dedifferentiation already occurs with expansion of the cells, specific factors may be obligatory for proper tissue maturation. These factors are even more essential for tissue engineering from precursor cells or mesenchymal stem cells. An increasing number of members of the TGF-β superfamily are known to direct morphogenesis and tissue formation during development. They are called bone morphogenetic proteins and their discovery has been initiated by Urist, using protein extracts from bone to induce ossification.[1] In another nomenclature called the "growth and differentiation factors (GDFs)", similar or identical molecules have been characterized within the last few years. Although their role is critical for the development of many organs beyond the skeletal tissues, these factors are precisely controlled[2] and may not only support maturation of tissues[3] but may also induce apoptosis of cells.[4] Very early, these proteins from demineralized bone have been shown to induce differentiation of mesenchymal precursors for example to form cartilage.[5] More recent work demonstrates that individual factors exhibit individual as well as overlapping effects. This may depend on the complexity of homo- and heterodimerisation of individual BMPs,[6] the promiscuity of binding to different types of receptors as well as the different pathways of intracellular signalling.[7]

Expression of certain BMPs has been associated with pathological ossification but may also support induction of ligaments[8] and cartilage. In adult tissue, a specific network of BMPs seems to be expressed for tissue homeostasis (Fig. 1) and maintenance of a differentiated phenotype.[9] Overwhelming stress and inflammation can deregulate this concert of morphogen action (Fig. 1) and may result in progressive deficiency of physiologic repair (unpublished observation). This riddle of specific effects, depending on the source of cells as well as the type, the concentration and the time of morphogen action is one of the most fascinating challenges for the engineering of human tissues.

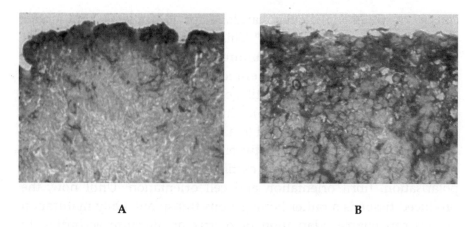

A   B

Fig. 1. Differential expression of BMP-7 MRNA in A normal and B inflamed synovial tissue, suggesting an important role in cartilage maintenance.

SYNTHESIS OF THE EXTRACELLULAR MATRIX

For the *in vitro* formation of an artificial tissue using a supporting structure with a high porosity, synthesis and aggregation of extracellular matrix, molecules are essential. In a monolayer or simple polymer fibre culture, these molecules dissolve in the surrounding media and distant remain distant to the cells.[10] This is completely different with gel cultures like agarose.[11] Large molecules like aggrecan only stay within the pericellular area so that intercellular matrix cannot develop.[12] To support the production of an extracellular matrix in a polymer fleece culture, the tissue is coated with a gel or, even better with a semipermeable membrane.[13] This way, macro molecular proteoglycan and collagens can be kept inside the artificial tissue. A high concentration or density of hygroscopic matrix molecules is of special importance for the pressure resistance of the tissues.[14] Areas in the joint cartilage that show the best mechanical resistance have the biggest proteoglycan concentration. The enormous water storage capacity of these giant molecules produces tissue swelling. About three-quarters of the extracellular cartilage matrix consist of water.[15] Articular cartilage for example is kept under pressure of two atmospheres.[16] This biomechanical quality cannot yet be achieved with *in vitro* cultured tissue. It is known that non-chondrogenic cells supplemented with hyaluronic acid and proteoglycan can produce a cartilage-like matrix if they have enough hyaluronic acid receptors (CD44).[17] In future, adding these substances to the cell suspension could lead to a far-improved matrix production.

An important difference between artificially made extracellular matrix and matrix of native cartilage tissue is the lack of any polarisation, fibril orientation and cell orientation. Until now, the produced tissue is a rather homogenous transplant. Only maturation *in vivo* can enable adaptation of matrix architecture according to natural forces and requirements.[18] Applying growth factors such as bone morphogenetic proteins along gradients could probably solve this problem.

BIOMATERIALS FOR TISSUE CONSTRUCTION
(see also Chapter 1, Hench *et al.*)

Specifically designed biomaterials are one of the fundamental and driving tools in tissue engineering. The formation of skeletal tissues derived from expanded cell cultures requires new techniques to engineer three-dimensional tissues (Fig. 2). Biomaterials may shape and guide the tissue development *in vitro* and *in vivo*. For cartilage and bone tissues, a suitable biomaterial provides or supports initial mechanical stability, and allows even cell distribution and a good biocompatibility.[19] The biomaterial has to promote rather than hinder the phenotypic development *in vitro* and *in vivo*. Many investigators frequently applied gels such as agarose or alginate. The major advantage of these gels is a good cell distribution and three-dimensional immobilisation, however their mechanical behaviour is usually not sufficient for transplantation in skeletal tissue repair. Initial biomechanical stability can be achieved mainly by solid resorbable fibre scaffolds or other porous structures. Possibly the

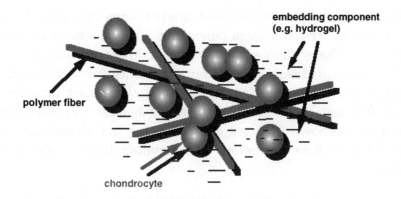

Fig. 2. *In vitro* tissue formation by scaffold fibers and embedding substances (Sittinger, 1994).

best solutions are techniques that combine the advantages of both fibre structures and gels. In general, the amount of any biomaterials should be kept at the possible minimum per volume to avoid negative effects such as toxic degradation products or inflammatory responses. Low concentrations of e.g. ultra low melting agarose, fibrin, alginate or hyaluronic acid are suitable substances for cell embedding that can be applied within scaffolding polymers like PGLA or PLLA fiber fleeces.[20] (Fig. 2). This technical approach allows an even cell distribution that is less dependent on uniformity or density of the scaffold structure. Rather convenient in handling is the loading of polymer fleeces with a cell suspension containing a low concentration of fibrin which is subsequently polymerized by thrombin.[21]

BIOREACTORS FOR BONE AND CARTILAGE ENGINEERING

In the last decade, isolated human cells have been grown as monolayers in flasks or petri dishes. Complex organs or tissues are often too difficult to engineer, but single human cells provide good method to understand development and functionality. Thus, human cells *in vitro* have allowed countless experiments to investigate the pathogenesis of diseases including approaches for their treatment.

Although our knowledge of cell biology increased considerably in recent years, the methods of handling human cells have hardly changed. As shown in recent experiments, conventional monolayer cultures have their limitations in generating highly differentiated structures.

i) The cells lack the typical extracellular matrix, so that the cells are cultured on an inappropriate support.
ii) The metabolic conditions within the culture medium are unstable.
iii) Finally, high density and long term cultures are at risk of bacterial or fungal contamination during frequent manual medium replacement procedures.[22,23]

Artificial tissue constructs can be cultured in perfusion culture systems (Fig. 3) to stabilise culture conditions. This is important due to the high nutrient consumption in high density cultures and acidic degradation products from the polymers.[24] Perfusion not only

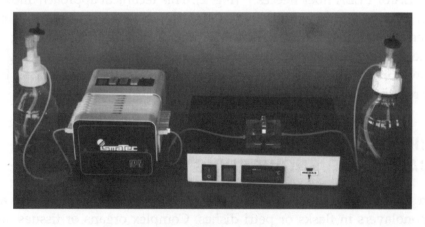

Fig. 3. Perfusion culture unit.

stabilises the components provided by the culture medium, it also, like *in vivo*, stabilises secreted autocrine factors such as morphogenetic signals and does not allow overreaction of paracrine factors similar to the situation *in vivo*. In addition, important advantages could be provided by novel gradient chambers to establish a concentration gradient of differentiating morphogenic factors across artificial tissues similar to conditions found during embryonic development.

IN VIVO DEVELOPMENT OF ENGINEERED CONNECTIVE TISSUES

1) Cartilage Transplants

Matrix formation and tissue morphology

Before engineered tissues could be made available for clinical application, extensive evaluation of tissue development *in vivo* is required. Frequently, the thymus aplastic nude mouse is used for a preliminary investigation to analyse the development of engineered transplants *in vivo* independent of future clinical indications.[25,26] These studies provide data on biomaterial degradation, matrix formation and morphological features as well as changes in tissue shape, size and also mechanical properties (Fig. 4). Depending on the applied experimental procedure and the cell source used, the engineered cartilage can be developed into histologically mature cartilage. However, collagen type II is not sufficiently produced when adult human cells are expanded through several passages in monolayer. A variety of different animal models has been used for joint cartilage regeneration by tissue engineering. The most frequently used and well accepted model is the rabbit. In those experiments articular cartilage defects have shown promising results when treated with *in vitro* engineered cartilage discs.[27,28] The macroscopical as well as the histological examination showed a cartilage-typical tissue development (Fig. 5).

Fig. 4. Histological appearance of transplanted cartilage after 14 weeks in nude mice. The transplant is usually surrounded by a fibrous tissue.

Fig. 5. Engineered cartilage tissues transplanted in a knee joint of a rabbit.

In some cases, areas of tissue or even the whole transplant changed to fibrotic tissue with stretched cell morphology. The reasons for that are not completely understood. It is thought that phenotypic differentiation might depend on the concentration of several growth factors near the transplant and possibly also on the surgical procedure.

Tissue surface and transplant protection

The most important problem currently faced in autologous bone and cartilage engineering is the longterm stability of the transplants *in vivo* against inflammatory and immunological processes.[29,30] There is a significant danger that the tissues could be invaded by surrounding host cells or by immune cells. Part of current research is therefore focused on projects dealing with tissue protection or encapsulation. Possible strategies are enclosing transplantable constructs with gels, membranes or cell layers[31,32] (Fig. 6). Ideally, such tissue protections should be resorbable and should not trigger any immune response. In native cartilage the perichondrium possibly

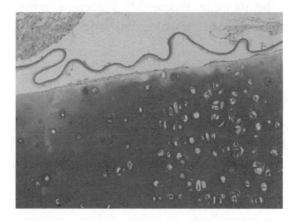

Fig. 6. Cartilage transplant in a nude mouse encapsulated by a semi-permeable membrane (polyelectrolyte complex membrane).

functions as a cartilage protecting tissue. Currently, engineered cartilage tissues usually lack surrounding cell layers similar to perichondrium. When artificial cartilage is transplanted in nude mice, a fibrous capsule is frequently formed possibly protecting the transplants from resorption or rejection processes.

2) Bone Transplants

Cells from the cambium layer of the periost may be the first contributors together with osteoblasts in driving the process of bone repair with the development of the initial fracture callus and subsequent remodelling.[33] Therefore, periosteal cells are expected to have important regenerative potential for bone tissue engineering. To analyse bone repair by engineered transplants, critical size bone defects in suitable animal models need to be chosen. In contrast to articular cartilage lesions, smaller bone defects are very likely to heal without treatment. Studies with critical size rabbit calvaria and ulna defects have shown that implanted cell-polymer tissue constructs are able to bridge the defects with osteoid tissue while carrier materials such as polymer fibers and fibrin alone were replaced by fibrous scar tissue.[34,35] Different scores have been established to evaluate histological and X-ray data in bone regeneration.

3) Biomechanical Properties

Recent studies on *in vitro* engineered cartilage show that prolonged culture times allow the growth of tissues with good mechanical stability and vitality.[36] *In vivo* tissue engineered cartilage achieved values in pressure resistance and stiffness comparable to native human septal cartilage.[37] However, so far it is not clear to what extent cartilage transplants in joint defects develop the natural mechanical properties, as the techniques for analysis are more complex. The histology of regenerated cartilage in joints usually reveals clear differences to neighbouring native cartilage: The distribution of cells appears at random and lacking a typical column formation with no indication of a typical collagen architecture. It is also questionable whether there is enough water binding capacity to provide the considerable hydroelastic tissue properties.

This is similarly true for engineered bone. Currently, most X-ray and histology is applied to evaluate potential healing of defects and fractures. In future, biomechanical parameters of regenerated bone tissue will have to be addressed.

4) Immunological Aspects

Two major immunological risks still exist even from autologous engineered bone and cartilage transplants. (a) Even when the amount of any biomaterial used is minimized, the presence of foreign body giant cells or granulocytes are seen invading the tissues, attracted by the scaffolds or other substances. (b) The engineered tissue is not completely mature and possibly exposes epitopes of cell surface or matrix proteins which are not usually "seen" by the immune system and may therefore be treated as "foreign". Immunological studies on patients receiving cartilage transplants have shown humoral reactions against mainly collagens type IX and XI, associated to collagen type II fibrils.[38,39] Even though tissue engineering is usually intended as an autologous therapy, it appears that solving immunological problems will be the major challenge to clinical application.

CLINICAL ASPECTS

1) Cell Preparation

Tissue Engineering requires extensive handling of autologous cells, which have been sampled from the patient and should finally be given back to the patient. This requires a careful manipulation at all stages to exclude possible infective agents or other types of contamination. The usual method of culturing cells in foetal calf cannot be recommended because of the possible prion contamination in the serum.[40] Up to now, it has not been possible to grow chondrocytes in serum-free culture media. Therefore, serum from the patient has to be used for all steps in tissue engineering.

The proliferation of chondrocytes of the human nasal septum has been investigated thoroughly. Proliferation rates up to 1000-times can be achieved.[41] A correlation between proliferation potential and patients age was not found. It seems more difficult to grow human articular chondrocytes. In contrast, articular cartilage cells of adult cattle or rabbit can be expanded easily.

2) Plastic Surgery and Shaping of Artificial Tissues

For reconstructive facial surgery, pre-shaping of artificial tissues is important. In general, there are two possibilities. Either the resorbable scaffold structure is being produced of the desired shapes, or a less rigid structure together with the cell suspension allows pre-shaping *in vitro*. Subcutaneously implanted structures in nude mice have proved to be useful in long-term experiments. Complex shapes such as an auricle can be prepared more accurately using pre-made silicon negatives (Fig. 7). However, it remains to be proven, whether such transplants can achieve convincing clinical results, especially due to a lack of elastic fibers in engineered cartilage which are an important component of the native external ear cartilage.

3) Repair of Injured Joint Cartilage

For joint cartilage replacement, pressure resistance and fixation of the transplant onto the bone is more important and difficult than in plastic surgery. Presently, tissue grown *in vitro* does not have the necessary stability for the treatment of larger joint surfaces. Animal studies with 5 mm tissues implanted in artificial defects in rabbit

Fig. 7. Artificial cartilage can be formed to require three-dimensional shapes.[42]

joints are not comparable with the replacement or regeneration of a whole joint surface. Two main problems have to be solved in the future: (a) the mechanical stability and (b) the fixation of the cartilage transplant in the joint. To achieve a sufficient mechanical stability, a technique has to be developed which increases the matrix density of the artificial cartilage tissue. Currently, studies are performed using heterotopic transplantation or extended *in vitro* cultures with hyaluronic acid to address the biomechanical problem.

In theory, the artificially grown cartilage layers could be attached directly onto the defect joint surface using fibrin glue, or it could be fixed using resorbable pins. However, so far it seems unlikely that by this method, an efficient connection between transplant and subchondrial bone can be achieved. Therefore, artificial cartilage tissue should be cultured directly on porous calcium carbonate.[43] This biomaterial is replaced by bone tissue after transplantation. The aim is to achieve a permanent, solid connection between cartilage and bone tissue.

Cartilage repair in chronic joint diseases by gene therapy

Inflammatory joint diseases are aggravated by cytokines like IL-1[44,45] and TNF-α, which activate metalloproteinases and thereby destroy extracellular matrix and the loss of cartilage. To regenerate cartilage and protect transplants from further destruction, artificial cartilage engineering can be combined with therapies to increase the regenerative potential and the matrix production as a new strategy in inflammatory and destructive joint diseases. A regenerative and anti-inflammatory potency of the transplant can be achieved by transferring genes of the TGF-β-superfamily (bone morphogenetic proteins).[46,47] The technique of *ex vivo* or *in vitro* gene therapy has the particular advantage that

  i) a defined population of cells is genetically modulated,
 ii) the effects of the gene therapy can be tested to control for a devastating outcome e.g. by accidental tumorigenesis,
iii) the dasage for optimal differentiation and protection can be controlled.

4) Bone Regeneration

Critical size bone defects whether caused by trauma, tumour excision, congenital malformation or aseptic loosening of prosthesis often require the transplantation of bone tissue or substitutes to restore bone integrity. So far, autologous bone grafts are considered as the "gold standard" in bone repair. However, autografts are available in limited amounts. Free bone grafts with microsurgical vascular anastomosis were used successfully for the repair of bone defects[48,49] but the limited availability of donor sites and patient morbidity are not advantageous. In contrast, the application of allogenic bone material avoids these problems. However, it carries a higher risk of infections and is also more expensive.[50,51]

    The implantation of biomaterials such as ceramics and biopolymers is based upon their osteoconductive properties, their structural flexibility and the potential to serve as a delivery system for bioactive molecules. Now the transplantation of osteogenic cells in these suitable carrier systems is an encouraging new approach to further enhance the process of bone reconstitution and remodelling (see further details in Chapter 1, Hench *et al.*).

Thus, such attempts have to focus on the synergistic interaction of the two key players:

i) Autologous cells, which have to possess a high osteogenic potential, and are easy to obtain and handle. Osteoblasts, periosteal cells, chondrocytes and pluripotent mesenchymal progenitor cells have been studied so far for this purpose.[52–55]

ii) The application of a biocompatible, osteoconductive, resorbable carrier substrate, which supports mesenchymal cell attachment and facilitates rapid vascularization. Furthermore, the ideal vehicle has to provide temporary mechanical strength and design flexibility for defined indications. Fibrin as well as resorbable polymers may be used as a cell delivery system although their properties are rather different. Fibrin can be considered as a cell embedding material, whereas resorbable polymers provide a more mechanically stable scaffold.[56] Fibrin was also used in combination with beta-Tri-Calcium-Phosphate (β-TCP),[57] hydroxyapatite[58] or demineralized bone[59] to fill bone defects. Biodegradable polymers have been applied as structural scaffolds, delivery systems for morphogenic proteins and osteogenic cells for the reconstruction of segmental bone defects, calvaria defects and in oral-maxillo-fascial surgery.[60–62]

IN VITRO APPLICATIONS

Tissue engineering approaches are mainly focused on the restoration of pathologically altered tissues and organs based on the transplantation of cells in combination with supportive matrices and biomolecules. However, the establishment of vital transplants was paralleled with new ways of cell culture systems: complex three-dimensional cell cultures in gels (collagen, agarose, alginate, matrigel, fibrin) or polymers (PLA, PGLA) within specific bioreactor modules. Thus, a complex cellular microenvironment was created mimicking the *in vivo* situation more closely than conventional cell cultures. Especially, mesenchymal cells such as chondrocytes and osteocytes undergo a process of phenotypic and functional dedifferentiation when cultured in monolayer systems lacking the essential influence of physiological cell–cell and cell-matrix interactions. A growing body of evidence indicates that these interactions, which directly influence cell signalling via cell adhesion molecules such as integrins or cadherins are of paramount importance for nearly all cell functions. Three-dimensional cell cultures provide the advantage of anchorage independent cell growth allowing cell motility, the synthesis of a specific pericellular or intercellular matrix and the physiological release and storage of bioactive molecules such as cytokines and morphogenic factors. These advanced cell culture systems open new perspectives for various experimental settings concerning the investigation of cell function under physiological and pathophysiological circumstances to supplement animal experiments. *In vitro* models and assay systems

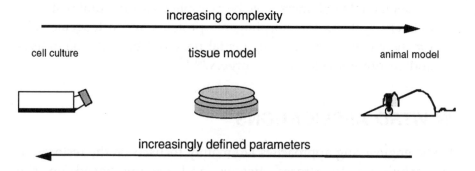

Fig. 8. Potential of tissue models to replace animal experiments such as in arthritis research and drug testing.

are currently used for the study of cellular differentiation or the influence of various biomolecules on cellular morphology and behaviour including the over-expression or deletion of defined genes (gain of function/loss of function) in inducible systems. Thus these techniques could have great potential to study induction processes in developmental biology, primarily the domain of animal models.

Newly developed culture models are designed for the investigation of interactions between different cell populations along with their specific extracellular matrix structures. This is advantageous since a defined set of cells can be investigated under reproducible and specific culture conditions. The experimental strategy can focus on various aspects of cell dynamics: structural and functional changes, cellular activation, migration, infiltration or degradation of the pericellular matrix (tumor models, models for destructive joint diseases), synthesis of specific proteins or apoptosis. The recently designed bioreactor systems allowing the establishment of separate cellular compartments depending on the experimental setting (serial culture, mixed culture, direct cell-cell contact). As a consequence, *in vitro* tests for drugs or bioactive molecules can be developed.[63]

MESENCHYMAL STEM CELLS AND FUTURE PERSPECTIVES

A basic tissue engineering principle of musculoskeletal tissue repair is the delivery and integration of functionally active cells within an appropriate carrier system into the original tissue to restore the pathologically altered architecture and function. This approach comprises the interactive triad of responsive cells, a supportive matrix and bioactive molecules promoting differentiation and regeneration. Within this interplay, mesenchymal cells are the major keyplayers driving the process of tissue remodelling, regeneration and maintenance via the synthesis of specific matrix proteins and the establishment of cell-cell and cell-matrix interactions. The ideal

Application	Example
in vitro-assay	test system for drugs, cytokines, morphogenic factors, inhibitors for proteolytic enzymes
infection model	investigation of infiltration and persistence of microorganism in a 3D-cell culture, interaction with mononuclear or other cells
morphogenesis-model	induction of proliferation and differentiation in an interactive 3D-culture
establishment of tissue transplants	combination with polymer scaffolds as supportive structures
angiogenesis-model	endothelial cells interacting with tumour cells, inflammatory cells etc.
cell migration	migration of mononuclear cells, fibroblasts etc. in an extracellular matrix, chemotaxis, cell adhesion, homing, cell infiltration
immunological studies	interaction of T-cells with macrophages, Antigen-Presenting-Cell (APC's) fibroblasts etc. in context of the extracellular matrix during the cellular immunological response
application of genetically altered cells	transfection of mesenchymal cells for the expression of morphogens, cytokines etc., interaction with resident cells (chondrocytes, osteocytes)

candidate for these cell therapies has to face several challenges: cells should be easy to obtain, autologous with the capacity of unlimited but controlled expansion and the potential to differentiate into various mesenchymal tissues depending on positional cues and the corresponding microenvironment. Since the availability of autologous differentiated cells such as osteocytes and chondrocytes is limited and their functional termination seems counter-productive for regeneration, interest has switched to uncommitted mesenchymal progenitor cells. Recent evidence indicates that even differentiated tissues contain populations of undifferentiated multipotent cells (stem cells, progenitor cells) that have the capacity for tissue regeneration after trauma, disease or ageing.[64-66]

Pluripotent embryonic stem cells, successfully cultured from human fetal tissue and able to differentiate into virtually every tissue and organ of the body, open fascinating perspectives for cell replacement strategies.[67-69] The rudimental knowledge about their differentiation pathway and the ethical debate concerning their origin remain an issue. Thus, tissue engineering approaches focused on progenitor cells of the corresponding tissues: Bone, cartilage, muscle and tendon are derived from mesenchymal stem cells (MSC), which are responsible for the morphogenesis, differentiation and homeostasis of these mesenchymal tissues. The major source of these pluripotent cells is the bone marrow, where they reside as supportive cells for hematopoiesis and possibly as a reservoir and regeneration pool for the various mesenchymal tissues.[70,71] In analogy to hematopoietic cells, MSCs commit to specific lineages, a process referred as mesengenesis.[72] During embryonic development, cells of the mesoderm differentiate in a series of signalling events into the various mesenchymal tissues. Regeneration and tissue repair seem to involve similar cellular transitional events observed during embryogenesis. However, the pace of these events in tissue homeostasis in the adult organism is much slower and the decline of uncommitted progenitor cells and the missing of corresponding signals prevents an effective rejuvenation or a complete regeneration of a complex tissue. Thus, local tissue turnover and repair activity of MSCs involves a multistep event: chemoattracted mesenchymal progenitor cells from resident sites as well as distant reservoirs group together and mitotically expand to form a repair blastema. Further differentiation along specific lineage progresses over several distinctive transitions and is determined by the local concentration of surrounding tissue-specific cytokines and growth factors. The process is based on the complex interaction between a set of defined bioactive molecules and responsive cells in a temporal and sequential manner.[73,74] These mesenchymal stem cells were characterised by their ability to proliferate in culture with a specific phenotype, by the expression of a consistent set of surface proteins and by their extensive differentiation to multiple mesenchymal lineages under specific controlled culture conditions.[75-78] As meso-dermal cells they are adherent and migratory, thus the purification and culture expansion of MSCs takes advantage of this capacity. Frequency of MSC in bone marrow varies between 1/104 and 1/106 and is decreasing with the age of the donor. The isolation from bone marrow is performed via density gradient. A small percentage of cells of a defined density interface (1.073g/ml) attach and grow as fibroblastic cells, developing into colonies after 5–7 weeks.[79]

In studies, cells comprised a single phenotypic population by flow cytometric analysis of expressed surface antigens: CD44, CD71, CD90, CD106, CD120a, CD124, but negative for markers of the hematopoietic lineage such as CD14, CD34, CD45. The functional characterisation of MSCs revealed beside the expression of surface molecules, a distinct pattern of secreted cytokines, which include Il-6, Il-11, LIF, GM-CSF.

No spontaneous differentiation was noted during expansion, cells maintained a normal kariotype and telomerase activity even at higher passages. The ability for specific differentiation pathways was studied in several *in vitro* assays by controlled culture conditions:

i) adipogenic pathway: addition of 1-methyl-3 isobutylxanthine, dexamethasone, insulin, indomethacin.

ii) chondrogenic pathway: pelleted micromass culture, without serum under influence of TGF beta3, cells developed a multilayered matrix-rich morphology (proteoglycans, collagen type II).

iii) osteogenic pathway: addition of dexamethasone, beta-glycerol phosphate, ascorbate, 10% FBS, cells exhibiting nodule aggregation and alkaline phosphatase expression.

Identification of MSCs *in situ* has remained elusive, in part, due to the relative paucity of specific molecular markers. Recently, several new antibodies against surface proteins of human MSC have been established. One of these referred as SB-10 is directed against surface proteins of uncommitted mesenchymal progenitor cells, which seems to be identical with ALCAM, a CD6 ligand. It is a type I membrane glycoprotein and member of the immunoglobulin-superfamily of cell adhesion molecules and can be found on lymphocytes, epithelial cells and monocytes. This molecule seems to be of functional importance since addition of SB-10 fragments to MSCs undergoing osteogenesis *in vitro* stimulates the differentiation process.[80]

Several experimental studies have been completed so far successfully to evaluate the potential of MSCs for their feasibility and efficacy to heal large osseous defects[81] or tendon defects[82] when embedded in appropriate carrier structures. In principle, transplantation of mesenchymal progenitor cells would attenuate or possibly correct genetic disorders of bone, cartilage and muscle, as shown in a study with children suffering from osteogenesis imperfecta.[83] Moreover, the functional properties of MSCs make them promising candidates for gene therapy strategies to enhance the process of tissue regeneration and repair and to deliver essential biological factors to restore and maintain tissue homeostasis. This will possibly include genetic disorders of these tissues, degenerative disorders such as osteoarthritis and osteoporosis or inflammatory diseases such as rheumatoid arthritis.[84,85]

REFERENCES

1. Urist MR. Bone: Formation by autoinduction. *Science* 1965; **150**:893–899.
2. DiLeone RJ, Russell LB, Kingsley DM. An extensive 3' regulatory region controls expression of Bmp5 in specific anatomical structures of the mouse embryo. *Genetics* 1998; **148**:401–408.

3. Wozney JM, Rosen V, Celeste AJ, Mitsock LM, Whitters MJ, Kriz RW, Hewick RM, Wang EA. Novel regulators of bone formation: Molecular clones and activities. *Science* 1988; **242**: 1528–1534.

4. Macias D, Ganan Y, Sampath TK, Piedra ME, Ros MA, Hurle JM: Role of BMP-2 and OP-1 (BMP-7) in programmed cell death and skeletogenesis during chick limb development. *Development* 1997; **124**:1109–1117.

5. Sampath TK, Nathanson MA, Reddi AH. *In vitro* transformation of mesenchymal cells derived from embryonic muscle into cartilage in response to extracellular matrix components of bone. *Proc Natl Acad Sci, USA*, 1984; **81**:3419–3423.

6. Israel DI, Nove J, Kerns KM, Kaufman RJ, Rosen V, Cox KA, Wozney JM. Heterodimeric bone morphogenetic proteins show enhanced activity *in vitro* and *in vivo*. *Growth Factors* 1996; **13**:291–300.

7. Heldin CH, Miyazono K, ten Dijke P. TGF-beta signalling from cell membrane to nucleus through SMAD proteins. *Nature* 1997; **390**:465–471.

8. Wolfman NM, Hattersley G, Cox K, Celeste AJ, Nelson R, Yamaji N, Dube JL, DiBlasio-Smith E, Nove J, Song JJ, Wozney JM, Rosen V. Ectopic induction of tendon and ligament in rats by growth and differentiation factors 5, 6, and 7, members of the TGF-beta gene family. *J Clin Invest* 1997; **100**:321–330.

9. Enomoto-Iwamoto M, Iwamoto M, Mukudai Y, Kawakami Y, Nohno T, Higuchi Y, Takemoto S, Ohuchi H, Noji S, Kurisu K. Bone morphogenetic protein signaling is required for maintenance of differentiated phenotype, control of proliferation, and hypertrophy in chondrocytes. *J Cell Biol* 1998; **140**:409–418.

10. Verbruggen G, Veys EM, Wieme N, Malfait AM, Gijselbrecht L, Nimmegeers J, Almquist KF, Broddelez C. The synthesis and immobilisation of cartilage — specific proteoglycan by human chondrocytes in different concentrations of agarose. *Clin Exp Rheumatol* 1990; **8**:371–378.

11. Bassleer C, Gysen P, Foidart JM, Bassleer R, Franchimont P. Human chondrocytes in tridimensional culture. *In Vitro Cell Dev Biol* 1986; **22**:113–119.
12. Verbruggen G, Veys EM, Malfait AM. Cornelissen M, Broddelez C, De Ridder L. Characterisation of the aggrecans synthesized by human chondrocytes cultured in suspension culture in agarose. XIIIth European Congress of Rheumatology, Amsterdam. 1995.
13. Sittinger M, Lukanoff B, Burmester GR, Dautzenberg H. Encapsulation of artificial tissues in polyelectrolyte complexes: Preliminary studies. *Biomaterials* 1996; **17**:1049–1051.
14. Hardingham TE, Muir H, Kwan MK, Lai WM, Mow VC. Viscoelastic properties of proteoglycan solutions with varying proportions present as aggregates. *J Orthop Res* 1987; **5**:36–46.
15. Lohmander S. Proteoglycans of joint cartilage: Structure, function, turnover and role as markers of joint disease. *Baillieres´s Clinical Rheumatology* 1988; **2**:37–57.
16. Burkhardt H, Hartmann F. Pathobiochemie und Pathobiomechanik. In *Rheumatologie Zeidler*, ed. Urban & Schwarzenberg, 1989.
17. Knudson W, Kundson CB. Assembly of a chondrocyte-like pericellular matrix on non-chondrogenic cells. Role of the cell surface hyaluronan receptors in the assembly of a pericellular matrix. *J Cell Sci* 1991; **99**:227–235.
18. Brittberg MA, Nilsson A, Lindahl C, Ohlsson, Peterson L. Rabbit articular cartilage defects treated with autologous cultured chondrocytes. *Clin Orthop* 1996; **326**:270–283.
19. Sittinger M, Bujia J, Rotter N, Reitzel D, Minuth WW, Burmester GR. Tissue engineering and autologous transplant formation: Practical approaches with resorbable biomaterials and new cell culture techniques. *Biomaterials* 1996; **17**:237–242.
20. Sittinger M, Bujia J, Hammer C, Minuth WW, Burmester GR. Engineering of cartilage tissue using bioresorbable polymer carriers in perfusion culture. *Biomaterials* 1994; **15**:451–456.

21. Haisch A, Schultz O, Perka C, Jahnke V, Burmester GR, Sittinger M. Tissue-engineering humanen Knorpelgewebes für die rekonstruktive Chirurgie unter Verwendung biokompatibler resorbierbarer Fibringel- und Polymervliesstrukturen. *HNO* 1996; **44**:624–629.

22. Minuth WW, Sittinger M, Kloth S. Tissue Engineering — generation of differentiated artificial tissues for biomedical applications. *Cell Tiss Res* 1998; **291**:1–11.

23. Minuth WW, Kloth S, Aigner J, Sittinger M, Röckl W. Approach to an organo-typical environment for cultured cells and tissues. *Biotechniques* 1996; **20**:498–501.

24. Sittinger M, Schultz O, Keyszer G, Minuth WW, Burmester GR. Artificial tissues in perfusion culture. *Int J Artif Org* 1997; **20**: 57–62.

25. Rotter N, Aigner J, Naumann A, Planck H, Hammer C, Burmester G, Sittinger M. Cartilage reconstruction in head and neck surgery: Comparison of resorbable polymer scaffolds for tissue engineering of human septal cartilage. *J Biomed Mater Res* 1998; **42**:347–356.

26. Haisch Rathert T, Jahnke V, Burmester GR, Sittinger M. *In vitro* engineered cartilage for auricular reconstruction. *Advances in Tissue Engineering and Biomaterials*, York, 1997.

27. Perka C, Sittinger M, Schultz O, Spitzer RS, Schlenzka D, Burmester GR. Tissue engineered cartilage repair using cryopreserved and noncryopreserved chondrocytes. *Clin Orthop Rel R* 2000; **378**:245–254.

28. Sittinger M, Perka C, Schultz O, Häupl T, Burmester G-R. Joint cartilage regeneration by tissue engineering. *Z Rheumatol* 1999; **58**:142–147.

29. Stoll W. Complications following implantation or transplantation in rhinoplasty. *Facial-Plast-Surg* 1997; **13**(1):45–50.

30. Bujia J, Alsalameh S, Naumann A, Wilmes E, Sittinger M, Burmester GR. Humoral immune response against minor collagens type IX and XI in patients suffering from cartilage

graft resorption after reconstructive surgery. *Ann Rheum Dis* 1994, **53**:229–234.

31. Haisch A, Gröger A, Radtke C, Ebmeyer J, Sudhoff H, Grasnick G, Jahnke V, Burmester GR, Sittinger M. Macroencapsulation of human cartilage implants: Polyelectrolyte complex membrane encapsulation. *Biomaterials* 2000; **21**:1561–1566.

32. Sittinger M, Lukanoff B, Burmester GR, Dautzenberg H. Encapsulation of artificial tissues in polyelectrolyte complexes: Preliminary studies. *Biomaterials* 1996, **17**:1049–1051.

33. Henricson A, Hulth A, Johnell O. The cartilaginous fracture callus in rats. *Acta Orthop Scand* 1987; **58**:244–248.

34. Breitbart AS, Grande DA, Kessler R, Ryaby JT, Fitzsimmons RJ, Grant RT. Tissue engineered bone repair of calvarial defects using cultured periosteal cells. *Plast Reconstr Surg* 1998; **101**: 567–574.

35. Redlich A, Perka C, Schultz O, Spitzer R, Häupl T, Burmester GR, Sittinger M. Bone engineering on the basis of periosteal cells cultured in polymer fleeces. *J Mat Sci* 1999; **10**:767–772.

36. Ma PX, Langer R. Morphology and mechanical function of long-term *in vitro* engineered cartilage. *J Biomed Mater Res* 1999; **44**:217–221.

37. Duda GN, Haisch A, Endres M, Geber C, Schroeder D, Hoffmann JE, Sittinger M. Mechanical quality of tissue engineered cartilage: Results after 6 and 12 *in vitro*. *J Biomed Mater Res* 2000; **53**: 673–677.

38. Sittinger M, Jerez R, Burmester GR, Krafft T, Spitzer W. Antibodies to collagens in sera from patients receiving bovine cartilage grafts. *Ann Rheum Dis* 1996; **55**:333–334.

39. Bujia J, Alsalameh S, Naumann A, Wilmes E, Sittinger M, Burmester GR. Humoral immune response against minor collagens type IX and XI in patients suffering from cartilage graft resorption after reconstructive surgery. *Ann Rheum Dis* 1994, **53**:229–234.

40. Gruber R, Sittinger M, Bujia J. Untersuchungen zur *in vitro* Kultivierung von Humanchondrozyten bei Einsatz FCS-freier Zuchtmedien: Minimierung des möglichen Risikos einer Infektion mit Erregern von Prionen-Erkrankungen. *Laryngo-Rhino Otol* 1996, 75:105–108.

41. Sittinger M, Bräunling J, Kastenbauer E, Hammer C, Burmester G, Bujia J. Untersuchungen zum Vermehrungspotential von Nasenseptum-Chondrozyten für die *in vitro* Züchtung von Knorpeltransplantaten. *Laryngo-Rhino Otol* 1997; 76:96–100.

42. Haisch A, Rathert T, Jahnke V, Sittinger M. *In vitro* engineered cartilage for external ear reconstruction. XVI World Congress of Otolaryngology Head and Neck Surgery, Sydney. 1997, 145.

43. Kreklau B, Sittinger M, Mensing M, Voigt C, Berger G, Burmester G-R, Rahmanzadeh R, Gross U. Tissue engineering of biphasic joint cartilage transplants. *Biomaterials* 1999; 20:1743–1749.

44. Goldring MB, Birkhead J, Sandell LJ, Kimura T, Krane SM: Interleukin 1 suppresses expression of cartilage-specific types II and IX collagens and increases types I and III collagens in human chondrocytes. *J Clin Invest* 1988; 82:2026–2037.

45. Campbell IK, Piccoli DS, Butler DM, Singleton DK, Hamilton JA. Recombinant human interleukin-1 stimulates human articular cartilage to undergo resorption and human chondrocytes to produce both tissue- and urokinase-type plasminogen activator. *Biochim Biophys Acta* 1988; 967:183–194.

46. Luyten FP, Chen P, Paralkar V, Reddi AH. Recombinant bone morphogenetic protein-4, transforming growth factor-beta 1, and activin A enhance the cartilage phenotype of articular chondrocytes *in vitro*. *Exp Cell Res* 1994; 210:224–229.

47. Häupl T, Sittinger M, Kaps C, Schultz O, Keyßer G, Gross G, Burmester G. Cartilage tissue engineering for implantation in inflammatory joint diseases: Applications for BMP gene therapy. 2nd Int. Conf. BMPs, Sacramento, 1997.

48. Dell PC, Burchardt H, Glowczewskie FP, Jr. A roentgenographic, biomechanical, and histological evaluation of vascularized and

non-vascularized segmental fibular canine autografts. *J Bone Joint Surg Am* 1985; **67**:105–112.

49. Taylor GI, Miller GD, Ham FJ. The free vascularized bone graft. A clinical extension of microvascular techniques. *Plast Reconstr Surg* 1975; **55**:533–544.

50. Buck BE, Malinin TI, Brown MD. Bone transplantation and human immunodeficiency virus. An estimate of risk of acquired immunodeficiency syndrome (AIDS). *Clin Orthop* **240**:129–136.

51. Buck BE, Malinin TI. Human bone and tissue allografts. Preparation and safety. *Clin Orthop* 1994; **303**:8–17.

52. Breitbart AS, Grande DA, Kessler R, Ryaby JT, Fitzsimmons RJ, Grant RT. Tissue engineered bone repair of calvarial defects using cultured periosteal cells. *Plast Reconstr Surg* 1998; **101**: 567–574.

53. Bruder SP, Kraus KH, Goldberg VM, Kadiyala S. The effect of implants loaded with autologous mesenchymal stem cells on the healing of canine segmental bone defects. *J Bone Joint Surg Am* 1998; **80**:985–996.

54. Kim WS, Vacanti CA, Upton J, Vacanti JP. Bone defect repair with tissue-engineered cartilage. *Plast Reconstr Surg* 1994; **94**: 580–584.

55. Vacanti CA, Vacanti JP. Bone and cartilage reconstruction with tissue engineering approaches. *Otolaryngol Clin North Am* 1994; **27**:263–276.

56. Perka C, Schultz O, Spitser RS, Lindenhayn K, Burmester GR, Sittinger M. Segmental bone repair by tissue engineered periosteal cell transplants with bioresorbable fleece and fibrin scaffolds in rabbits. *Biomaterials* 2000; **21**:1145–1153.

57. Reck R, Bernal Sprekelsen M. Fibrinkleber und Tricalciumphosphat-Implantate in der Mittelohrchirurgie. Eine tierexperimentelle Studie. *Laryngorhinootologie* 1989; **68**: 152–156.

58. Oberg S, Kahnberg KE. Combined use of hydroxy-apatite and Tisseel in experimental bone defects in the rabbit. *Swed Dent J* 1993; **17**:147–153.

59. Lasa C, Jr, Hollinger J, Drohan W, MacPhee M. Delivery of demineralized bone powder by fibrin sealant. *Plast Reconstr Surg* 1995; **96**:1409–1417.

60. Meikle MC, Mak WY, Papaioannou S, Davies EH, Mordan N, Reynolds JJ. Bone-derived growth factor release from poly(alpha-hydroxy acid) implants *in vitro*. *Biomaterials* 1993; **14**:177–183.

61. Meikle MC, Papaioannou S, Ratledge TJ, Speight PM, Watt Smith SR, Hill PA, Reynolds JJ. Effect of poly DL-lactide — coglycolide implants and xenogeneic bone matrix-derived growth factors on calvarial bone repair in the rabbit. *Biomaterials* 1994; **15**:513–521.

62. Renier ML, Kohn DH. Development and characterization of a biodegradable polyphosphate. *J Biomed Mater Res* 1997; **34**: 95–104.

63. Schultz O, Keyszer G, Zacher J, Sittinger M, Burmester G-R. Development of *in vitro* model systems for destructive joint diseases. Novel strategies to establish inflammatory pannus. *Arthritis and Rheumatism* 1997; **40**:1420–1429.

64. Prockop DJ. Marrow stromal cells as stem cells for nonhematopoietic tissues. *Science* 1997; **276**:71–74.

65. Bruder SP, Jaiswal N, Ricalton NS, Mosca JD, Kraus KH, Kadiyala S. Mesenchymal stem cells in osteobiology and applied bone regeneration. *Clin-Orthop* 1998; **355**:S247–256.

66. Yoo JU, Johnstone B. The role of osteochondral progenitor cells in fracture repair. *Clin-Orthop* 1998; **355**:S73–81.

67. Shamblott MJ, Axelman J, Wang S, Bugg EM, Littlefield JW, Donovan PJ, Blumenthal PD, Huggins GR, Gearhart JD. Derivation of pluripotent stem cells from cultured human primordial germ cells. *Proc Natl Acad Sci USA* 1998; **95**: 13726–13731.

68. Thomson JA, Itskovitz Eldor J, Shapiro SS, Waknitz MA, Swiergiel JJ, Marshall VS, Jones JM. Embryonic stem cell lines derived from human blastocysts. *Science* 1998; **282**:1145–1147.

69. Rathjen PD, Lake J, Whyatt LM, Bettess MD, Rathjen J. Properties and uses of embryonic stem cells: Prospects for application to human biology and gene therapy. *Reprod Fertil Dev* 1998; **10**: 31–47.

70. Gerson SL. Mesenchymal stem cells: No longer second class marrow citizens. *Nat-Med* 1999; **5**:262–264.

71. Young HE, Mancini ML, Wright RP, Smith JC, Black AC Jr, Reagan CR, Lucas PA. Mesenchymal stem cells reside within the connective tissues of many organs. *Dev-Dyn* 1995; **202**: 137–144.

72. Caplan AI. The mesengenic process. *Clin-Plast-Surg* 1994; **21**: 429–435.

73. Pittenger MF, Mackay AM, Beck SC, Jaiswal RK, Douglas R, Mosca JD, Moorman MA, Simonetti DW, Craig S, Marshak DR. Multilineage potential of adult human mesenchymal stem cells. *Science* 1999; **284**:143–147.

74. Bruder SP, Fink DJ, Caplan AI. Mesenchymal stem cells in bone development, bone repair, and skeletal regeneration therapy.[j] *Science* 1999; **284**:143–147.

75. Reddi AH. Bone morphogenetic proteins, bone marrow stromal cells, and mesenchymal stem cells. *Clin-Orthop* 1995; **313**: 115–119.

76. Mackay AM, Beck SC, Murphy JM, Barry FP, Chichester CO, Pittenger MF. Chondrogenic differentiation of cultured human mesenchymal stem cells from marrow. *Tissue-Eng* 1998; **4**: 415–428.

77. Kadiyala S, Young RG, Thiede MA, Bruder SP. Culture expanded canine mesenchymal stem cells possess osteochondrogenic potential *in vivo* and *in vitro*. *Cell-Transplant* 1997; **6**:125–134.

78. Majumdar MK, Thiede MA, Mosca JD, Moorman M, Gerson SL. Phenotypic and functional comparison of cultures of marrow-derived mesenchymal stem cells (MSCs) and stromal cells. *J-Cell-Physiol* 1998; **176**:57–66.

79. Pittenger MF, Mackay AM, Beck SC, Jaiswal RK, Douglas R, Mosca JD, Moorman MA, Simonetti DW, Craig S, Marshak DR. Multilineage potential of adult human mesenchymal stem cells. *Science* 1999; **284**:143–147.

80. Bruder SP, Ricalton NS, Boynton RE, Connolly TJ, Jaiswal N, Zaia J, Barry FP. Mesenchymal stem cell surface antigen SB-10 corresponds to activated leukocyte cell adhesion molecule and is involved in osteogenic differentiation. *J-Bone-Miner-Res* 1998; **13**:655–663.

81. Nevo Z, Robinson D, Horowitz S, Hasharoni A, Yayon A. The manipulated mesenchymal stem cells in regenerated skeletal tissues. *Cell-Transplant* 1998; **7**:63–70.

82. Young RG, Butler DL, Weber W, Caplan Al, Gordon SL, Fink DJ. Use of mesenchymal stem cells in a collagen matrix for Achilles tendon repair. *J-Orthop-Res* 1998; **16**:406–413.

83. Horwitz EM, Prockop DJ, Fitzpatrick LA, Koo WW, Gordon PL, Neel M, Sussman M, Orchard P, Marx JC, Pyeritz RE, Brenner MK. Transplantability and therapeutic effects of bone marrow-derived mesenchymal cells in children with osteogenesis imperfecta. *Nat-Med* 1999; **5**:309–313.

84. Riew KD, Wright NM, Cheng S, Avioli LV, Lou J. Induction of bone formation using a recombinant adenoviral vector carrying the human BMP-2 gene in a rabbit spinal fusion model. *Calcif Tissue Int* 1998; **63**:357–360.

85. Gordon EM, Skotzko M, Kundu RK, Han B, Andrades J, Nimni M, Anderson WF, Hall FL. Capture and expansion of bone marrow-derived mesenchymal progenitor cells with a transforming growth factor-beta1-von Willebrand's factor fusion protein for retrovirus-mediated delivery of coagulation factor IX. *Hum Gene Ther* 1997; **8**:1385–1394.

Chapter 6

Engineering the Liver

Clare Selden & Humphrey Hodgson

INTRODUCTION

A combination of biological and clinical facts combines to render the development of a functioning replacement system for the liver a demanding and complex task. These reflect first the complexity of the functions of the liver, and secondly the clinical circumstances under which liver replacement would be required.

Unlike many other organs whose function has been successfully replaced, the functions of the liver cannot be simply represented in physical or mechanical terms such as a pump, gas-exchanger, or a filter. The liver has been likened to a multi-functional chemical factory and reprocessing plant, with a sophisticated command and control system and multi servo mechanisms to ensure coordination of function. Many hundreds of distinct but often inter-related chemical reactions underlie the processing of carbohydrate and fat metabolism, the synthesis and secretion of many proteins, and the detoxification of both endogenous products and xenobiotics. Detoxification pathways are generally two-step, culminating in the production of water-soluble molecules excreted via the hepatocyte canaliculi into the biliary system, or excreted via the kidney.

From the clinical perspective, the diseases that severely impair liver function fall into two groups. First, acute insults that affect previously healthy people — generally viral or drug or toxin induced hepatitis, and second the chronic conditions such as cirrhosis. In chronic liver disease, the function of the liver is slowly imperiled by inflammatory processes that disturb the normal relationships between liver blood flow and liver cells, so that the normal architecture of the liver is obliterated; scarring and fibrosis in the liver precipitate the various manifestations of liver decompensation. Thus in advanced cirrhosis, the defective liver function may manifest as clouding of consciousness (reflecting diminished removal of endogenous toxins), bleeding into the gut (reflecting diversion of blood away from the liver in response to scarring) and fluid accumulation reflecting in part diminished plasma oncotic pressure from poor hepatic synthetic function.

A striking characteristic of the liver is, however, its ability to respond to an insult in which liver cells have died, by regeneration based on division of pre-existing cells. In the most extreme experimental example of this, the model of surgical removal of 70% of the liver in rodents, the normal liver size is restored in ten days by successive cycles of cell replication involving all the cell types of the liver. Even in chronic liver disease, there is however an increase above normal levels in the proliferative rate of hepatocytes and other liver cells.[1]

Currently the potential role of any tissue engineering project seeking to replicate liver function, and particularly the aims of an extracorporeal device to replicate liver function, are circumscribed. Fortunately, 98% of patients with viral hepatitis have an acute illness which is not life threatening. Of the 2% in whom it is, who go into acute hepatic failure, artificial liver replacement might buy time to allow the normal liver regenerative processes to take place. In addition, as we know substances that can inhibit liver function and repair accumulate in the blood of patients with liver failure, an artificial liver could also provide a more favourable milieu for the sick liver to repair and regenerate. A less ambitious but more immediately achievable aim for an artificial liver in acute liver failure is to buy time until a liver becomes available from a donor for transplantation. An artificial liver could similarly be used in chronic liver disease to buy time after an acute decompensation, either for some recovery of function to take place or for a donor organ to become available.

There has been no critical definition of the functions that need to be replaced to allow the above aims to be achieved, and therefore no firm definition of the minimal requirements. There is a school of thought that maintains that the critical 'failed' functions in acute liver failure are those of detoxification, and thus physical-mechanical approaches to detoxification, such as the use of adsorptive resins, may be adequate. However, there is no indication as yet that this approach will prove effective. Early experiments in the 1970s failed to prove efficacy of charcoal adsorption columns in patients with acute hepatic failure.[2] It is likely that the multi-step processes of detoxification in the liver cannot be adequately replaced by physical systems, and most workers have concluded that an artificial liver will need a major biological component.[3] The synthetic function of the liver has been thought to be less critical in the therapy of liver failure, as many liver-derived proteins can be replaced clinically using plasma-derived products. However if an artificial liver has a functional liver cell component, the synthetic function is unlikely to be disadvantageous.

Current investigation is concentrating on the problems of providing a bio-artificial liver for use as an extracorporeal liver support. In principle, this is a liver cell culture in a suitable chamber through which blood or plasma from a patient with liver failure can be passed. This article will first address progress in this field. The later portion of this will consider issues of *in vivo* liver cell engineering. This latter topic concerns the transplantation of hepatocytes into ectopic sites such as the spleen or the peritoneum, or into native liver, sometimes in association with natural or artificial support matrices. This is being investigated both in the context of the treatment of acute liver failure, and to replace individual functions in inborn errors of liver metabolism.

DEVELOPING THE BIOARTIFICIAL LIVER

The problems of development of a bioartificial liver include:

  i) the cell source.
 ii) the extracellular support system.
iii) oxygenation and nutrition of the system.
 iv) design of extracorporeal circuits to exchange with the patient's circulation.

The Cell Source

The liver contains many distinct cell sub-populations. Most liver-specific metabolic functions reside in hepatocytes, which are therefore the prime constituent of a bio-artificial liver. It is currently assumed that the supporting liver cells — sinusoidal endothelial cells, the macrophage-allied Kupffer cells, the stellate cells producing extracellular matrix, and biliary epithelial cells providing drainage, are not essential for a functioning bio-artificial liver. Whether that is true or not remains speculative, but for the time-being, the problems of providing functioning hepatocytes are sufficiently formidable for those considerations to take second place!

Accessing hepatocytes for a bio-artificial liver is a major challenge. In the adult, the vast majority of the functioning hepatocytes are non-dividing. Less than 1 in 300 are in mitosis at any one time. Thus, sourcing human cells to fill an extracorporeal liver is a major challenge. As a rough estimate of the order of 2×10^{11}, cells would be required to replace 100% of liver function. If whole human livers become available, they are in general and appropriately used for whole organ transplantation. One challenge is therefore to provide human liver cells that can divide and replicate in culture, to provide a constant and available cell source. The hepatocytes isolated from adult human liver have only limited potential to divide using current technologies.

To overcome this problem, there are a number of potential approaches. These include design of culture conditions and growth factor combinations that will allow adult hepatocytes to proliferate. Adult hepatocytes do maintain their potential to divide, demonstrable if the liver is subjected to major surgical resection to reduce mass, following which cells will go through cycles of cell division to restore hepatocyte mass and number. The use of hormonally defined media and growth factor stimulation has allowed cells to go through a number of cycles in culture, but thus far has not created conditions in which these vast cell numbers are achieved. In future, hepatocytes from human foetal sources,[4,5] or from neonatal liver,[6] have the potential to provide proliferating cell cultures. Fetal cells for example have high spontaneous mitotic rates, independent of growth factor stimulation. There are of course ethical and practical issues in accessing these cell sources.

The second approach is to manipulate hepatocytes genetically to create a proliferating cell population. This can be achieved in adult human hepatocytes by for example in our hands, the transfection of a cDNA expressing the SV40 large T antigen, which binds to the growth control protein P53 and initiates replicative cycling[7,8] Other immortalized cell lines (eg Hep Z) have been achieved by interfering with the cell cycle proteins Retinoblastoma gene product and P53 directly, as well as with the transcription factors E2F and D1.[9] In general, the transformation of cells by this sort of approach tends to reduce the expression of normal differentiated function. Hep Z maintains differentiated hepatocyte functions at lower levels than *in vivo*.[10] Even using a conditionally immortalised cell line which ceases replication at a non-permissive temperature — at which the large T antigen dissociates from the P53 — the level of differentiated function expressed is significantly below that expressed *in vivo* by hepatocytes. A report using foetal hepatocytes for transfection initially produced SV40 transformed hepatocytes with a poorly differentiated phenotype;[11] with the use of differentiating agents such as sodium butyrate albumin secretion increased 10-fold, and urea synthesis more than five-fold However, even those levels are several orders of magnitude less than observed *in vivo.*[12]

There are, however, recent reports of SV40 transfection of foetal human hepatocytes leading to clonal hepatocyte cell lines maintaining differentiated function. One such line is OUMS-29, and its potential utility has been demonstrated, not yet in a bio-artificial liver, but in cell transplant experiments. Intrasplenic transplantation of 20 million OUMS-29 cells protected rats from liver failure induced by 90% hepatectomy, affecting both hyperammonaemia and hepatic encephalopathy to prolong survival.[13]

An analogous approach is to use naturally transformed cells, and clones of the human hepatoblastoma-derived cell line Hep G2 have been used, and indeed formed the basis of one of the devices tried in clinical practice.[14] The performance of such cells is undoubtedly deficient in various respects from normal hepatocytes — and in general the limited detoxification capacity is most likely to prove problematic. However, there are potential means of enhancing the performance of such cells — by transfection of transcription factors for example,[15] or as discussed below by manipulation of their environment.

There is an additional and exciting approach that is under study — the exploitation of liver stem cells. In recent years, two probably distinct stem cell populations potentially capable of developing into hepatocytes have been demonstrated — most convincingly in rodents but with indications that similar systems pertain in man. It has been known for some time that if the normal proliferative potential for hepatocytes is impaired, for example by the administration of agents such as acetyl-aminofluorene, and a stimulus to hepatocyte proliferation then applied in vivo, a cell population proliferates — the oval cells, derived at the interface between hepatocytes and the biliary epithelium, and these cells are potentially capable of dividing into either hepatocytes or biliary cells.[16] Recent work suggests that a similar population may exist in man,[17] and be capable of proliferating under conditions of extreme stress such as acute severe hepatitis. Whether it will in the future be possible to isolate such cells to provide the starting point for developing large numbers of hepatocytes is a question that is beginning to be addressed. Rodent experiments suggest that purification procedures aimed at identifying small periportal hepatocytes (i.e. adjacent to the biliary apparatus) may provide hepatocytes with substantial replicative potential even from adult liver.[18,19]

Evidence has also accumulated from bone marrow transplantation experiments that bone marrow may contain cells that are capable of hepatocytic differentiation.[20-22] In animals treated by bone marrow ablation and subsequent marrow transplantation, subsequent analysis of the liver has shown than hepatocytes genetically identifiable as being of donor origin can be identified. In different systems, these have been found both when the recipient liver was subjected to damage — and therefore received a stimulus to repopulation — and in animals subjected to manoeuvres to damage the liver. Whether bone marrow harvesting can be harnessed to provide a source of stem cells for hepatocytes is one of the most intriguing avenues currently under investigation.

Faced with the difficulties of providing human hepatocytes in adequate quantities, some groups have turned to the use of primary porcine hepatocytes.[23-26] Anxieties about this strategy fall into two areas — firstly the bio-compatibility of the xenogeneic cells and the human immune system — which could both initiate antibody-mediated attack on the cells, or lead to deleterious consequences from incompatible pig proteins entering the human circulation.[27-29] These problems can in part be addressed by using diffusion barriers between cells and a patient's circulation to prevent high molecular weight molecule transfer, although it is likely this will diminish the efficiency of cell function provided. There are also safety concerns due to the presence of endogenous porcine retroviruses in pig cells, and these are under active debate. Thus far the objections are at the theoretical level, and such procedures as have been performed in man using porcine tissues (discussed in part below) have not shown infection of human cells.[30,31]

Extracellular Support and Oxygenation

Early experience with hepatocytes demonstrated that they rapidly lose liver specific function and then die in culture.[32,33] The last decade has documented many techniques which can maintain liver specific function.[34,35] These include co-culture with other cells of the liver, micro-gravity systems, and culturing in gels. A common feature of all these systems is that they maintain hepatocytes in a near-cuboidal shape reminiscent of that found *in vivo*. The link between cell shape and cell function involves interaction between surface integrins on the hepatocyte surface and extracellular matrix proteins.[36-38] The latter express repetitive integrin-binding domains, which by a combination of geometrical interaction, activation of intracellular signalling and activation of transcription factors, mediates the expression of adult differentiated function. The major role of non-hepatocytes in co-cultures, in which hepatocyte function is maintained, is probably to provide a source of extracellular matrix proteins.[39-41] In designing extra-corporeal liver cultures, extracellular matrix can conceivably be derived from a number of sources — non-parenchymal cells such as stellate cells,[42,43] exogenous sources,[44-46] or cooperation of non-parenchymal cells with hepatocytes themselves.[43]

A number of sources have been found for exogenous matrix proteins. The normal matrix proteins of the liver are a complex mixture with regional variation even in the liver parenchyma, and furthermore the synthesis and distribution of extracellular matrix can undergo rapid change, as a consequence of both degradation and synthesis, both during development and repair. The most prominent components are laminins, fibronectins and collagens, with complex and often multiple integrin-binding domains. Some workers have indicated that liver-derived matrix is superior to matrix proteins from other sources in maintaining differentiated function,[47,48] but others have found equivalent activity from sources such as the EHS tumour of mice.[49,50] This latter, which is the source of the most readily available matrix mixture for experimental purposes, is certainly capable or markedly up-regulating the expression of differentiated function by cultured hepatocytes. Issues such as the density of matrix, and thus the density of integrin-receptor interactions, are also critical.[51]

Our own experiments have indicated that if a three dimensional structure is imposed — in those experiments by culture in a gel formation alginate — hepatocyte-derived cell lines will produce a variety of matrix proteins, which are laid down intercellularly in patterns similar to those adjacent to hepatocytes in the adult liver (Fig. 1). As a strategy, persuading hepatocytes to synthesise their own extracellular support, rather then isolating matrix proteins and designing a structure in which cells and matrix become apposed, may be a helpful simplification.

However, an appropriate extracellular matrix providing the correct structure at the cellular level is not in itself sufficient to solve the issue of allowing up to 2×10^{11} cells to function adequately. The liver *in vivo* has a very large blood flow, the diffusion distances between liver cell and blood are small, and oxygen consumption is high. Providing an analogous system extracorporeally has only just begun to be addressed, using for example scaffolding systems such as a polyester matrix,[52,53] polypropylene membranes,[54] or polymers of poly-lactic, Polycaprolactone or polyglycolic acid. The scaffold design requires not only attention to the matrix itself but also to the void size,[55] a feature which has a dramatic effects on both cell adhesion, viability, and 3D aggregation. Micromachining on silicon and pyrex surfaces using photolithography can create trench patterns which mimic the architecture of natural vascular network.[56] Whilst

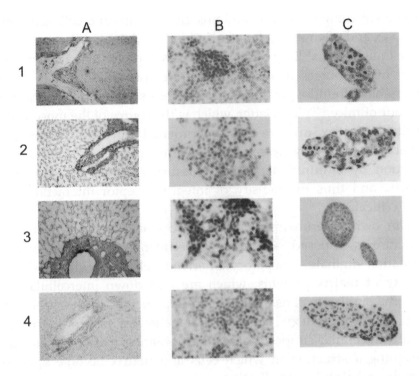

Fig. 1. Immunocytochemical demonstration of extracellular matrix proteins in human hepatocyte cell lines cultured as monolayers or in 3-dimensional spheroids cultured in alginate. A = normal human liver; B = monolayer HepG2 cells; C = 3D spheroids. 1 = collagen I; 2 = collagen III; 3 = collagen VI; 4 = negative control, no primary antibody.

small scale experiments show that these can support hepatocyte function, designing the micro-architecture to allow high volume blood flow, oxygenation and so on is a formidable task.

The Extracorporeal Circuit

Choices lie between directly exposing blood to the artificial liver device, or using a plasmapheresis step so that cells are exposed to plasma.[57] The latter has the disadvantage of added complexity, but it is more straightforward at the cell-culture end to bathe the cells in plasma than in blood. Most workers have used, for either system, some modification of a hollow fibre cartridge system.[26,58–63] At the most straightforward, hepatocytes are placed in the extra-capillary space, and plasma perfused in via the capillaries. By varying pressures across the system, plasma may be forced out of the capillaries and over the cells to increase exchange. In some systems to enhance turnover, a secondary circuit permits re-circulation of the effluent from the culture chamber several times before return to the patient.

There are however much more complicated systems. Nyberg *et al.* cultured hepatocytes entrapped in collagen gel within capillary membranes with a 100 kDa molecular mass cut-off, with perfusion fluid through the extracapillary space.[64] Gerlach *et al.* constructed a complex bioreactor with cells in an extracapillary space, but four different interwoven hollow fibre systems for plasma inflow, oxygen supply and carbon dioxide removal, plasma outflow, sinusoidal endothelial co-culture.[65] Time is allowed for the cells within the extracapillary space to establish cell to cell contact and alignment. Others have cultured hepatocytes on a polyester sheet wound in a "swiss-roll" shape to increase surface area.[66]

These systems have currently been used in either animal or clinical experiments or both. In clinical application, both human hepatoblastoma-derived cell lines (the C3A clone of Hep G2 cells), and primary pig hepatocytes have been tried. In brief summary, it has been shown that such bio-artificial machines can be utilised even in sick patients, although problems such as defibrination in the extracorporeal system do present risks.[62,67] Unequivocal evidence of benefit is not apparent, although there are some rapid changes in physiological status reported at the time of initial application — notably a fall in intracranial pressure.[25] Evidence of other replacement of function, either detoxificatory or synthetic, is scant. A number of patients have been "bridged to transplantation". It has been difficult to establish controlled trials to prove benefit, but these are now in progress.[62] These equivocal results contrast with a series of animal experiments in which evidence of useful hepatic detoxificatory function in the extracorporeal component, and prolongation of survival, has been reported with a number of bio-artificial liver circuits.[23,68-70] In summary, this should be regarded as a field which is attracting immense interest and experimental work; it has not yet yielded machines of proven clinical benefit; providing an appropriate cell source for use in patients is probably the biggest hurdle to be overcome.

IN VIVO LIVER CELL ENGINEERING

In contrast to bio-artificial systems intent on recreating useful liver function outside the body, *in vivo* replacement of liver function provides a different set of challenges. Orthotopic liver transplantation, now a routine procedure, provides the crudest as well as the most effective paradigm for *in vivo* liver engineering. In addition to whole organ transplantation for liver failure there have now been several developments in the field of cut-down and auxiliary liver transplants which have made a significant contribution to improving the lot of patients with hepatic failure in a society where the shortage of donor organs is ever increasing. These techniques rely on the ability of the liver to regenerate and re-create sufficient liver mass if a cut-down liver is used, or to take the pressure off the host liver if an auxiliary liver is implanted to allow time for the host liver to repair itself. These are however still major surgical techniques with the inevitable associated risk of mortality.

A more elegant approach is the use not of the liver organ, but of the liver cell. Since the demonstration that implanted liver cells can survive, proliferate and function, for months or years, harnessing this approach for genuine therapeutic gain has been a major challenge In principle, the technique could have wide application — implantation of normal cells to substitute for a single missing enzyme defect; re-implantation of autologous cells after they have been genetically modified as a more sophisticated approach to such inborn errors of metabolism; supplementing liver function short or long term in liver failure. As for the bio-artificial liver, the problem of the initial source of liver cells remains problematic in all except the autologous experiments, and even that implies the need for harvesting cells by surgery if hepatocytes are to be genetically altered. For some indications, only relatively small numbers of cells would be required — particularly inborn errors of metabolism, in which only a relatively small number of cells would suffice to replace missing function,[71-73] and furthermore as discussed below they might be under selective pressure to proliferate. In severe liver failure however, there would probably be a requirement for a very large number of cells both to replace global liver function and to do so within the hostile context of liver failure plasma.

Whilst the majority of specific liver function derives from the hepatocyte, non-parenchymal cells also play an interactive and essential part, and in fact such is the difficulty of obtaining 100% pure preparations of hepatocytes that most experiments have in fact been done with *de facto* mixed cell populations. Hepatocyte implantation for therapeutic gain raises a wide variety of issues which this section will address — the site of implantation and encouragement of engraftment; selective growth of implanted cells; application in animal models of acute and chronic disease, the early clinical experience in man; and finally future refinements of the approach — the use of cryopreserved cells, the use of stem cells, the prospect of xenogeneic transplantation, and adaptions of the technique for gene therapy.

SITE OF IMPLANTATION

There are numerous potential sites for implantation. Initial reports concerned spleen, fat pads, kidney capsule, peritoneum, lung, and the liver itself.[74-80] Whilst from a clinical standpoint the liver may prove the best site long term, there were initial concerns of induction of portal hypertension with the introduction of hepatocytes,[81] and perhaps more importantly during this developmental experimental phase, the problem of identifying transplanted cells on a background of host hepatocytes. The latter has now been solved by various cell-marking techniques, of which the most elegant is the Fisher rat system in which inbred animals lacking the dipeptidylpeptidase (DPPIV) enzyme are implanted with wild-type cells in which the normal enzyme can be stained histochemically.[82–84]

Hepatocytes are adherent cells, which normally lie within the portal circulation, constantly bathed in a growth factor and nutrient-rich blood supply; in the liver they are closely apposed to the sinusoidal cells and other non-parenchymal cells, generally in cords only a single cell thick, and closely related to extracellular matrix proteins as has already been discussed. Of the ectopic sites tried, the spleen was the most readily exploited. Hepatocytes transplanted by direct injection into the spleen invade the red, but not the white pulp of the spleen, and form cords reminiscent of the native liver[85,86] (Fig. 2). The proliferative rate of hepatocytes within the spleen is several fold greater than that in the liver of the same animal.[87] The extracellular matrix structure of the spleen presumably plays a significant role, both as a provider of scaffolding, and by interaction with hepatocyte integrins. Given the tight controls exerted on the liver to maintain sufficient mass to achieve normal liver function, and to respond appropriately to removal of mass by regeneration of only that same amount of mass, the fact that hepatocytes survive ectopically in the presence of a normal fully functional liver is remarkable. It is also of *importance* that this good survival and performance is seen in a milieu which is not receiving portal venous blood, as organ transplant experiments had emphasised the value of maintaining portal blood to prevent atrophy. However, a good blood supply alone is inadequate to allow cells to thrive. Hepatocytes survived poorly in the lung and did not proliferate in spite of good perfusion in pulmonary capillaries.[77]

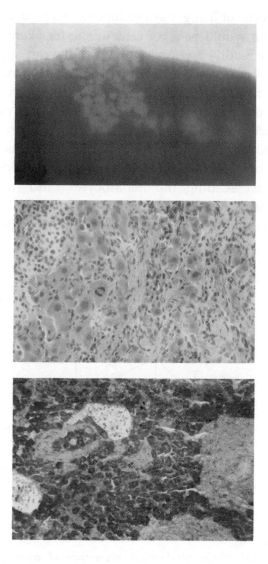

Fig. 2. Survival of rat hepatocytes transplanted intrasplenically in syngeneic rats, 10 months after implant. (a) macroscopic evidence of hepatocyte survival and proliferation in the spleen, (b) H&E of hepatocytes in spleen; note the hepatocyte cords, and occasional mitoses in hepatocytes lying within the red pulp of the spleen, (c) Glucose-6-phosphatase activity specific for hepatocytes (brown stain) as an indication of functional hepatocyte colonies on a negative spleen background.

Initial reports utilising the peritoneum were disappointing;[88-91] the peritoneum would be a convenient site for eventual clinical use since it can be readily accessed with a needle, can take a large mass of cells, and access could be repeated regularly if necessary. There might of course be problems from invading the peritoneal space, the potential for obstructing organs, if cells proliferate excessively and become space-occupying. Since hepatocytes are adherent, it is perhaps not surprising that the peritoneum was not readily a successful ectopic site. Attempts were made to provide the adherent milieu by first seeding the hepatocytes onto microcarriers or into polymeric sponges. The former was not successful, and the latter only marginally so if the pre-implantation culture conditions

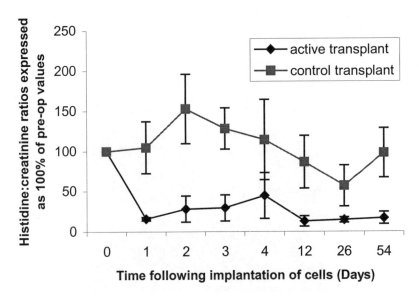

Fig. 3. Amelioration of the biochemical phenotype of murine histidinaemia in histidinaemic mice transplanted with a mixed population of liver cells isolated from wild type mice. Urinary histidine levels in histidase-deficient animals were monitored in two groups of mice. Active transplant group (n = 5) received histidase competent cells, control transplant group received cells from histidase-deficient donors (n = 6). Results, normalised to pre-transplantation values, are expressed as mean ± SEM.

were ideal. We explored the hypothesis that provision of the non-parenchymal cell population together with the hepatocytes would provide the necessary microclimate for hepatocytes to survive, function and proliferate within the peritoneal cavity. In a mouse model of histidinaemia, an inborn error of metabolism resulting from a single base mutation leading to inactive hepatic histidase, transplantation of a mixed population of liver cells led to complete amelioration of the biochemical phenotype, restoring urinary histidine levels to within the normal range (Fig. 3). This was associated with the appearance of ectopic islands of hepatocytes expressing histidase immunoreactivity, in contrast to the histidase deficient host liver.[72] These islands were distributed throughout the peritoneal cavity. A recent report using hepatocytes transfected with genes expressing Hepatocyte growth factor, TNF alpha and Vascular endothelial cell growth fector prior to implantation into the peritoneum demonstrated a significant improvement not only of tissue growth but of blood vessels, in particular, when the cells were transplanted in a restrictive mesh bag within the peritoneum; the blood vessels invaded the parenchyma within the bag.[92]

ENHANCEMENT OF ENGRAFTMENT AND ENCOURAGEMENT OF PROLIFERATION OF IMPLANTS

The survival of an implant must depend upon the immediate possibility of hepatocytes attaching, and whether they find an extracellular matrix compatible with survival, maintenance of differentiation, and able to support proliferation. It is reasonable to hypothesise that a stimulus to proliferation may enhance engraftment. This can be approached in a variety of ways experimentally, although the clinical usefulness of each approach is rather more limited. For example the strongest stimulus to hepatocyte proliferation *in vivo* is that of 70% partial hepatectomy, particularly in rats, and in that species the procedure increases proliferation of ectopically situated cells — for example a tenfold increase in proliferation of ectopic cells in the spleen, although this contrasted with a >100% stimulation of the remaining hepatocytes within the residual liver. However, in man such a surgical procedure is a major risk. Likewise, induction of damage to the recipient liver by a chemical insult such as galactosamine is also a stimulus to growth, but not readily applicable clinically. More potentially useful are infusions of growth factors such as Hepatocyte growth factor (HGF),[93-95] Keratinocyte growth factor (KGF)[96] and tri-iodothyronine,[97,98] all potent stimulators of hepatocyte DNA synthesis. There are potential deleterious consequences of long-term exposure to growth factors, such as initiation of neoplasia, but short-term stimulation of the duration envisaged is unlikely to be problematic.

Furthermore, there are differences in the response of transplanted hepatocytes dependent on the stimulus to growth given. In a study comparing the engraftment and proliferation response after 70% partial hepatectomy, with that after carbon tetrachloride liver injury, or after an infusion of HGF, the transplanted hepatocytes behaved differently. If CCl_4 was administered 24 hours before transplant, the transplanted cells were injured and cleared; if transplantation had preceded CCl_4 treatment by some weeks, both the host and transplanted hepatocytes responded with identical increases of DNA synthesis. In contrast, HGF had no effect on the DNA synthesis of transplanted hepatocytes.[99]

A recent experimental approach is co-transplantation of hepatocytes with pancreatic islets based on the hypothesis that the mitogenic potential of islets for hepatocytes will stimulate the hepatocyte engraftment. Islets have been either co-injected with hepatocytes or seeded as co-cultures onto polymeric matrices.[100] Hepatocyte engraftment was enhanced, and in comparison with a portacaval shunt, seeding with islets was superior, and optimal at a ratio of 40 islets per million hepatocytes. Portacaval shunting was in fact additive.[101,102] In the Gunn rat model of hyperbilirubinemia, cotransplantation of hepatocytes with islets was compared with transplantation of hepatocytes alone in the splenic parenchyma. Both caused a significant decrease in unconjugated bilirubin, and the effect was more marked in the co-transplantation group (47% reduction in bilirubin with hepatocytes alone, 65% with islets).[103] In these experiments, the stimulus that the islets provide to hepatocyte proliferation and engraftment has not been defined, but both insulin and glucagon have been demonstrated to be useful hepatocyte co-mitogens in culture.

SELECTIVE ADVANTAGE

Engraftment and proliferation of transplanted cells should be stimulated if the recipient's liver function is inadequate, or if the implanted cells have a greater proliferative potential than the native cells and a stimulus to growth is provided, or if the transplanted cells have additional properties encouraging survival — for example the ability to metabolise an endogenous toxin. Each of these situations can be created experimentally, and the proliferation of selectively advantaged cells provides an intriguing example of natural selection *in vivo*.

The first experimental indication of this came from studies of the murine urokinase inhibitor knockout mouse.[104] In this model, some cells undergo somatic mutation and express normal function. These cells gradually repopulate the liver, replacing the metabolically abnormal population. Subsequently this experimental system has been modified to demonstrate the tremendous doubling potential of transplanted hepatocytes so that in principle, a very low number of implanted cells has the potential to repopulate the liver. Subsequently, Overturf *et al.* investigated the murine inborn error of metabolism tyrosinaemia. This model, like the human counterpart, can be treated by the exogenous agent NTBC (2-(2-nitro-4-trifluoro-methylbenzoyl)1,3-cyclohexedione), which prevents the build-up of tyrosine. In the presence of NTBC, implanted wild-type hepatocytes have no particular survival advantage within the liver of tyrosinaemic mice. However, in the absence of NTBC, as few as 100 implanted hepatocytes with competent tyrosine metabolising enzymes can repopulate the liver. This growth of the advantaged population can be seen not only with wild-type implanted hepatocytes, but with deficient hepatocytes in which the enzyme defect had been treated by retroviral gene transfer.[105]

Other experimental systems have depended on disadvantaging the normal liver cells, with or without additional stimulation to encourage growth of implanted cells. After hepatic damage due to irradiation, transplanted hepatocytes proliferate more rapidly than the previously irradiated cells. The pyrrolidine alkaloid retrorsine inhibits hepatocyte proliferation, with a long lasting effect over months. Animals previously treated with retrorsine, and receiving implants of untreated hepatocytes, can demonstrate virtually complete repopulation of the liver if there is a stimulus to growth.[106,107] This can be done by partial hepatectomy, but more elegantly by specific mitogens. The growth proliferating effect of T3, which as a "primary mitogen" bypasses tyrosine kinase pathways and acts at a nuclear level, has been used to initiate virtually total repopulation of the liver after prior administration of retrorsine. 60–80% of the hepatocyte mass could be restored from transplanted hepatocytes within 60 days under repeated stimulation with T3. The approach has been used in the analbuminaemic rat to encourage the proliferation of albumin secreting transplanted hepatocytes.[108]

THE SCOPE OF POTENTIAL CLINICAL APPLICATION

Experimentally, hepatocyte transplantation can improve survival. This has been demonstrated in both acute and chronic models. A variety of experiments in rodents, many in the 1970s, demonstrated that survival after acute damage could be enhanced by administration of hepatocytes, intrasplenically or in other sites.[89,109-112] Interpretation of these experiments required caution, as it was demonstrated that dead hepatocytes or cytosolic extracts of hepatocytes,[113,114] could also enhance survival in severe acute liver damage, and in these circumstances, the effect of the manoeuvre may be to enhance regeneration of the recipient liver.

Chronic liver disease presents different problems, and in this condition enhanced regeneration of the native liver is a less likely mechanism for any favourable effect on survival, as the cirrhotic liver appears to be under a high level of stimulation to regenerate already. Cirrhosis is an attractive target for transplantation clinically, since a slow deterioration might allow transplanted cells sufficient time to implant, proliferate and function. In cirrhotic rats, hepatocyte transplants into the spleen improved clotting parameters, serum ammonia, and clinical encephalopathy scores. The number of cells required was fifty million. Interestingly implantation into the peritoneum of either hepatocytes or hepatocyte homogenate or indeed bone marrow cells were not effective.[115]

A much more substantial body of evidence supports the amelioration of inborn errors of metabolism in animal models by hepatocyte transplantation. These include improvements in tyrosinaemia,[116] analbuminaemia,[117] and histidinaemia,[72] and as already discussed, these experiments are also being harnessed to investigate the results of *ex vivo* genetic cell manipulation. In Dalmatian dogs with hyperuricosuria, repeated intrasplenic transplantation with hepatocytes either fresh or cryopreserved, taken from mongrel dogs led to significant decreases in uric acid secretion in urine. Although the decrease was transient and at no time reduced values to within the normal range, the effectiveness of cryopreserved cells has important implications for clinical hepatocyte transplantation.[118] A study looking specifically at the ability of cryopreserved hepatocytes to repopulate the liver by clonal replication in mice demonstrated that in spite of a loss of viability, individual viable frozen-thawed hepatocytes had identical replicative potential to that of fresh hepatocytes.[119]

HUMAN STUDIES

One message of the various reports in human hepatocyte transplantation is that, if this has been performed to supplement global function — in either acute or chronic liver disease, it is currently more or less impossible to assess whether there has been any functional benefit. It is only in inborn errors of metabolism, where there is a stable background level of one specific function to be investigated, that it has been possible to demonstrate useful transfer of function — and currently in little more than a handful of patients.

In acute liver failure, if a patient survives without requiring conventional orthotopic liver transplantation after receiving an experimental hepatocyte transplant, there will always be a strong possibility that the improvement would have occurred anyway. One patient received 800 million hepatocytes via the portal vein when they had fulminant hepatic failure; liver biopsy at three months, exploiting HLA Class I non-identity between donor and recipient, showed no donor DNA in the liver at three months. The patient's clinical parameters at the time of implantation confirmed very severe hepatitis, but did not infallibly predict death so the experience remains essentially uninterpretable (other than as a safety/feasibility study).[120] Strom et al. transplanted hepatocytes intraarterially into the spleen in five children with acute hepatic failure.[121] Three survived, but for clear ethical reasons all underwent orthotopic transplantation, so evidence of benefit was only indicated by alterations in notoriously unstable parameters e.g. serum ammonia.

Another study used intrasplenic or intrahepatic hepatocyte transplantation in five patients with grade III or IV encephalopathy, who were ventilator dependent, but were not candidates for organ transplantation. Three patients survived more than 48 hours and in those, there were substantial improvements in encephalopathy scores, arterial ammonia and prothrombin times, after a 24–72 hour delay. Whilst these three patients lived longer than would have been expected with only clinical management, all patients eventually died. At post-mortem, transplanted hepatocytes were still surviving, recognised by fluorescent *in situ* hybridisation with Y-chromosome markers to detect transplanted male hepatocytes in female livers.[112]

In metabolic defects, unequivocal evidence of transfer of function has now emerged. Fox *et al.* treated a child with congenital hyperbilirubinaemia (Crigler-Najjar syndrome) and demonstrated a fall in serum bilirubin following intrasplenic infusion of hepatocytes.[123] Whilst the recipient subsequently required immunosuppression, and bilirubin levels halved rather than returning to normal, useful amelioration was achieved, with the number of daily hours of phototherapy required to prevent bilirubin-mediated neurological damage decreasing. In a far more complex experiment, hepatocyte implantation was combined with gene therapy in a small number of patients.[124,125] In familial hypercholesterolaemia (a hepatocyte receptor defect) hepatocytes were harvested from a patient by partial liver resection; the cells were separated, transduced *in vitro* with a retroviral vector construct expressing the missing receptor, and the transduced cells then re-implanted into the liver via the portal vein. The partial correction of the defect, whilst disappointing in terms of overall efficacy, and obtained at the cost of a complex and potentially dangerous procedure, is nevertheless a landmark study pointing the way towards successful hepatic gene therapy. However, whether in the future hepatocyte gene therapy will rely on *ex vivo* transduction is unclear, and current advances suggest that *in vivo* approaches are much more likely to be used.

FUTURE HUMAN STUDIES — RETURN TO THE PROBLEM OF THE CELL SOURCE!

As discussed in considering cells for extracorporeal support — human hepatocytes are scarce and poorly proliferative. A major encouragement in the field of transplantation rather than extracorporeal utilisation of hepatocytes is that if appropriately implanted, both the *in vivo* environment and selective pressure may allow progressive expansion of cells — so that the problem is theoretically less marked. However, two alternative cell sources in addition to differentiated human hepatocytes are being investigated.

Xenogeneic hepatocytes. A xenogeneic source for hepatocytes is being investigated, despite the species difference which in whole organ transplants precipitated hyper-acute rejection. That process however is largely explained by antibodies to antigens on endothelial cells which should be absent from highly purified hepatocyte preparations. This has been explored in cirrhotic models transplanting porcine hepatocytes intra-splenically into rats. Providing cyclosporine immunosuppression was maintained porcine hepatocytes integrated into host hepatic cords and continued to express porcine albumin for more than 50 days.[126] A likely problem is that xenogeneic cells will induce not only immune reactions within the host against the cells themselves, but also against the proteins synthesised by the xenogeneic hepatocytes. A study using human hepatoma cells (HepG2) implanted into the peritoneum of Lewis rats illustrated the expected antibody response, but also demonstrated that short-term cyclosporine immunosuppression (two weeks) could reduce both antibody production and hepatocyte injury. In rats without immunosuppression, human albumin was evident only for a month, whilst cyclosporine treated rats demonstrating human albumin for up to 60 days. The indication that a short course of immunosuppression is sufficient to prolong graft survival and function may be of particular importance in the setting of acute liver failure, as short-term xenogeneic support might buy sufficient time and clinical improvement for the host liver to repair and regenerate.[127] The field of xenogeneic transplantation raises many issues in addition to those of immunological compatibility: the safety issues, particularly the risk of transmission of endogenous retroviruses, and ethical issues, are all under current debate (see further details in chapters dealing with Xenotransplantation, Chapters 9–12).

Stem cells. The prospects for culturing, expanding and maturing liver cell stem cells are enormously exciting. During foetal life, the liver is a haemopoietic organ as well as the site of development of the future hepatocytic and biliary epithelial cell series. Experiments based on surface markers, such as cytokeratins, and integrin expression, as well as functional identification of hepatocyte products such as alpha-fetoprotein and specific biliary enzymes demonstrate that in mid-foetal life, there are pluripotential hepatoblasts. Culture experiments on such populations after isolation demonstrate that with appropriate manipulation of environment, particularly extracellular matrix and addition of differentiation factors such as retinoids, cells can be persuaded to take on phenotypic characteristics of one or other cell type.[128–131]

In adult liver, including observations on human liver, if there is severe liver damage — for example in the context of the slow development of sub-fulminant hepatic failure in viral hepatitis — a population of oval cells with similar bipolar potential develops near the portal tract. They can experimentally be stimulated to grow in rats if normal hepatocyte proliferation is inhibited with various organic chemicals, and stimuli to proliferation applied. Recent work has focused on the cells making up the canals of Hering, which in normal liver represents only a minor component; and these cells may well represent a resting stem cell-population.[17] Molecular markers such as a c-kit may identify these cells prominently in neonatal liver.[132,133] Whether these cells can be usefully garnered from livers for use in transplantation is currently speculative but under active investigation.

An alternative approach to finding, if not stem cells in the liver, at least cells with greater capacity to repopulate the liver is to identify differentiated hepatocyte cells with the greatest proliferative potential. Foetal and neonatal hepatocytes are one source, as they have high proliferative rates, but they are prone to slow down on harvesting and culture. The adult hepatocyte population itself is heterogeneous comprising small (~16 μm), medium (~21 μm) and large hepatocytes (~27 μm). The evidence concerning the mitotic potential of these cells is conflicting. Some reports indicate a higher proliferative rate for small hepatocytes.[134] However, in a recent study by Overturf's group in rat liver, after separation by centrifugal elutriation and subsequent transplantation, contrary to expectation the small hepatocytes had a lower repopulation ability than the medium or larger hepatocytes.[135]

Finally, there is now evidence for a stem cell within the bone marrow that under specialised circumstances can repopulate the liver.[20,22] This was first demonstrated experimentally using rodents subjected to bone marrow transplantation and liver damage, and subsequently shown — although numbers of cells are small in animals receiving bone marrow transplants without any additional liver damage being inflicted. In the latter experiments, Y-chromosome probes were positive using FISH techniques in hepatocytes in female mice that had received male bone marrow cells. This latter observation raises many fascinating conjectures — is bone marrow normally the source of some of the new cells that replace effete hepatocytes as part of normal wear and tear? Therapeutically, identifying and culturing such cells would provide a source of hepatocyte function of awesome potential, capable of revolutionising the treatment of acute and chronic liver disease and many metabolic conditions.

REFERENCES

1. Kaita KD, Pettigrew N, Minuk GY. Hepatic regeneration in humans with various liver disease as assessed by Ki-67 staining of formalin-fixed paraffin-embedded liver tissue. *Liver* 1997; **17**(1):13–16.
2. Hughes RD, Williams R. Use of sorbent columns and haemofiltration in fulminant hepatic failure. *Blood Purif* 1993; **11**(3):163–169.

3. Sussman NL, Kelly JH. Extracorporeal liver support: Cell-based therapy for the failing liver. *Am J Kidney Dis* 1997; **30**(5 Suppl 4):S66–S71.

4. Lilja H, Arkadopoulos N, Blanc P, Eguchi S, Middleton Y, Meurling S, *et al.* Fetal rat hepatocytes: Isolation, characterization, and transplantation in the Nagase analbuminemic rats. *Transplantation* 1997; **64**(9):1240–1248.

5. Gruppuso PA, Awad M, Bienieki TC, Boylan JM, Fernando S, Faris RA. Modulation of mitogen-independent hepatocyte proliferation during the perinatal period in the rat. *In Vitro Cell Dev Biol Anim* 1997; **33**(7):562–568.

6. Caperna TJ, Failla ML, Kornegay ET, Richards MP, Steele NC. Isolation and culture of parenchymal and nonparenchymal cells from neonatal swine liver. *J Anim Sci* 1985; **61**(6): 1576–1586.

7. Smalley M, Selden C, O'Hare M, Hodgson HJF. Cell lines derived from normal human liver by infection with a retrovirus containing the SV40 large T antigen. Gut 38s1, A57. 1996 (Ref Type: Abstract).

8. Smalley MJ, McCloskey P, Leiper K, O'Hare MJ, Hodgson HJF. Cell strains derived from normal human hepatocytes by infection with a retrovirus containing the SV40 large-T-antigen. *Hepatology* **24**(4): 261A. 1996. (Ref Type: Abstract).

9. Werner A, Duvar S, Müthing J, Büntemeyer H, Kahmann U, Lünsdorf H, *et al.* Cultivation and characterization of a new immortalized human hepatocyte cell line, HepZ, for use in an artificial liver support system. *Ann NY Acad Sci* 1999; **875**: 364–368.

10. Werner A, Duvar S, Muthing J, Buntemeyer H, Lunsdorf H, Strauss M, *et al.* Cultivation of immortalized human hepatocytes HepZ on macroporous CultiSpher G microcarriers. *Biotechnol Bioeng* 2000; **68**(1):59–70.

11. Park J, Lee H-S, Park JB, Kim CY. Establishment of continuous human fetal hepatocyte line by transfection of simian virus 40 T gene. *Mol Cells* 1996; **6**:534–540.

12. Yoon JH, Lee HV, Lee JS, Park JB, Kim CY. Development of a non-transformed human liver cell line with differentiated-hepatocyte and urea-synthetic functions: applicable for bioartificial liver. *Int J Artif Organs* 1999; **22**(1):769–777.

13. Kobayashi N, Miyazaki M, Fukaya K, Inoue Y, Sakaguchi M, Uemura T, *et al.* Transplantation of highly differentiated immortalized human hepatocytes to treat acute liver failure. *Transplantation* 2000; **69**(2):202–207.

14. Sussman NL, Gislason GT, Conlin CA, Kelly JH. The Hepatix extracorporeal liver assist device: initial clinical experience. *Artif Organs* 1994; **18**(5):390–396.

15. Spath GF, Weiss MC. Hepatocyte nuclear factor 4 provokes expression of epithelial marker genes, acting as a morphogen in dedifferentiated hepatoma cells. *J Cell Biol* 1998; **140**(4): 935–946.

16. Alison M, Golding M, Lalani EN, Nagy P, Thorgeirsson S, Sarraf C. Wholesale hepatocytic differentiation in the rat from ductular oval cells, the progeny of biliary stem cells. *J Hepatol* 1997; **26**(2):343–352.

17. Theise ND, Saxena R, Portmann BC, Thung SN, Yee H, Chiriboga L, *et al.* The canals of Hering and hepatic stem cells in humans. *Hepatology* 1999; **30**(6):1425–1433.

18. Yin L, Lynch D, Sell S. Participation of different cell types in the restitutive response of the rat liver to periportal injury induced by allyl alcohol. *J Hepatol* 1999; **31**(3):497–507.

19. Sell S. Comparison of liver progenitor cells in human atypical ductular reactions with those seen in experimental models of liver injury. *Hepatology* 1998; **27**(2): 317–331.

20. Petersen BE, Bowen WC, Patrene KD, Mars WM, Sullivan AK, Murase N, *et al.* Bone marrow as a potential source of hepatic oval cells. *Science* 1999; **284**(5417):1168–1170.

21. Strain AJ. Changing blood into liver: Adding further intrigue to the hepatic stem cell story. *Hepatology* 1999; **30**(4):1105–1107.

22. Theise ND, Badve S, Saxena R, Henegariu O, Sell S, Crawford JM, *et al.* Derivation of hepatocytes from bone marrow cells in mice after radiation-induced myeloablation. *Hepatology* 2000; **31**(1):235–240.

23. Naka S, Takeshita K, Yamamoto T, Tani T, Kodama M. Bioartificial liver support system using porcine hepatocytes entrapped in a three-dimensional hollow fiber module with collagen gel: an evaluation in the swine acute liver failure model. *Artif Organs* 1999; **23**(9):822–828.

24. Naruse K, Nagashima I, Sakai Y, Harihara Y, Jiang GX, Suzuki M, *et al.* Efficacy of a bioreactor filled with porcine hepatocytes immobilized on nonwoven fabric for *ex vivo* direct hemoperfusion treatment of liver failure in pigs. *Artif Organs* 1998; **22**(12): 1031–1037.

25. Chen SC, Hewitt WR, Watanabe FD, Eguchi S, Kahaku E, Middleton Y, *et al.* Clinical experience with a porcine hepatocyte-based liver support system. *Int J Artif Organs* 1996; **19**(11):664–669.

26. Demetriou AA, Rozga J, Podesta L, Lepage E, Morsiani E, Moscioni AD, *et al.* Early clinical experience with a hybrid bioartificial liver. *Scand J Gastroenterol Suppl* 1995; **208**:111–117.

27. Baquerizo A, Mhoyan A, Kearns-Jonker M, Arnaout WS, Shackleton C, Busuttil RW, *et al.* Characterization of human xenoreactive antibodies in liver failure patients exposed to pig hepatocytes after bioartificial liver treatment: An *ex vivo* model of pig to human xenotransplantation. *Transplantation* 1999; **67**(1):5–18.

28. Baquerizo A, Mhoyan A, Shirwan H, Swensson J, Busuttil RW, Demetriou AA, *et al.* Xenoantibody response of patients with severe acute liver failure exposed to porcine antigens following treatment with a bioartificial liver. *Transplant Proc* 1997; **29**(1–2/01):964–965.

29. te Velde AA, Flendrig LM, Ladiges NC, Chamuleau RA. Immunological consequences of the use of xenogeneic

hepatocytes in a bioartificial liver for acute liver failure. *Int J Artif Organs* 1997; **20**(4):229–233.

30. Pitkin Z, Mullon C. Evidence of absence of porcine endogenous retrovirus (PERV) infection in patients treated with a bioartificial liver support system. *Artif Organs* 1999; **23**(9):829–833.

31. Nyberg SL, Hibbs JR, Hardin JA, Germer JJ, Persing DH. Transfer of porcine endogenous retrovirus across hollow fiber membranes: significance to a bioartificial liver. *Transplantation* 1999; **67**(9):1251–1255.

32. Gregory PG, Connolly CK, Toner M, Sullivan SJ. *In vitro* characterization of porcine hepatocyte function. *Cell Transplant* 2000; **9**(1):1–10.

33. Guguen-Guillouzo C, Guillouzo A. Modulation of functional activities in cultured rat hepatocytes. *Mol Cell Biochem* 1983; **53–54**(1–2):35–56.

34. Mitaka T. The current status of primary hepatocyte culture. *Int J Exp Pathol* 1998; **79**(6):393–409.

35. Bhandari RN, Riccalton LA, Lewis AL, Fry JR, Hammond AH, Tendler SJ, *et al.* Liver tissue engineering: a role for co-culture systems in modifying hepatocyte function and viability. *Tissue Eng* 2001; **7**(3):345–357.

36. Clark EA, Brugge JS. Integrins and signal transduction pathways: the road taken. *Science* 1995; **268**(5208):233–239.

37. Iredale JP, Arthur MJ. Hepatocyte-matrix interactions. *Gut* 1994; **35**(6):729–732.

38. Stamatoglou SC, Hughes RC. Cell adhesion molecules in liver function and pattern formation. *FASEB J* 1994; **8**(6):420–427.

39. Bhatia SN, Balis UJ, Yarmush ML, Toner M. Effect of cell-cell interactions in preservation of cellular phenotype: cocultivation of hepatocytes and nonparenchymal cells. *FASEB J* 1999; **13**(14):1883–1900.

40. Michalopoulos GK, Bowen WC, Zajac VF, Beer Stolz D, Watkins S, Kostrubsky V, *et al.* Morphogenetic events in

mixed cultures of rat hepatocytes and nonparenchymal cells maintained in biological matrices in the presence of hepatocyte growth factor and epidermal growth factor. *Hepatology* 1999; 29(1):90–100.

41. Mitaka T, Sato F, Mizuguchi T, Yokono T, Mochizuki Y. Reconstruction of hepatic organoid by rat small hepatocytes and hepatic nonparenchymal cells. *Hepatology* 1999; 29(1): 111–125.

42. Arenson DM, Friedman SL, Bissell DM. Formation of extracellular matrix in normal rat liver: lipocytes as a major source of proteoglycan. *Gastroenterology* 1988; 95(2):441–447.

43. Loreal O, Levavasseur F, Fromaget C, Gros D, Guillouzo A, Clement B. Cooperation of Ito cells and hepatocytes in the deposition of an extracellular matrix *in vitro*. *Am J Pathol* 1993; 143(2):538–544.

44. Elçin YM, Dixit V, Gitnick G. Hepatocyte attachment on biodegradable modified chitosan membranes: *in vitro* evaluation for the development of liver organoids. *Artif Organs* 1998; 22(10):837–846.

45. Bissell DM, Arenson DM, Maher JJ, Roll FJ. Support of cultured hepatocytes by a laminin-rich gel. Evidence for a functionally significant subendothelial matrix in normal rat liver. *J Clin Invest* 1987; 79(3):801–812.

46. Oda H, Nozawa K, Hitomi Y, Kakinuma A. Laminin-rich extracellular matrix maintains high level of hepatocyte nuclear factor 4 in rat hepatocyte culture. *Biochem Biophys Res Commun* 1995; 212(3):800–805.

47. Ponce ML, Rojkind M. Rat hepatocytes attach to laminin present in liver biomatrix proteins by an Mg(++)-dependent mechanism. *Hepatology* 1995; 22(2):620–628.

48. Rojkind M, Gatmaitan Z, Mackensen S, Giambrone MA, Ponce P, Reid LM. Connective tissue biomatrix: its isolation and utilization for long-term cultures of normal rat hepatocytes. *J Cell Biol* 1980; 87(1):255–263.

49. Runge D, Runge DM, Bowen WC, Locker J, Michalopoulos GK. Matrix induced re-differentiation of cultured rat hepatocytes and changes of CCAAT/enhancer binding proteins. *Biol Chem* 1997; **378**(8):873–881.

50. Nagaki M, Shidoji Y, Yamada Y, Sugiyama A, Tanaka M, Akaike T, *et al.* Regulation of hepatic genes and liver transcription factors in rat hepatocytes by extracellular matrix. *Biochem Biophys Res Commun* 1995; **210**(1):38–43.

51. Smalley M, Leiper K, Floyd D, Mobberley M, Ryder T, Selden C, *et al.* Behavior of a cell line derived from normal human hepatocytes on non-physiological and physiological-type substrates: Evidence for enhancement of secretion of liver-specific proteins by a three-dimensional growth pattern. *In Vitro Cell Dev Biol Anim* 1999; **35**(1):22–32.

52. Flendrig LM, te Velde AA, Chamuleau RA. Semipermeable hollow fiber membranes in hepatocyte bioreactors: A prerequisite for a successful bioartificial liver? *Artif Organs* 1997; **21**(11):1177–81.

53. Flendrig LM, la Soe JW, Jorning GG, Steenbeek A, Karlsen OT, Bovee WM, *et al. In vitro* evaluation of a novel bioreactor based on an integral oxygenator and a spirally wound nonwoven polyester matrix for hepatocyte culture as small aggregates. *J Hepatol* 1997; **26**(6):1379–92.

54. Gerlach J, Kloppel K, Stoll P, Vienken J, Muller C. Gas supply across membranes in bioreactors for hepatocyte culture. *Artif Organs* 1990; **14**(5):328–33.

55. Ranucci CS, Moghe PV, Polymer substrate topography actively regulates the multicellular organization and liver-specific functions of cultured hepatocytes. *Tissue Eng* 1999; **5**(5): 407–420.

56. Kaihara S, Borenstein J, Koka R, Lalan S, Ochoa ER, Ravens M, *et al.* Silicon micromachining to tissue engineer branched vascular channels for liver fabrication. *Tissue Eng* 2000; **6**(2): 105–117.

57. Dixit V. Development of a bioartificial liver using isolated hepatocytes. *Artif Organs* 1994; **18**(5):371–384.
58. Chen SC, Mullon C, Kahaku E, Watanabe F, Hewitt W, Eguchi S, *et al.* Treatment of severe liver failure with a bioartificial liver. *Ann N Y Acad Sci* 1997; **831**:350–360.
59. Rozga J, Podesta L, Lepage E, Morsiani E, Moscioni AD, Hoffman A, *et al.* A bioartificial liver to treat severe acute liver failure. *Ann Surg* 1994; **219**(5):538–544.
60. Shatford RA, Nyberg SL, Meier SJ, White JG, Payne WD, Hu WS, *et al.* Hepatocyte function in a hollow fiber bioreactor: a potential bioartificial liver. *J Surg Res* 1992; **53**(6):549–557.
61. Sheil AG, Sun J, Mears DC, Waring M, Woodman K, Johnston B, *et al.* Positive biochemical effects of a bioartificial liver support system (BALSS) in a porcine fulminant hepatic failure (FHF) model. *Int J Artif Organs* 1998; **21**(1):43–48.
62. Ellis AJ, Hughes RD, Wendon JA, Dunne J, Langley PG, Kelly JH, *et al.* Pilot-controlled trial of the extracorporeal liver assist device in acute liver failure. *Hepatology* 1996; **24**(6):1446–1451.
63. Sajiki T, Iwata H, Paek HJ, Tosha T, Fujita S, Ueda Y, *et al.* Transmission electron microscopic study of hepatocytes in bioartificial liver. *Tissue Eng* 2000; **6**(6):627–640.
64. Nyberg SL, Shatford RA, Payne WD, Hu WS, Cerra FB. Primary culture of rat hepatocytes entrapped in cylindrical collagen gels: An *in vitro* system with application to the bioartificial liver. Rat hepatocytes cultured in cylindrical collagen gels. *Cytotechnology* 1992; **10**:205–215.
65. Gerlach JC, Encke J, Hole O, Muller C, Ryan CJ, Neuhaus P. Bioreactor for a larger scale hepatocyte *in vitro* perfusion. *Transplantation* 1994; **58**(9):984–988.
66. Flendrig LM, la Soe JW, Jörning GG, Steenbeek A, Karlsen OT, Bovée WM, *et al. In vitro* evaluation of a novel bioreactor based on an integral oxygenator and a spirally wound nonwoven

polyester matrix for hepatocyte culture as small aggregates. *J Hepatol* 1997; **26**(6):1379–1392.

67. Hughes RD, Nicolaou N, Langley PG, Ellis AJ, Wendon JA, Williams R. Plasma cytokine levels and coagulation and complement activation during use of the extracorporeal liver assist device in acute liver failure. *Artif Organs* 1998; **22**(10): 854–858.

68. Arnaout WS, Moscioni AD, Barbour RL, Demetriou AA. Development of bioartificial liver: bilirubin conjugation in Gunn rats. *J Surg Res* 1990; **48**(4):379–382.

69. Flendrig LM, Calise F, Di Florio E, Mancini A, Ceriello A, Santaniello W, et al. Significantly improved survival time in pigs with complete liver ischemia treated with a novel bioartificial liver. *Int J Artif Organs* 1999; **22**(10):701–709.

70. Cuervas-Mons V, Colas A, Rivera JA, Prados E. *In vivo* efficacy of a bioartificial liver in improving spontaneous recovery from fulminant hepatic failure: A controlled study in pigs. *Transplantation* 2000; **69**(3):337–344.

71. Lee B, Dennis JA, Healy PJ, Mull B, Pastore L, Yu H, et al. Hepatocyte gene therapy in a large animal: A neonatal bovine model of citrullinemia. *Proc Natl Acad Sci, USA*, 1999; **96**(7): 3981–3986.

72. Selden C, Calnan D, Morgan N, Wilcox H, Carr E, Hodgson HJ. Histidinemia in mice: A metabolic defect treated using a novel approach to hepatocellular transplantation. *Hepatology* 1995; **21**(5):1405–1412.

73. Nakazawa F, Onodera K, Kato K, Sawa M, Kino Y, Imai M, et al. Multilocational hepatocyte transplantation for treatment of congenital ascorbic acid deficiency rats. *Cell Transplant* 1996; **5**(5 Supp 1):S23–S25.

74. Cobourn CS, Makowka L, Falk JA, Falk RE. Allogeneic intrasplenic hepatocyte transplantation in the Gunn rat using cyclosporine A immunosuppression. *Transplant Proc* 1987; **19**: 1002–1003.

75. Ponder KP, Gupta S, Leland F, Darlington G, Finegold M, DeMayo J, *et al.* Mouse hepatocytes migrate to liver parenchyma and function indefinitely after intrasplenic transplantation. *Proc Natl Acad Sci, USA*, 1991; **88**:1217–1221.

76. Rosenthal RJ, Chen SC, Hewitt W, Wang CC, Eguchi S, Geller S, *et al.* Techniques for intrasplenic hepatocyte transplantation in the large animal model. *Surg Endosc* 1996; **10**(11):1075–1079.

77. Selden C, Gupta S, Johnstone R, Hodgson HJ. The pulmonary vascular bed as a site for implantation of isolated liver cells in inbred rats. *Transplantation* 1984; **38**:81–83.

78. de Roos WK, von Geusau BA, Bouwman E, van Dierendonck JH, Borel R, I, Terpstra OT. Monitoring engraftment of transplanted hepatocytes in recipient liver with 5-bromo-2′-deoxyuridine. *Transplantation* 1997; **63**(4):513–518.

79. Johnson LB, Aiken J, Mooney D, Schloo BL, Griffith-Cima L, Langer R, *et al.* The mesentery as a laminated vascular bed for hepatocyte transplantation. *Cell Transplant* 1994; **3**(4):273–281.

80. Jirtle RL, Michalopoulos G. Effects of partial hepatectomy on transplanted hepatocytes. *Cancer Res* 1982; **42**(8):3000–3004.

81. Gupta S, Yerneni PR, Vemuru RP, Lee CD, Yellin EL, Bhargava KK. Studies on the safety of intrasplenic hepatocyte transplantation: Relevance to *ex vivo* gene therapy and liver repopulation in acute hepatic failure. *Hum Gene Ther* 1993; **4**(3):249–257.

82. Thompson NL, Hixson DC, Callanan H, Panzica M, Flanagan D, Faris RA, *et al.* A Fischer rat substrain deficient in dipeptidyl peptidase IV activity makes normal steady-state RNA levels and an altered protein. Use as a liver-cell transplantation model. *Biochem J* 1991; **273**:497–502.

83. Gupta S, Vasa SR, Rajvanshi P, Zuckier LS, Palestro CJ, Bhargava KK. Analysis of hepatocyte distribution and survival in vascular beds with cells marked by 99mTC or endogenous dipeptidyl peptidase IV activity. *Cell Transplant* 1997; **6**(4): 377–386.

84. Sigal SH, Rajvanshi P, Reid LM, Gupta S. Demonstration of differentiation in hepatocyte progenitor cells using dipeptidyl peptidase IV deficient mutant rats. *Cell Mol Biol Res* 1995; **41**(1):39–47.

85. Selden AC, Darby H, Hodgson H. Further observations on the survival, proliferation and function of ectopically implanted syngeneic and allogeneic liver cells in rat spleen. *Eur J Hepatol Gastro* 1991; **3**:607–611.

86. Darby H, Gupta S, Johnstone R, Selden C, Hodgson HJ. Observations on rat spleen reticulum during the development of syngeneic hepatocellular implants. *Br J Exp Pathol* 1986; **67**: 329–339.

87. Gupta S, Johnstone R, Darby H, Selden C, Price Y, Hodgson HJ. Transplanted isolated hepatocytes: Effect of partial hepatectomy on proliferation of long-term syngeneic implants in rat spleen. *Pathology* 1987; **19**:28–30.

88. Demetriou AA, Whiting J, Levenson SM, Chowdhury NR, Schechner R, Michalski S, *et al*. New method of hepatocyte transplantation and extracorporeal liver support. *Ann Surg* 1986; **204**(3):259–271.

89. Gupta S, Vemuru RP, Lee CD, Yerneni PR, Aragona E, Burk RD. Hepatocytes exhibit superior transgene expression after transplantation into liver and spleen compared with peritoneal cavity or dorsal fat pad: Implications for hepatic gene therapy. *Hum Gene Ther* 1994; **5**(8):959–967.

90. Wong H, Chang TM. The viability and regeneration of artificial cell microencapsulated rat hepatocyte xenograft transplants in mice. *Biomater Artif Cells Artif Organs* 1988; **16**(4):731–739.

91. te Velde AA, Bosman DK, Oldenburg J, Sala M, Maas MA, Chamuleau RA. Three different hepatocyte transplantation techniques for enzyme deficiency disease and acute hepatic failure. *Artif Organs* 1992; **16**(5):522–526.

92. Ajioka I, Nishio R, Ikekita M, Akaike T, Sasaki M, Enami J, *et al*. Establishment of heterotropic liver tissue mass with direct

link to the host liver following implantation of hepatocytes transfected with vascular endothelial growth factor gene in mice. *Tissue Eng* 2001; **7**(3):335–344.

93. Kaibori M, Kwon AH, Nakagawa M, Wei T, Uetsuji S, Kamiyama Y, *et al.* Stimulation of liver regeneration and function after partial hepatectomy in cirrhotic rats by continuous infusion of recombinant human hepatocyte growth factor. *J Hepatol* 1997; **27**(2):381–390.

94. Roos F, Ryan AM, Chamow SM, Bennett GL, Schwall RH. Induction of liver growth in normal mice by infusion of hepatocyte growth factor/scatter factor. *Am J Physiol* 1995; **268**(2 Pt 1):G380–G386.

95. Kobayashi Y, Hamanoue M, Ueno S, Aikou T, Tanabe G, Mitsue S, *et al.* Induction of hepatocyte growth by intraportal infusion of HGF into beagle dogs. *Biochem Biophys Res Commun* 1996; **220**(1):7–12.

96. Forbes SJ, Themis M, Alison MR, Sarosi I, Coutelle C, Hodgson HJ. Synergistic growth factors enhance rat liver proliferation and enable retroviral gene transfer via a peripheral vein. *Gastroenterology* 2000; **118**(3):591–598.

97. Short J, Brown RF, Husakova A, Gilbertson JR, Zemel R, Lieberman I. Induction of deoxyribonucleic acid synthesis in the liver of the intact animal. *J Biol Chem* 1972; **247**(6): 1757–1766.

98. Oren R, Dabeva MD, Karnezis AN, Petkov PM, Rosencrantz R, Sandhu JP, *et al.* Role of thyroid hormone in stimulating liver repopulation in the rat by transplanted hepatocytes. *Hepatology* 1999; **30**(4):903–913.

99. Gupta S, Rajvanshi P, Aragona E, Lee CD, Yerneni PR, Burk RD. Transplanted hepatocytes proliferate differently after CCl4 treatment and hepatocyte growth factor infusion. *Am J Physiol* 1999; **276**(3 Pt 1):G629–G638.

100. Kneser U, Kaufmann PM, Fiegel HC, Pollok JM, Rogiers X, Kluth D, *et al.* Interaction of hepatocytes and pancreatic islets

cotransplanted in polymeric matrices. *Virchows Arch* 1999; **435**(2):125–132.

101. Kaufmann PM, Sano K, Uyama S, Breuer CK, Organ GM, Schloo BL, *et al.* Evaluation of methods of hepatotrophic stimulation in rat heterotopic hepatocyte transplantation using polymers. *J Pediatr Surg* 1999; **34**(7):1118–1123.

102. Kaufmann PM, Kneser U, Fiegel HC, Pollok JM, Kluth D, Izbicki JR, *et al.* Is there an optimal concentration of cotransplanted islets of Langerhans for stimulation of hepatocytes in three dimensional matrices? *Transplantation* 1999; **68**(2): 272–279.

103. Genin B, Andereggen E, Rubbia-Brandt L, Birraux J, Morel P, Le Coultre C. Improvement of the effect of hepatocyte isograft in the Gunn rat by cotransplantation of islets of Langerhans. *J Pediatr Surg* 1999; **34**(2):321–324.

104. Sandgren EP, Palmiter RD, Heckel JL, Daugherty CC, Brinster RL, Degen JL. Complete hepatic regeneration after somatic deletion of an albumin-plasminogen activator transgene. *Cell* 1991; **66**(2):245–256.

105. Overturf K, Al Dhalimy M, Tanguay R, Brantly M, Ou CN, Finegold M, *et al.* Hepatocytes corrected by gene therapy are selected *in vivo* in a murine model of hereditary tyrosinaemia type I. [published erratum appears in *Nat Genet* 1996 Apr; **12**(4):458]. *Nat Genet* 1996; **12**(3):266–273.

106. Dabeva MD, Laconi E, Oren R, Petkov PM, Hurston E, Shafritz DA. Liver regeneration and alpha-fetoprotein messenger RNA expression in the retrorsine model for hepatocyte transplantation. *Cancer Res* 1998; **58**(24): 5825–5834.

107. Laconi E, Oren R, Mukhopadhyay DK, Hurston E, Laconi S, Pani P, *et al.* Long-term, near-total liver replacement by transplantation of isolated hepatocytes in rats treated with retrorsine. *Am J Pathol* 1998; **153**(1):319–329.

108. Oren R, Dabeva MD, Petkov PM, Hurston E, Laconi E, Shafritz DA. Restoration of serum albumin levels in Nagase

analbuminemic rats by hepatocyte transplantation. *Hepatology* 1999; **29**(1):75–81.

109. Vogels BA, Maas MA, Bosma A, Chamuleau RA. Significant improvement of survival by intrasplenic hepatocyte transplantation in totally hepatectomized rats. *Cell Transplant* 1996; **5**(3):369–378.

110. Zhang H, Miescher Clemens E, Drugas G, Lee SM, Colombani P. Intrahepatic hepatocyte transplantation following subtotal hepatectomy in the recipient: A possible model in the treatment of hepatic enzyme deficiency. *J Pediatr Surg* 1992; **27**:312–315.

111. Kasai S, Sawa M, Hirai S, Nishida Y, Onodera K, Yamamoto T, *et al.* Beneficial effect of hepatocyte transplantation on hepatic failure in rats. *Transplant Proc* 1992; **24**(6):2990–2992.

112. Kasai S, Sawa M, Kondoh K, Ebata H, Mito M. Intrasplenic hepatocyte transplantation in mammals. *Transplant Proc* 1987; **19**(1 Pt 2):992–994.

113. Baumgartner D, LaPlante-O'Neill PM, Sutherland DE, Najarian JS. Effects of intrasplenic injection of hepatocytes, hepatocyte fragments and hepatocyte culture supernatants on D-galactosamine-induced liver failure in rats. *Eur Surg Res* 1983; **15**(3):129–135.

114. Woods RJ, Parbhoo SP. An explanation for the reduction in bilirubin levels in congenitally jaundiced Gunn rats after transplantation of isolated hepatocytes. *Eur Surg Res* 1981; **13**(4):278–284.

115. Kobayashi N, Ito M, Nakamura J, Cai J, Gao C, Hammel JM, *et al.* Hepatocyte transplantation in rats with decompensated cirrhosis. *Hepatology* 2000; **31**(4):851–857.

116. Fabrega AJ, Bommineni VR, Blanchard J, Tetali S, Rivas PA, Pollak R, *et al.* Amelioration of analbuminemia by transplantation of allogeneic hepatocytes in tolerized rats. *Transplantation* 1995; **59**(9):1362–1364.

117. Wu GY, Wilson JM, Shalaby F, Grossman M, Shafritz DA, Wu CH. Receptor-mediated gene delivery *in vivo*. Partial correction of genetic analbuminemia in Nagase rats. *J Biol Chem* 1991; **266**:14338–14342.

118. Dunn TB, Kumins NH, Raofi V, Holman DM, Mihalov M, Blanchard J, *et al.* Multiple intrasplenic hepatocyte transplantations in the dalmatian dog. *Surgery* 2000; **127**(2):193–199.

119. Jamal HZ, Weglarz TC, Sandgren EP. Cryopreserved mouse hepatocytes retain regenerative capacity *in vivo*. *Gastroenterology* 2000; **118**(2):390–394.

120. Fisher RA, Bu D, Thompson M, Tisnado J, Prasad U, Sterling R, *et al.* Defining hepatocellular chimerism in a liver failure patient bridged with hepatocyte infusion. *Transplantation* 2000; **69**: 303–307.

121. Strom SC, Fisher RA, Thompson MT, Sanyal AJ, Cole PE, Ham JM, *et al.* Hepatocyte transplantation as a bridge to orthotopic liver transplantation in terminal liver failure. *Transplantation* 1997; **63**(4):559–569.

122. Bilir BM, Guinette D, Karrer F, Kumpe DA, Krysl J, Stephens J, *et al.* Hepatocyte transplantation in acute liver failure. *Liver Transpl* 2000; **6**(1):32–40.

123. Fox IJ, Chowdhury JR, Kaufman SS, Goertzen TC, Chowdhury NR, Warkentin PI, *et al.* Treatment of the Crigler-Najjar syndrome type I with hepatocyte transplantation. *N Engl J Med* 1998; **338**(20):1422–1426.

124. Grossman M, Rader DJ, Muller DW, Kolansky DM, Kozarsky K, Clark BJ, III *et al.* A pilot study of *ex vivo* gene therapy for homozygous familial hypercholesterolaemia. *Nat Med* 1995; **1**(11):1148–1154.

125. Grossman M, Raper SE, Kozarsky K, Stein EA, Engelhardt JF, Muller D, *et al.* Successful *ex vivo* gene therapy directed to liver in a patient with familial hypercholesterolaemia. *Nat Genet* 1994; **6**(4):335–341.

126. Stefan AM, Coulter S, Gray B, LaMorte W, Nikelaeson S, Edge AS, *et al.* Xenogeneic transplantation of porcine hepatocytes into the CCl4 cirrhotic rat model. *Cell Transplant* 1999; **8**(6):649–659.

127. Kobayashi N, Miyazaki M, Fukaya K, Inoue Y, Sakaguchi M, Noguchi H, *et al.* Establishment of a highly differentiated immortalized human hepatocyte cell line as a source of hepatic function in the bioartificial liver. *Transplant Proc* 2000; **32**(2): 237–241.

128. Sanchez A, Alvarez AM, Pagan R, Roncero C, Vilaro S, Benito M, *et al.* Fibronectin regulates morphology, cell organization and gene expression of rat fetal hepatocytes in primary culture. *J Hepatol* 2000; **32**(2):242–250.

129. Fiorino AS, Diehl AM, Lin HZ, Lemischka IR, Reid LM. Maturation-dependent gene expression in a conditionally transformed liver progenitor cell line. *In Vitro Cell Dev Biol Anim* 1998; **34**(3):247–258.

130. Rogler LE. Selective bipotential differentiation of mouse embryonic hepatoblasts *in vitro. Am J Pathol* 1997; **150**(2): 591–602.

131. Sanchez A, Pagan R, Alvarez AM, Roncero C, Vilaro S, Benito M, *et al.* Transforming growth factor-beta (TGF-beta) and EGF promote cord-like structures that indicate terminal differentiation of fetal hepatocytes in primary culture. *Exp Cell Res* 1998; **242**(1):27–37.

132. Fujio K, Hu Z, Evarts RP, Marsden ER, Niu CH, Thorgeirsson SS. Coexpression of stem cell factor and c-kit in embryonic and adult liver. *Exp Cell Res* 1996; **224**(2):243–250.

133. Blakolmer K, Jaskiewicz K, Dunsford HA, Robson SC. Hematopoietic stem cell markers are expressed by ductal plate and bile duct cells in developing human liver. *Hepatology* 1995; **21**(6):1510–1516.

134. Rajvanshi P, Liu D, Ott M, Gagandeep S, Schilsky ML, Gupta S. Fractionation of rat hepatocyte subpopulations with varying

metabolic potential, proliferative capacity, and retroviral gene transfer efficiency. *Exp Cell Res* 1998; **244**(2):405–419.
135. Overturf K, Al Dhalimy M, Finegold M, Grompe M. The repopulation potential of hepatocyte populations differing in size and prior mitotic expansion. *Am J Pathol* 1999; **155**(6): 2135–2143.

Chapter 7

Tissue Engineering of the Genitourinary System

A Atala

INTRODUCTION

Tissue engineering follows the principles of cell transplantation, materials science and engineering towards the development of biological substitutes which would restore and maintain normal function. Tissue bioengineer involves both the use of matrices alone, or matrices with cells.[1]

Using recently developed cell culture techniques, it is now possible to expand a cell strain from a single specimen which initially covers a surface area of 1 cm^2 to one covering a surface area of 4202 m^2 (the equivalent area of one football field) within eight weeks.[1,2]

Both synthetic (Polyglycolic and poly-l-lactic acid) and natural biodegradable materials (processed collagen derived from donor bladder submucosa, intestinal submucosa, and dermis) have been used as cell delivery and regeneration vehicles at our centre.[1] The cell-support matrix becomes vascularized in concert with expansion of the cell mass following implantation.

Initial experiments were designed in order to explore the possibility of engineering urologic tissue components *ex-situ* (outside the urinary tract) in an *in vivo* animal model.[3,4] These experiments demonstrated that newly isolated human bladder urothelial and muscle cells would attach to artificial and natural matrices *in vitro* and that, when implanted into animals, these constructs could survive, proliferate and reorganize into newly formed multi-layered structures which exhibit spatial orientation and a normal histomorphology *in vivo*. These experiments also demonstrated for the first time that composite tissue engineered structures could be created *de novo*.[4]

BLADDER

Bladder engineering experiments were initially performed using animal models of augmentation. Partial cystectomies were performed in dogs and both urothelial and smooth muscle cells were harvested and expanded separately.[5] A collagen matrix obtained from processed allogenic bladder submucosa was seeded with muscle cells on one side and urothelial cells on the opposite side. Augmentation cystoplasty was performed with either cell seeded or unseeded matrices. Bladders augmented with the cell seeded-matrix scaffolds showed a 99% increase in capacity compared to bladders augmented with the unseeded matrices, which showed only a 30% increase in capacity. Contracture and resorption of the graft was evident in the implanted cell-free matrices. Matrices implanted without cells showed a marked paucity of muscle. Similar results have been observed with the use of SIS without cells for bladder augmentation, where diminished muscle was seen histologically.[6] Recently, homologous bladder acellular matrix grafts used in canine bladders were shown to be morphologically and functionally adequate.[7]

In order to better address the functional parameters of tissue engineered bladders, an animal model was designed which required a subtotal cystectomy with subsequent replacement with either a cell-seeded or unseeded polymer scaffold.[8] The cystectomy only controls and polymer-only grafts maintained low capacities and poor compliance while the tissue-engineered bladder replacements had adequate capacities and compliance. The polymer-only bladders presented a pattern of normal urothelial cells with a thickened fibrotic submucosa and a thin layer of muscle fibers. The retrieved tissue engineered bladders showed a normal cellular organization, consisting of a tri-layer of urothelium, submucosa and muscle. The results from this study showed that unseeded artificial matrices, without cells, are not adequate for the formation of functionally adequate bladder reservoirs. However, it is possible to engineer anatomically and functionally normal bladders using cell-seedeed matrices. Due to the above studies, clinical trials are in progress for patients requiring augmentation cystoplasty, wherein autologous engineered bladder tissue is used instead of bowel.

URETHRA

A similar strategy as described above has been used in trying to engineer urethral tissue in rabbits, using polymer scaffolds, SIS, and acellular bladder grafts.[9-12] The patch grafts promoted urethral regeneration in all experiments.

After our experimental experience with the collagen-based acellular matrix, we used the material clinically for urethral reconstruction in four patients with a history of prior hypospadias.[13] The neourethras were created by anastomosing the matrix in an onlay fashion to the urethral plate. The size of the created neourethra ranged from five to fifteen centimetres. After a three-year follow-up, all patients had a successful outcome with regards to their function. One patient who had a 15 cm neourethra created developed a subglanular fistula. The acellular collagen based matrix eliminated the necessity of performing additional surgical procedures for graft harvesting, and operative time, as well as the potential morbidity due to the harvest procedure, were decreased. We obtained similar results in pediatric and adult patients with urethral stricture disease.[14] Over 60 patients have been successfully treated in this manner using these techniques.

KIDNEY

Renal transplantation is severely limited by a critical donor shortage. Augmentation of either isolated or total renal function with kidney cell expansion *in vitro* and subsequent autologous transplantation may be a feasible solution. The possibility of harvesting and expanding renal cells *in vitro* and implanting them *in vivo* in a three dimensional organization in order to achieve a functional artificial renal unit wherein urine production could be achieved was explored.[15] Studies demonstrated that renal cells can be successfully harvested, expanded in culture and transplanted *in vivo* where the single suspended cells form and organize into functional renal structures which are able to excrete high levels of uric acid and creatinine through a yellow urine-like fluid. This work is being continued in our laboratory.

INJECTABLE THERAPIES

Long-term studies were conducted by our group in order to determine the effect of injectable chondrocytes *in vivo* as a potential bulking agent for urinary incontinence and vesicoureteral reflux.[16,17] Studies in mini-swine showed that alginate, a liquid solution of glucuronic and mannuronic acid, embedded with chondrocytes, could serve as a synthetic substrate for the injectable delivery and maintenance of cell architecture *in vivo*. These experiments showed, for the first time in the field of tissue engineering, that injectable cells could reform into tissue structures *in vivo*.

Using the same line of reasoning as with the chondrocyte technology, the possibility of using autologous smooth muscle cells

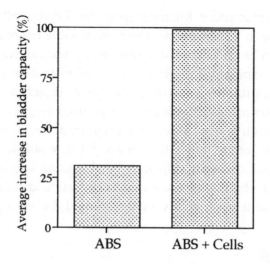

Fig. 1. Bladders augmented with a matrix seeded with urothelial and smooth muscle cells showed a 100% increase in capacity compared to bladders augmented with the cell-free unseeded matrix, which showed only a 30% increase in capacity within three months after implantation.

was investigated.[18] *In vivo* experiments were conducted in mini-pigs and reflux was successfully corrected. The use of myoblasts, derived from skeletal muscle has also been used in animal studies, either alone, or in combination with genes, and is very promising.

The first human application of a tissue engineering technology for urologic applications occurred with the injection of cells for the correction of vesicoureteral reflux in children at our institution and the treatment of urinary incontinence in adults.[20,21] Phase 3 clinical trials are being completed for vesicoureteral reflux and phase 2 trials are being initiated for urinary incontinence.

TESTIS

A system was designed wherein Leydig cells were microencapsulated in an alginate solution coated with 0.1% poly-L-lysine, for controlled testosterone replacement. Microencapsulated Leydig cells offer several advantages, such as serving as a semipermeable barrier between the transplanted cells and the host's immune system, as well as allowing for the long term physiological release of testosterone.

The encapsulated Leydig cells were injected in castrated animals and serum testosterone was measured serially. The castrated animals receiving the microencapsulated cells were able to maintain testosterone levels long term.[22] These studies suggest that microencapsulated Leydig cells may be able to replace or supplement testosterone in situations were anorchia or testicular failure is present. A similar system is currently being studied for estrogen.

GENITALIA

Primary cultures of human corpus cavernosum smooth muscle cells were derived from operative biopsies obtained during penile prosthesis surgery. Cells were maintained in continuous multilayered cultures, seeded onto polymers of nonwoven polyglycolic acid, and implanted subcutaneously in athymic mice, either alone, or with corporal endothelial cells.[23-25] At retrieval, by six weeks, there was tissue organization similar to normal corpora.

To apply tissue engineering techniques to reconstruct corporal tissue clinically, further studies are being performed in our laboratory. These include the further development of cell delivery vehicles identical to that of native corpus cavernosal architecture, and functional and biomechanical studies of the neo-corpora.

The possibility of creating a natural phallic prosthesis consisting of autologous chondrocytes was investigated which if biocompatible and elastic, could be used in patients who require genital reconstruction. Chondrocytes were isolated, grown, expanded *in vitro*, and seeded onto pre-formed cylinders which were implanted *in vivo*.[26] Gross examination showed the presence of well formed milky-white rod shaped solid cartilage structures. Biomechanical analyses demonstrated that the retrieved rods were able to withstand

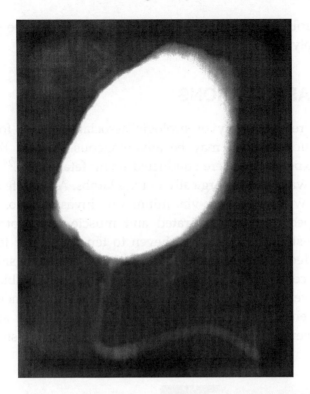

Fig. 2. Post-operative urethrogram of a patient with severe hypospadias recon-structed with an off-the-shelf collagen based matrix. Maintenance of a normal urethral caliber without any evidence of stricture is observed one year after surgery.

high degrees of compression, tension and bending and were readily elastic.

In a subsequent study, autologous cartilage seeded rods were implanted into rabbit corpora. The scaffolds were able to form cartilage rods *in vivo*, in the corpora. The engineered penile prostheses were stable, without any evidence of infection or erosion.[27] Subsequent studies were performed assessing the functionality of the cartilage penile rods *in vivo* long term. To date, the animals have done well, and can copulate and impregnate their female partners without problems. Further functional studies need to be completed before applying this technology to the clinical setting.

FETAL APPLICATIONS

Having a ready supply of urologic associated tissue for surgical reconstruction at birth may be advantageous. Toward this end, a series of experiments were conducted using fetal lambs.[28,29] Bladder exstrophy was created surgically in fetal lambs. A small fetal bladder specimen was harvested via minimally invasive fetoscopy. The bladder specimen was separated and muscle and urothelial cells were harvested and expanded. Seven to ten days prior to delivery, the expanded bladder muscle cells were seeded on one side and the urothelial cells on the opposite side of a biodegradable scaffold. After delivery, the lambs had surgical closure of their bladder using the engineered tissue. Molecular and functional analyses of the engineered tissue showed a pattern indistinguishable from

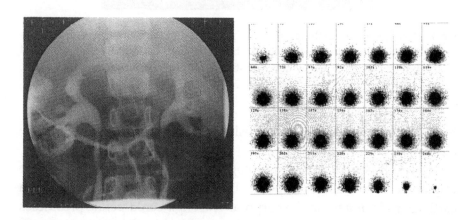

Fig. 3. Cystograms of pre- and post-endoscopic treatment of reflux in a patient. A chondrocyte-alginate suspension was injected endoscopically in the subureteral region. Preoperative fluoroscopic cystogram (left) shows bilateral reflux. Postoperative radionuclide cystogram (right) shows resolution of reflux.

native bladder.[41] Similar prenatal studies were performed in lambs, engineering skin, for reconstruction at birth.[42] Other fetal tissues and organs such as skeletal muscle and trachea have also been harvested and engineered in the same manner in our laboratory.

GENE THERAPY

Based on the feasibility of tissue engineering techniques in which cells seeded on biodegradable polymer scaffolds form tissue when implanted *in vivo*, the possibility was explored of developing a neo-organ system for *in vivo* gene therapy.[30] Human urothelial cells seeded on biodegradable scaffolds were transfected with PGL3-luc, pCMV-luc and pCMVβ-gal promoter-reporter gene constructs. The transfected cell-polymer scaffolds were then implanted *in vivo*. The results indicated that successful gene transfer could be achieved. The transfected cell/polymer scaffold formed organ-like structures with functional expression of the transfected genes.[30] This technology may be applicable throughout the spectrum of diseases which may be manageable with tissue engineering.

STEM CELLS FOR TISSUE ENGINEERING

Most current strategies for engineering urologic tissues involve harvesting of autologous cells from the host diseased organ. However, in situations wherein extensive end-stage organ failure is present, a tissue biopsy may not yield enough normal cells for expansion. Under these circumstances, the availability of pluripotent stem cells may be beneficial. Pluripotent embryonic stem cells are known to form teratomas *in vivo*, which are composed of a variety of differentiated cells. However, these cells are immunocompetent, and would require immunosuppression if used clinically.

We investigated the possibility of deriving stem cells from post-natal mesenchymal tissue from the same host, and inducing their differentiation *in vitro* and *in vivo*. Stem cells were isolated from human foreskin derived fibroblasts. Stem cell derived chondrocytes were obtained through a chondrogenic lineage process. The cells were grown, expanded, seeded onto biodegradable scaffolds, and implanted *in vivo*, where they formed mature cartilage structures. This was the first demonstration that stem cells can be derived from postnatal connective tissue and can be used for engineering tissues *in vivo ex situ*.[31]

A second approach which we have pursued involves the isolation of stem cells from individual organs. For example, daily female hormone supplementation is used widely, most commonly in post-menopausal women. A continuous and unlimited hormonal supply produced from ovarian granulosa cells would be an attractive alternative. We examined the feasibility of isolating functional human ovarian granulosa stem cells, which unlike primary cells, may have the ability to proliferate and function indefinitely.

Granulosa stem cells were selectively isolated from post-menopausal human ovaries and their phenotype was confirmed with the stem cell marker antibodies, CD 34, CD 105, and CD 90. The granulosa stem cells in culture showed steady state progesterone (5–7 ng/ml) and estradiol (2500–3000 pg/ml) production either with or without hCG stimulation.[32]

THERAPEUTIC CLONING FOR TISSUE ENGINEERING

Recent advances with the cloning of embryos and newborn animals have expanded the possibilities of this technology for tissue engineering and organ transplantation. There are many ethical concerns with cloning in terms of creating humans for the sole purpose of obtaining organs. However, the potential for retrieving cells from early-stage cloned embryos for subsequent regeneration is being proposed as an ethically viable benefit of therapeutic cloning.

We investigated the feasibility of engineering syngeneic tissues *in vivo* using cloned cells.

Unfertilized donor bovine eggs were retrieved and the nuclear material was removed. Bovine fibroblasts from the skin of a steer were obtained. The nuclear material was removed from the fibroblast and microinjected into the donor egg shell (nuclear transfer). A short burst of energy was delivered, initiating neo-embryogenesis. After eight days, the embryos were placed in the same steer uterus from which the fibroblasts had been obtained. The cloned embryo, with identical genetic material as the steer, was retrieved after 40 days for tissue harvest. Various cell types were harvested, expanded *in vitro*, and seeded on biodegradable scaffolds. The cell-polymer scaffolds were implanted into the back of the same steer from

<u>Therapeutic Cloning Strategies</u>

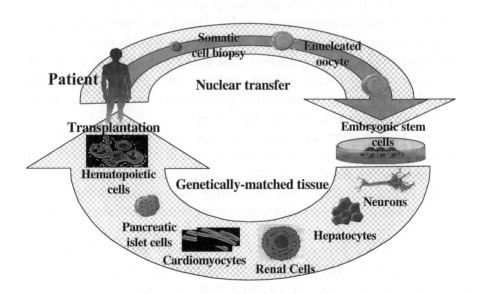

Fig. 4. Outline of therapeutic cloning.

which the cells were cloned. The implants were retrieved at various time points for analyses. Renal tissue, cartilage, cardiac, skeletal and smooth muscle were engineered successfully using therapeutic cloning.

These studies demonstrated that cells obtained through nuclear transfer can be successfully harvested, expanded in culture and transplanted *in vivo* with biodegradable scaffolds where the single suspended cells form and organize into tissue structures, which are the same genetically as the host. These studies were the first demonstration of the use of therapeutic cloning for the regeneration of tissues *in vivo*.[33]

CONCLUSION

Tissue engineering efforts are currently being undertaken for every type of tissue and organ within the urinary system. Primary autologous cells, stem cells and therapeutic cloning are being applied for the creation of tissues and organs. Tissue engineering techniques require expertise in growth factor biology, a cell culture facility designed for human application, and personnel who have mastered the techniques of cell harvest, culture and expansion. Polymer scaffold design and manufacturing resources are essential for the successful application of this technology.

The first human application of cell based tissue engineering technology for urologic applications occurred at our institution with the injection of autologous cells for the correction of vesicoureteral reflux in children. The same technology has been recently expanded to treat adult patients with urinary incontinence. Trials involving urethral tissue replacement using processed collagen matrices are in progress at our centre for both hypospadias and stricture repair. Bladder replacement using tissue engineering techniques is being explored. Recent progress suggests that engineered urologic tissues may have a wider clinical applicability in the future.

REFERENCES

1. Atala A. Tissue engineering in the genitourinary system. In *Tissue Engineering*, eds. A Atala and D Mooney, Boston, Birkhauser Press, 1997; 149.

2. Cilento BG, Freeman MR, Schneck FX, Retik AB, Atala A. Phenotypic and cytogenetic characterization of human bladder urothelia expanded *in vitro*. *J Urol* 1994; **152**:655–670.

3. Atala A, Vacanti JP, Peters CA, Mandell J, Retik AB, Freeman MR. Formation of urothelial structures *in vivo* from dissociated cells attached to biodegradable polymer scaffolds *in vitro*. *J Urol* 1992; **148**:658–662.

4. Atala A, Freeman MR, Vacanti JP, Shepard J, Retik AB. Implantation *in vivo* and retrieval of artificial structures consisting of rabbit and human urothelium and human bladder muscle. *J Urol* 1993; **150**:608–612.

5. Yoo JJ, Meng J, Oberpenning F, Atala A. Bladder augmentation using allogenic bladder submucosa seeded with cells. *Urol* 1998; **51**:221–222.

6. Kropp BP, Rippy MK, Badylak SF, Adams MC, Keating MA, Rink RC, Thor KB. Small intestinal submucosa: Urodynamic and histopathologic evaluation in long term canine bladder augmentations. *J Urol* 1996; **155**:2098–2104.

7. Probst M, Piechota HJ, Dahiya R, Tanagho EA. Homologous bladder augmentation in dog with the bladder acellular matrix graft. *BJU Int* 2000; **85**(3):362–371.

8. Oberpenning FO, Meng J, Yoo J, Atala A. *De novo* reconstitution of a functional urinary bladder by tissue engineering. *Nature Bio* 1999; **17**:149–155.

9. Cilento BG, Retik AB, Atala A. Urethral reconstruction using a polymer scaffolds seeded with urothelial and smooth muscle cells. *J Urol* 1996; **155**:5 (supp).

10. Kropp BP, Ludlow JK, Spicer D, Rippy MK, Badylak SF, Adams MC, Keating MA, Rink RC, Birhle R, Thor KB. Rabbit

urethral regeneration using small intestinal submucosa onlay grafts. *Urology* 1998; **52**(1):138–142.

11. Chen F, Yoo J, Atala A. Acellular collagen matrix as a possible "off the shelf" biomaterial for urethral repair. *Urol* 1999; **54**:407–410.

12. Sievert KD, Bakircioglu ME, Nunes L, Tu R, Dahiya R, Tanagho EA. Homologous acellular matrix graft for urethral reconstruction in the rabbit: histological and functional evaluation. *J Urol* 2000; **163**(6):1958–1965.

13. Atala A, Guzman L, Retik A. A novel inert collagen matrix for hypospadias repair. *J Urol* 1999; **162**:1148–1151.

14. Kassaby EA, Yoo J, Retik A, Atala A. A novel inert collagen matrix for urethral stricture repair. *J Urol* 2000; **308S**:70.

15. Amiel GE and Atala A. Current and future modalities for functional renal replacement. *Urol Clin* 1999; **26**:235–246.

16. Atala A, Cima LG, Kim W, Paige KT, Vacanti JP, Retik AB, Vacanti CA. Injectable alginate seeded with chondrocytes as a potential treatment for vesicoureteral reflux. *J Urol* 1993; **150**:745–747.

17. Atala A, Kim W, Paige KT, Vacanti CA, Retik AB. Endoscopic treatment of vesicoureteral reflux with chondrocyte-alginate suspension. *J Urol* 1994; **152**:641–643.

18. Cilento BG, Atala A. Treatment of reflux and incontinence with autologous chondrocytes and bladder muscle cells. *Dialogues in Pediatric Urol* 1995; **18**:11.

19. Yokoyama T, Huard J, Chancellor MB. Myoblast therapy for stress urinary incontinence and bladder dysfunction. *World J Urol* 2000; **18**(1):56–61.

20. Kershen RT, Atala A. Advances in injectable therapies for the treatment of incontinence and vesicoureteral reflux. *Urol Clin* 1999; **26**:81–94.

21. Diamond DA, Caldamone AA. Endoscopic correction of vesicoureteral reflux in children using autologous chondrocytes: Preliminary results. *J Urol* 1999; **162**:1185–1188.

22. Machluf M, Boorjian S, Caffaratti J, Kershen R, Atala A. Microencapsulation of Leydig Cells: A new system for the therapeutic delivery of testosterone. *Pediatrics* 1998; **102S**:32.

23. Kershen RT, Yoo JJ, Moreland RB, Krane RJ, Atala A. Novel system for the formation of human corpus cavernosum smooth muscle tissue *in vivo*. *Tissue Eng*, in press, 2000.

24. Park HJ, Yoo JJ, Kershen R, Atala A. Reconstitution of corporal tissue using human cavernosal smooth muscle and endothelial cells. *J Urol* 1999; **162**:1106–1109.

25. Falke G, Yoo J, Machado M, Moreland R, Atala A. Penile reconstruction using engineered corporal tissue. *J Urol* 2000; **980S**:221.

26. Yoo JJ, Lee I, Atala A. Cartilage rods as a potential material for penile reconstruction. *J Urol* 1998; **160**:1164–1168.

27. Yoo JJ, Park HJ, Lee I, Atala A. Autologous engineered cartilage rods for penile reconstruction. *J Urol* 1999; **162**:1119–1121.

28. Fauza DO, Fishman S, Mehegan K, Atala A. Videofetoscopically assisted fetal tissue engineering: Bladder augmentation. *J Ped Surg* 1998; **33**:7.

29. Fauza DO, Fishman S, Mehegan K, Atala A. Videofetoscopically assisted fetal tissue engineering: Skin replacement. *J Ped Surg* 1998; **33**:357–361.

30. Yoo JJ, Atala A. A novel gene delivery system using urothelial tissue engineered neo-organs. *J Urol* 1997; **158**:1066–1070.

31. Bartsch G, Yoo J, Kim B, Atala A. Stem cells in tissue engineering applications for incontinence. *J Urol* 2000; **1009S**;227.

32. Raya-Rivera A, Yoo J, Atala A. Hormone producing granulosa stem cells for intersex disorders. American Academy of Pediatrics Meeting, Urology section, Chicago, abs 29, 2000.

33. Atala A, Yoo J, Cibelli JB, Blackwell C, West N, Lanza. Therapeutic cloning applications for the engineering of kidney tissues. American Academy of Pediatrics Meeting, Urology section, Chicago, abs 9, 2000.

Chapter 8

Implantation of Kidney Rudiments

Marc R Hammerman

INTRODUCTION

End-stage chronic renal failure afflicts many hundreds of thousands of individuals worldwide.[1] Left untreated, end-stage disease would lead to death within days–weeks. Treatment consists of dialysis, a modality with considerable morbidity,[2] or renal transplantation using human donor organs (allografts). Transplantation has many advantages relative to dialysis. However, its use is limited by the number of organs available.[3]

One possible solution to the lack of organ availability is the use of kidneys from animals (renal xenografts). The clinical renal xenografts performed to date have utilized primate donors because the closer the species are phylogenetically, the more easily xenografts are accepted.[4] The clinical experience with the use of primates as kidney donors dates from the 1960s. However, the results of xenografting of kidneys have been unsatisfactory, and this technique has remained an experimental one.[4]

A second possible solution to the problem of limited human organ availability is the transplantation of kidney rudiments or developing kidneys to replace diseased renal tissue. Such transplantations have been performed by several investigators in animals.[5-14]

WHY RENAL RUDIMENTS?

There are four theoretical reasons why the transplantation of renal rudiments or developing kidneys (metanephroi) into adult animals might be advantageous relative to the transplantation of developed kidneys.

First, if obtained as a sufficiently early stage, such as prior to embryonic day 15 (E15) in the rat, a developing metanephros would be expected to be depleted of dendritic antigen-presenting cells (APCs)[15] that mediate "direct" host recognition of alloantigen or xenoantigen.[16] Dendritic cells would be absent from developing metanephroi by virtue of the absence of a vasculature from which APCs can enter the organ[5,12] and the absence of mature APCs themselves at this stage of rat development.[15]

It has been suggested that APC depletion from rat pancreatic islets resulting in prolongation of islet survival after injection into the portal vein of host mice, can be achieved by culturing the islets *in vitro* prior to injection.[17]

It is more difficult to achieve the same results with whole vascularized organs.[18,19] Dendritic cells can be depleted from developed kidneys by the combination of total body irradiation and cyclophosphamide pretreatment of donors. APC depletion in this manner prolongs graft survival in MHC (RT1) incompatible rat hosts.[19] Also, long surviving "immunologically enhanced" (dendritic cell depleted) kidneys transplanted into MHC incompatible hosts do not elicit strong primary T cell-dependent alloimmunity after transplantation into a secondary recipient of the same genotype as the original host.[18] "Immunological enhancement" is accomplished by injecting hosts with donor-strain spleen cells prior to transplantation and host anti-donor antiserum prior to transplantation and at the time of transplantation.[18]

Second, the transplanted metanephros becomes a chimeric organ in that it is vascularized in part by blood vessels originating from the host.[5,12] Hyperacute rejection of xenografts, that is initiated by circulating preformed natural antibodies which activate complement after binding to glycoprotein antigens on endothelial cell surfaces,[20,21] should be circumvented to the extent that the transplanted organ is supplied by host vessels.[4]

Insight into the origin of the renal blood supply is provided by experiments in which developing kidneys are transplanted to ectopic sites. In the case of 11 day-old mouse or chick metanephroi grafted onto the chorioallantoic membrane of the quail, the vasculature is derived entirely from the host.[12] In the case of 11–12 day-old mouse metanephroi grafted into the anterior chamber of the eye, the glomerular microvascular endothelium derives from both donor and host.[5] In either case, large external vessels derive from the host.[11,12]

Third, the expression of class II MHC heterodimers on non lymphoid, renal proximal tubule cells without co-expression of costimulatory receptors such as B7,[22] may serve as an extra thymic mechanism for the maintenance of immune tolerance.[20,21]

Fourth, the trafficking of host T cells through transplanted neonatal nonlymphoid tissue may represent a mechanism by which hosts can be rendered tolerant to that tissue.[23]

TRANSPLANTATION OF RENAL RUDIMENTS INTO HOST KIDNEYS

Woolf *et al.* implanted pieces of sectioned metanephroi originating from E13–E16 mice into tunnels fashioned in the cortex of kidneys of newborn outbred mice. Differentiation and growth of donor nephrons occurred in the host kidney. Glomeruli were vascularized, mature proximal tubules were formed and extensions of metanephric tubules into the renal medulla were observed. Glomerular filtration was demonstrable in donor nephrons using fluorescently-labeled dextran as a marker of filtration into the proximal tubules. However, connection of donor nephrons to the collecting system of hosts that would be required for plasma clearance to occur could not be demonstrated.[7]

Abrahamson *et al.* implanted metanephroi from E17 rat embryos beneath the renal capsule of adult rat hosts. Within 9–10 days post-implantation, every graft became vascularized, new nephrons were induced to form and glomerular and tubular cytodifferentiation occurred. Intravenous injection of antilaminin IgG into hosts resulted in labeling of glomerular basement membranes of grafted kidneys, confirming perfusion of the grafts by the host's vasculature.[6]

Robert *et al.* grafted metanephroi from E12 mouse embryos into kidney cortices of adult and newborn ROSA26 mouse hosts. ROSA26 mice bear an ubiquitously expressed β-galactosidase transgene that can be identified by staining in histological sections, permitting differentiation of transplanted from host tissue. Grafts into both newborn and adult hosts examined for 7 days post-transplantation were vascularized by components originating from both donor and host.[8]

Koseki *et al.* transplanted rat nephrogenic mesenchymal cells that had been transfected with a Lac Z reporter gene by a retrovirus, underneath the capsule of kidneys of neonatal rats. Transplanted mesenchymal cells were integrated into functioning host nephron segments.[9]

We transplanted E15 rat metanephroi beneath the renal capsule of adult Sprague Dawley rat kidneys. Within several weeks, transplanted metanephroi had developed glomeruli, and mature proximal and distal tubules characteristic of mature nephrons. In addition, collecting ducts of donor origin grew towards the papilla of the host.[10] However, as in the studies of Woolf *et al.*[6] we could not demonstrate any connection between the collecting systems of the donor and host kidneys.[10]

Four or six weeks following renal subcapsular transplantation when host kidneys were examined, cysts containing clear fluid surrounded the sites where metanephroi were transplanted under the renal capsule.[10] Levels of urea nitrogen and creatinine were concentrated in cyst fluid relative to blood. The concentrations of urea, nitrogen and creatinine in cyst fluid were significantly less than the concentrations in bladder urine.[10] These observations raise the possibility that the cyst fluid represents urine originating from the transplant. That cyst fluid is diluted relative to bladder urine and is consistent with the reduced ability of a four week-old kidney (transplanted kidney) to clear the blood of urea nitrogen and creatinine relative to a ten week-old kidney (host kidney).[10]

Armstrong *et al.* reported the formation of cysts in metanephroi transplanted under the kidney capsule of mice.[13] They suggested that the presence of cysts in developed donor metanephroi, coupled with their inability to demonstrate any connection between the donor and host collecting systems, indicated that transplanted metanephroi become obstructed in the subcapsular site. Our findings[10] are consistent with theirs.[13]

TRANSPLANTATION OF DEVELOPING METANEPHROI INTO THE OMENTUM

Barakat and Harrison showed that sections of rat metanephroi transplanted into a subcutaneous site in the abdominal wall of host rats remain vascularized for several days.[14] To provide alternative conditions for transplantation of metanephroi into adult hosts, we implanted whole metanephroi from embryonic day 15 (E15) outbred Sprague Dawley rats into the omentum of uni-nephrectomized non-immunosuppressed adult Sprague Dawley rat hosts.[10]

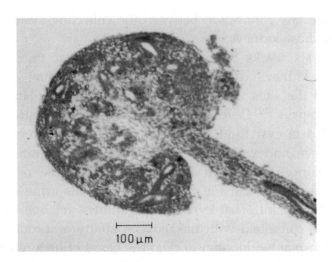

Fig. 1. Photomicrographs of an E15 rat metanephros stained with hematoxylin and eosin. Magnification is shown. Reproduced with permission.[10]

E15 metanephroi contained only metanephric blastema, segments of ureteric bud, and primitive nephrons with no glomeruli (Fig. 1). Four to six weeks post-implantation, metanephroi from E15 rats had enlarged, attracted a vasculature originating from the host's omentum, had formed mature tubules and glomeruli and a ureter that could be anastomosed to a ureter of the host (Fig. 2).

Four weeks following ureteroureterostomy, after the removal of both host kidneys, transplanted developed metanephroi were shown to function (to clear inulin infused into the host's circulation).

Electron microscopy of metanephroi revealed normal renal structures (Fig. 3). Shown in Fig. 3(a) is a glomerular capillary loop. A mesangial cell is labeled (m). A higher-power view of another glomerular capillary loop with an endothelial cell labeled (en) is shown in Fig. 3(b). A still higher-power view showing an epithelial cell (ep), an endothelial cell (en) podocytes (pd) and a basement membrane of normal thickness and appearance (arrows) is provided in Fig. 3(c). A proximal tubule (pt) with a brush border membrane

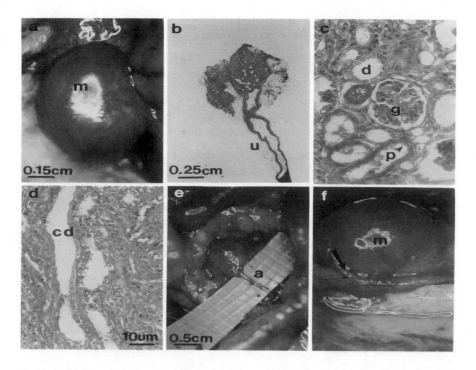

Fig. 2. Photographs of rat metanephroi, six weeks post-transplantation into the omentum of unilaterally nephrectomized host rats (a, e, f) and photomicrographs of hematoxylin and eosin-stained sections of metanephroi (b, c, d): (a) Developed metanephros (m) in abdominal cavity. (b) Section of developed metanephros with ureter (u). (c) Section of cortex from developed metanephros. Glomerulus (g), proximal tubule (p), brush border (arrowhead) and distal tubule (d) are shown. (d) Section of medulla from developed metanephros. Collecting duct (cd) is shown. (e) An artery originating from the omentum (a) is shown. (f) Anastomosis is shown (arrow) between host ureter and ureter from implanted metanephros (m). Magnifications are shown for a and f (a), for b, for c and d (d) and for e. Reproduced with permission.[10]

(arrowhead) is labeled in Fig. 3(d). A proximal tubule (pt), distal tubule (dt) and collecting duct (cd) are labeled in Fig. 3(e).

Our findings[10] establish that functional chimeric kidneys develop from metanephroi transplanted in adult hosts.

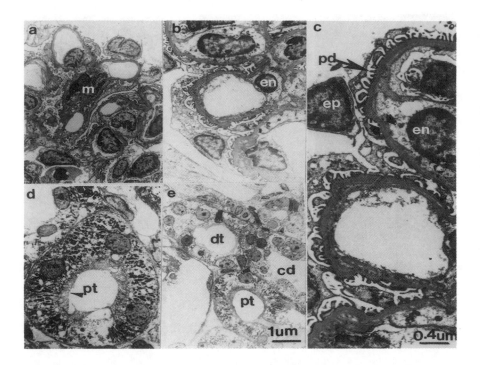

Fig. 3. Electron micrographs of transplanted rat metanephroi: (a) A glomerular capillary loop. A mesangial cell is labeled (m). (b) A glomerular capillary loop. An endothelial cell labeled (en) is labeled. (c) A glomerular capillary loop. An epithelial cell (ep), an endothelial cell (en) and podocytes (pd) are labeled and a basement membrane is delineated (arrows). (d) A proximal tubule (pt) with a brush border membrane (arrowhead). (e) A proximal tubule (pt), distal tubule (dt) and collecting duct (cd). Magnifications are shown for c and e.

THE USE OF GROWTH FACTORS TO ENHANCE THE GROWTH AND FUNCTION OF TRANSPLANTED METANEPHROI

Experiments utilizing metanephric organ culture have shown that kidney development *in vitro* is dependent upon the production of a number of polypeptide growth factors within the developing organ. Blocking the expression or action of any of the growth factors, transforming growth factor-α (TGF-α), hepatocyte growth factor (HGF), insulin-like growth factor I (IGF I) or insulin-like growth factor II (IGF II), inhibits growth and development *in vitro*.[24] Vascular endothelial growth factor (VEGF) is also produced by developing kidneys. Blocking VEGF activity *in vivo* inhibits renal vascularization.[25] Exposure of developing metanephroi to vitamin A stimulates glomerulogenesis *in vitro*.[26]

To ascertain whether growth factors can be used to enhance the growth and function of developing metanephroi, we performed two types of experiments. First, we determined whether the administration to transplanted hosts of IGF I, a growth factor known to enhance renal growth and function in a variety of settings,[27] can increase inulin clearance of transplanted metanephroi.

In preliminary experiments, we demonstrated that neither weights nor inulin clearances of metanephroi transplanted into rats increase between weeks 12–32 post transplantation.[28] In contrast, infusion of IGF I into hosts enhances inulin clearances of metanephroi measured at 12–16 weeks post-transplantation.[28]

Second, we exposed metanephroi to growth factors in two ways:

i) We incubated metanephroi with growth factors prior to transplantation into the omentum (pre-treatment).
ii) We bathed metanephroi in growth factors four weeks post-transplantation at the time of ureterouretostomy between the transplanted metanephros and the host (post-treatment).

Through the use of growth factors, we have increased inulin clearances to 300-times those measured in our first experiments. In some rats, clearances were about 10% of normal, approximately the level achieved by dialysis in humans.[27]

SUMMARY AND CONCLUSIONS

The studies described above demonstrate the feasibility of transplantation of allograft metanephroi subcapsularly in the kidney and into the omentum of adult outbred rat hosts. When implanted in the omentum of non-immunosuppressed hosts, metanephroi grow, differentiate into kidneys with normal renal structures, become vascularized (Figs. 2–3) and clear inulin from the circulation.[10] In contrast, transplanted developed kidneys are rejected within a week.[9] Studies performed subsequently demonstrate the feasibility of:

i) transplanting rat metanephroi into the mouse using co-stimulatory blockade, in which case the vasculature of the transplant originates from the mouse host.[29]

ii) transplanting metanephroi across the rat MHC (RT1) without immunosuppression, in which case peripheral tolerance results from T cell ignorance.[30]

iii) preserving metanephroi *in vitro* for as long as three days prior to transplantation.[31]

iv) transplanting pig metanephroi into the rat.[32]

Inulin clearances in transplanted metanephroi are very low.[10] However, clearances can be enhanced by growth promoting stimuli[10] including the use of exogenous growth factors.[27]

Although the rat provides a good model with which to characterize the growth of transplanted metanephroi, the applicability of this technique to humans will require that its feasibility be demonstrated in a larger mammal. Therefore, additional studies will be undertaken using the pig, the kidneys of which are more similar to those of man, and which has been proposed as a potential source for xenografts into humans.[33]

ACKNOWLEDGEMENTS

MRH is supported by grants DK-45181 and DK53487 from the National Institutes of Health, and by a grant from Intercytex Ltd., Manchester U.K.

REFERENCES

1. US Renal Data System, USRDS Annual Data Report. N.I.H., N.I.D.D.K. Bethesda MD *Am J Kidn Dis* 1999; **34**:S144–S151.
2. US Renal Data System, USRDS Annual Data Report, N.I.H., N.I.D.D.K. Bethesda MD *Am J Kidn Dis* 1999; **34**:S40–S50.
3. US Renal Data System, USRDS Annual Data Report, N.I.H., N.I.D.D.K. Bethesda MD *Am J Kidn Dis* 1999; **34**:S74–S86.
4. Auchincloss H Jr. Xenogeneic Transplantation. *Transplantation* 1988; **46**:1–20.
5. Hyink DP, et al. Endogenous origin of glomerular endothelial and mesangial cells in grafts of embryonic kidneys. *Am J Physiol* 1996; **270**:F886–F889.
6. Abrahamson DR, St John PL, Pillion DL, Tucker DC. Glomerular development in intraocular and intra renal grafts of fetal kidneys. *Lab Invest* 1991; **64**:629–639.
7. Woolf AS, Palmer SJ, Snow ML, Fine LG. Creation of a functioning mammalian chimeric kidney. *Kidney International* 1990; **38**:991–997.
8. Robert B, St John PL, Hyink DP, Abrahamson DR. Evidence that embryonic kidney cells expressing flk-1 are intrinsic, vasculogenic angioblasts. *Am J Physiol* 1996; **271**:F744–F753.
9. Kosecki C, Herzlinger D, Al-Awqati Q. Integration of embryonic nephrogenic cells carrying a reporter gene into functioning nephrons. *Am J Physiol* 1991; **261**:C550–C554.
10. Rogers SA, Lowell JA, Hammerman NA, Hammerman MR. Transplantation of developing metanephroi into adult rats. *Kidney International* 1998; **54**:27–37.
11. Wolf AS, Loughna S. Origin of glomerular capillaries: Is the verdict in? *Experimental Nephrology* 1998; **6**:17–21.
12. Sariola H, Ekblom P, Lehtonen E, Saxen L. Differentiation and vascularization of the metanephric kidney grafted on the chorioallantoic membrane. *Developmental Biology* 1983; **96**:427–435.

13. Armstrong JF, Kaufman MH, van Heyningen V, Bard JBL. Embryonic kidney rudiments grown in adult mice fail to mimic the Wilms' phenotype but show strain-specific morphogenesis. *Exp Nephrol* 1983; **1**:168–174.

14. Barakat TL, Harrison RG. The capacity of fetal and neonatal renal tissues to regenerate and differentiate in a heterotopic allogeneic subcutaneous tissue site in the rat. *J Anat* 1971; **110**:393–407.

15. Naito M. Macrophage heterogeneity in development and differentiation. *Archives of Histology and Cytology* 1993; **56**: 331–351.

16. Murphy B, Sayegh M. Why do we reject a graft? Mechanisms for recognition of transplantation antigens. *Transplantation Reviews* 1996; **10**:150–159.

17. Lacy PE, Davie JM, Finke FH. Prolongation of islet xenograft survival (rat to mouse). *Diabetes* 1981; **30**:285–291.

18. Lechler RI, Batchelor JR. Restoration of immunogenicity to passenger cell-depleted kidney allografts by the addition of donor strain dendritic cells. *J Exp Med* 1962; **155**:31–41.

19. McKenzie JL, Beard MEJ, Hart DNJ. Depletion of donor kidney dendritic cells prolongs graft survival. *Transplantation Proc* 1984; **16**:948–951.

20. Charlton B, Auchincloss H Jr, Fathman CG. Mechanisms of transplantation tolerance. *Annu Rev Immunol* 1994; **12**:707–734.

21. Kaufman CL, Gainse BA, Ildstad SY. Xenotransplantation. *Annu Rev Immunol* 1995; **13**:339–267.

22. Hagerty DT, Evavold DB, Allen PM. Regulation of the costimulator B7, not class II major histocompatibility complex, restricts the ability of murine kidney tubule cells to stimulate CD4+ T cells. *J Clin Invest* 1994; **93**:1208–1215.

23. Alferink J. *et al.* Control of neonatal tolerance to tissue antigens by peripheral T cell trafficking. *Science* 1998; **282**:1338–1341.

24. Hammerman MR. Growth factors in renal development. *Seminars in Nephrology* 1995; **15**:291–299.

25. Kitamano Y, Tokunaga H, Tomita H. Vascular endothelial growth factor is an essential molecule for mouse kidney development: Glomerulogenesis and nephrogenesis. *J Clin Invest* 1997; **99**:2351–2357.

26. Vilar J, Gilbert T, Moreau E, Merlet-Binichou C. Metanephros organogenesis is highly stimulated by vitamin A derivatives in organ culture. *Kidney International* 1995; **49**:1478–1487.

27. Hammerman MR. Recapitulation of phylogeny by ontogeny in nephrology. *Kidney International* 2000; **57**:742–755.

28. Rogers SA, Powell-Braxton L, Hammerman MR. Insulin-like growth factor I regulates renal development in rodents. *Developmental Genetics* 1999; **24**:293–298.

29. Rogers SA, Hammerman MR. Transplantation of rat metanephroi into mice. *Am J Physiol* 2001; **280**:R1865–R1869.

30. Rogers SA, Liapis H, Hammerman MR. Transplantation of metanephroi acress the major histocompatibility complex in rats. *Am J Physiol* 2001; **280**:R132–R136.

31. Rogers SA, Hammerman MR. Transplantation of metanephroi after preservation *in vitro*. *Am J Physiol* 2001; **281**:R661–665.

32. Hammerman MR. Growing kidneys. *Current Opinion in Nephrology and Hypertension* 2001; **10**:13–17.

33. Sachs DH. The pig as a potential xenograft donor. *Veterinary Immunology and Immunopathology* 1994; **43**:185–191.

Section 4

Xenotransplantation

Chapter 9

Xenotransplantation: Will Pigs Fly?

Andrew JT George & Robert I Lechler

INTRODUCTION

Xenotransplantation, or the transplantation of animal organs, to treat patients with end-stage organ failure has aroused considerable scientific, medical, ethical and public interest. To some extent, xenotransplantation is acting as a trail blazer for other novel forms of organ replacement therapy, causing the biomedical community not only to produce solutions to the technical problems of xenotransplantation, but also to engage with a wider audience in debates on the ethics and desirability of this new form of treatment. In this chapter, we will ask the question "Will pigs fly?" — in other words, whether pig (the favoured source) organs will be used to treat patients. This single question has a number of components; questioning the need for xenotransplantation, the ethics and safety of this approach and the scientific problems faced.

WHY SHOULD WE WANT PIGS TO FLY?

The first question that must be addressed is whether there is a need for xenotransplantation. Conventional allotransplantation (between genetically disparate individuals) is one of the medical success stories of the 20th century, and can be used in treatment of kidney, heart, lung, liver or corneal failure (Fig. 1). Organ transplantation is the only routine treatment available for patients with life threatening end-stage failure of heart, liver or lungs. Even when not life saving, renal transplantation offers a considerable improvement in the quality of life of patients with renal failure, at a reduced cost when compared to that of long term dialysis. Corneal transplantation can restore sight when the cornea has failed. In addition, the more experimental approaches of bowel and islet/pancreas transplantation offer considerable promise to appropriate patient groups. Finally, cellular transplantation in the form of bone marrow transplantation is routine for patients with some forms of haematopoietic malignancy or

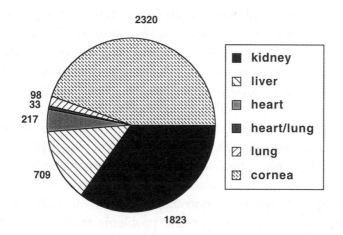

Fig. 1. Transplant numbers in the UK and Republic of Ireland, 2000. This demonstrates the number of transplants carried out (major types, not including rarer, experimental forms of transplant) as reported to UK Transplant in year 2000. Data from Ref. 1.

congenital abnormalities of the bone marrow — and transplantation of cells is under active clinical investigation for conditions such as Parkinson's disease.

However, there are two considerable limitations in our current transplantation practice. The first is that the immune system recognises the transplanted tissue as foreign and mounts a vigorous response that can destroy the organ. Advances in immunosuppressive therapy, using a variety of pharmaceutical agents, have had a major impact in preventing acute rejection, and graft survival of organs transplanted is approaching 90% at one year. However, this is at a cost. Patients on long-term immunosuppressive therapy are prone to increased rates of infection and malignancy because of a generalised blunting of the immune response. In addition, many of the pharmaceutical agents have toxic effects, such as nephrotoxicity. It should also be noted that the improvements in immunosuppression have had little effect on the incidence or progression of chronic transplant rejection. There is therefore a need to develop new ways of preventing the immune system from destroying the graft, if possible by inducing tolerance to the foreign organ. As we will discuss later, xenotransplantation may offer unique opportunities to do this.

The second major problem is the shortage of organs for transplantation. This is a problem for nearly all forms of organ transplantation. For example, in 1999, 289 heart transplants were performed in the United Kingdom and Republic of Ireland.[1] This is a rate of five transplants per million of the population. However, the demand for organs has been estimated to be between 20 and 60 per million of the population.[2] This estimate may be conservative; patient survival following cardiac transplantation is superior to that of patients with moderate to high severity chronic cardiac failure treated more conventionally, indicating that transplantation might be an appropriate option for many more patients. It is also important to note that the quality of life of patients with transplants is very high. In one report, 90% of cardiac graft recipients considered there was no limitation to their activity.[3]

While the demand for organs is increasing, the supply of organs is decreasing. The number of transplants performed has plateaued or decreased since the early 1990s (Fig. 2). This shortage of donors is the major limitation on transplantation becoming a much more common and routine form of therapy. In some parts of the world, the problem is particularly severe owing to religious and cultural beliefs about the way the body is treated after death,[4] and in these areas, transplantation programmes are relatively very underdeveloped when compared to Europe, Australasia or the USA.

There are several solutions to this problem. One is to increase the donor base by a process of education. This may be helped by

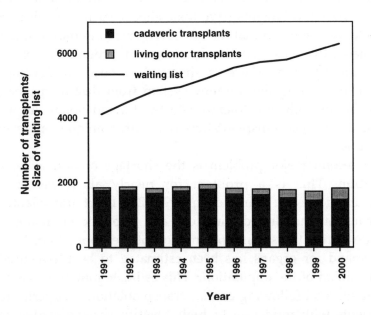

Fig. 2. Number of renal transplants and waiting list, 1991–2000. The number of renal transplants using grafts from cadaveric or living donors is shown. The number of transplants has been static over the 10 years, with the number of grafts performed using cadaveric donors showing a substantial decline. However, the number of patients on the waiting list has risen inexorably over this time. Data from Ref. 1.

changes in the way consent is obtained for donation (possibly moving to presumed consent for donation rather than the current need to obtain consent). In addition, there can be more efficient use of potential donors, with multiple organs being harvested from donors (for example in 1998, 2965 solid organs were retreived from just 847 donors in the UK and Republic of Ireland[5]). There is also an increased use of "marginal donors" whose organs would previously not have been accepted for transplantation,[6,7] and the use of "non heart beating" donors.[8-10] A final approach is to increase the use of living donors. Living donation is routinely in use for renal transplantation (Fig. 2), and is being developed for liver and lung transplantation. However, while all these approaches will be useful in increasing the number of organs available for donation, they will not be able to provide enough material to satisfy the demand. In addition, there are considerable problems with some of these approaches. These include public concern over donation policies, with potential loss of confidence in the medical system. The use of living donors raises important issues,[11] as there may be mortality and morbidity associated with donation (indeed, living lung transplantation, in which lobes from two donors are transplanted into the recipient, can be described as the only operation with a potential 300% mortality).

These considerations indicate that the increase of the number of organs available for transplantation would have an enormous impact on health. This is the major motivating factor for xenotransplantation. There are other strategies being developed, as discussed elsewhere in this book. However, at present animal organs are the most developed alternative to human organs available.

SHOULD PIGS FLY?

Xenotransplantation has aroused considerable public debate concerning the ethics and safety of xenotransplantation. While these two issues are related, we will consider some of the ethical points in this section.

There are several grounds for objecting to xenotransplantation. The first set of objections concerns the use of animals. Some would consider that it is morally wrong to kill or inflict suffering on animals. This can be because animals are seen to have rights or a moral status equivalent to that of human beings. It has been argued that genetic engineering and the breeding and raising of animals in conditions suitable for transplantation will inflict suffering on animals, possibly in an unforeseen way. In addition, some believe that the genetic engineering of animals interferes with either the naturalness of an animal or species or intrudes into the autonomy of their being. For a balanced overview and discussion of some of these objections, see Ref. 12.

These objections are, of course, not restricted to xenotransplantation but to the use of animals in agriculture, research and entertainment. The standard response of scientists is to point out that humans have been altering creatures for their own use for millennia, through a process of selective breeding.[13] In addition, the use of animals for food and clothing indicates that, in general, there is public acceptance that animals can be killed and harmed for human use (it should be noted that more than 100 million pigs are slaughtered for food in the USA each year[14]). Of course, while persuasive, such arguments do not really counter those of the objectors; just because we have carried out certain practices for many years and that they are accepted by the public does not make them necessarily ethically or morally correct actions.

The alternative is to argue that humans do have a moral status that is different from animals. This can then be used to justify the killing or harming of animals (which have a lower status) to benefit humans (which have a higher status). Individuals might see the moral status of humans being conferred by a god who, in Judeo-Christian belief can be seen to have placed humans in charge of different orders of creation. Alternatively it might be argued that humans have an increased capacity to suffer, or are self conscious, and so can be seen as different from other animals. There are some problems with such arguments; if moral status is conferred by a god, then the argument will only have weight with others who share the same beliefs! The use of the self-conscious nature or capacity for suffering to draw a distinction between us and other animals may suggest that it would be acceptable to harm individuals who do not have such attributes. Modern genetics can also suggest that, as there is considerable similarity between the DNA of humans and animals, the differences may not be as great as once imagined. However, it is an argument that is instinctively shared by the majority of society, even if the basis for the distinction is difficult to determine. In general, when faced between saving the life of an animal, even a much-loved family pet, and a human being, most people would recognise a difference and save the person.

Alternative arguments against xenotransplantation include general objections to the way in which medical science is going. It can be argued that too much resource is being used to marginally prolong life, which might be better spent in preventing disease. We live in a society that is spending more and more on healthcare, with less and less benefit. Such an argument is unlikely to have much of an impact on someone faced with death due to organ failure! It is also one not shared by the majority of the population, who continue to support medical research through charitable giving. However, it may be that the cost benefit ratio of new treatments is not always considered early in their development, leading to the unfortunate situation that funders of health care (either government or insurance companies) may increasingly have to prevent access to expensive but effective forms of treatment. In general, however, xenotransplantation will probably come out of any cost benefit analysis quite well, the long term costs are likely to be similar to allotransplantation (with the addition of an "up front" purchase of organs, but a reduced need to maintain typing laboratories or organ sharing schemes). At present, transplantation offers a high quality prolongation of life that is often cheaper than current maintenance therapy (e.g. dialysis).

The final set of objections rests in the effect of transplantation on the recipient. At one level, these objections can be very emotional; people can feel there is something wrong with putting animal material into humans, the so called "yuck factor". Some worry that insertion of an animal organ will cause psychological harm to the recipient, but since the alternative is probably death that might appear the lesser of two evils. In addition, there is little evidence of major psychiatric morbidity as a result of allotransplantation. Other objections are those surrounding any novel form of therapy; the risk to the patient and their ability to give fully informed consent to the procedure. This might be particularly important if there is a need to monitor the first patients for a long period for safety reasons. However, it should be able to produce protocols that overcome these problems.

In general it would appear that, with the exception of safety issues as will be discussed below, for most individuals there are no insurmountable ethical barriers to xenotransplantation. The majority of reports, such as that of the Nuffield Council of Bioethics have endorsed this.[15] In addition the Pope, at the 2000 Transplantation Congress in Rome gave his qualified support to xenotransplantation.[16]

WILL FLYING PIGS BE DANGEROUS?

There are safety issues concerning xenotransplantation. In particular the risk of zoonotic infection of the recipient by the donor organ. It should be possible, by careful husbandry and microbiological screening, to prevent the transmission of viruses causing an acute infection of the pigs.[17] Far more problematic is the existence of latent retroviruses in the porcine genome. These porcine endogenous retroviruses (PERVs) were first described 30 years ago.[18-20] Recently it has been shown that these PERV are capable of infecting human cells,[21-23] and so are a potential risk to the recipient. More recent experiments have highlighted the concern, by showing the transmission of PERV *in vivo* following islet cell transplantation to either *scid*/NOD or *scid* mice.[24,25] These data are balanced by other experiments, in which no evidence of PERV transmission has been seen in humans exposed to porcine tissue.[26-30] However, the danger of transmission of a potentially infectious agent to a relatively large number of immunosuppressed patients, as well as the potential for recombination with human retroviruses to create novel pathogens, has stopped any clinical developments with solid organ xenotransplantation until the issue is satisfactorily resolved.

Sequence analysis of PERV show them to be similar to type C retroviruses of mice, cats and gibbons. The viral genome is around 8.1 kb and contains conventional *gag*, *pol* and *env* genes. The viruses are divided into three families: A, B and C.[31,32] The amino acid homology between families of the products of the *gag* and *pol* open reading frames is very high (>95%), but there is considerable differences in the envelope proteins.[31-33] These differences in the coat proteins may affect the tropism of the viruses. The number of copies of PERV in the porcine genome varies between animals, with an estimated 50 loci in some strains.[22] However, the number of complete, infectious, genomes is not generally known for each strain. In one study, 11 full length PERV transcripts were found with at least 4 with functional envelope protein.[33] In addition in inbred strains, individual animals contain additional proviral loci, that might have arisen by recombination, retrotransposition of active PERV or new infections, suggesting a plasticity in the numbers and position of PERV in the genome.[31]

The mRNA for PERV are expressed in a wide range of pig tissues. Infectious PERV-A and PERV-B virions are released by the porcine kidney epithelial line PK-15[22] and primary endothelial cell lines.[21] PERV-C can be released by porcine peripheral blood mononuclear cells after mitogen stimulation.[23]

A number of strategies could be employed in order to prevent PERV spread into recipients. These include the use of conventional transcriptase or protease inhibitors, though the long term use of such agents cannot be optimal. In addition PERV show a lower susceptibility to transcriptase inhibitors, and no sensitivity to current protease inhibitors.[34] Vaccination of recipients has been suggested as an alternative,[35] but would seem not to be advisable in a setting in which one is trying to reduce the anti-graft immune response. An alternative strategy would be to breed out the PERV in the genome. Given the large number of copies of PERV, and the potential plasticity of the PERV copies, this would be extremely difficult to achieve. It is likely that such breeding would be beneficial in reducing the number of PERV, though not in abolishing it. Finally, it might be feasible to "knock out" the PERV in the genome, possibly in a cell line followed by cloning. Again the large number of PERV makes this a challenging task.

There is no doubt that the risk of infection is a major hurdle that needs to be overcome if xenotransplantation is to be successful. At present, we do not have the necessary information to let us decide whether xenotransplantation will be safe. Even when more data is available it will be difficult to decide how to make the decision. Normally in the case of a novel therapy, the risk to the individual of the treatment must be balanced with the potential benefit to that person. In the case of xenotransplantation, the majority of the patients, at least initially, will die without treatment. Therefore the slight risk of infection by PERV might be reasonable, especially as there is a risk of infection (for example with CMV, HIV, prions etc) following conventional allotransplantation. The problem with xenotransplantation is that there is a risk to society that a new and virulent virus might be unleashed. In the worst Doomsday scenario, this could threaten all human life. How does one balance what is thought to be an extremely remote risk, but with potentially devastating consequences, with the benefit of prolongation of life for a limited number of individuals. At present, there is no consensus on this.

It is clear that the first recipients of pig organs, as well as those in close contact with them, will have to be monitored and screened for long periods — quite possibly for life. If there does turn out to be a risk of transmission, then contingency plans will need to be drawn up, and included in the information given to the recipient when they consent to the trial.

It must be noted that this problem is not confined to xenotransplantation of organs from animals such as pigs. The US Public Health Service has recently amended their definition of xenotransplantation to be: "*Any procedure that involves the transplantation, implantation or infusion into a human recipient of either (a) live cells, tissues or organs from a nonhuman animal source, or (b) human body fluids, cells, tissues or organs that have had ex vivo contact with nonhuman cells, tissues or organs*". As many strategies that involve tissue engineering or cell replacement may include the use of animal serum or cells or other material for the growth of the tissue, such safety considerations must be addressed more widely. At present, the UK Xenotransplantation Interim Regulatory Authority is considering whether to recommend a modification of the definition used in the UK.[36]

HOW TO MAKE PIGS FLY?

The debate above about the ethics and safety of xenotransplantation will be irrelevant if it cannot be made to work! Xenotransplantation has been performed since at least the 1800s on an occasional basis. However, these attempts were all doomed to failure as there was no realisation of the role of the immune system in rejection of organs. In more modern times (since the advent of immunosuppressive drugs), kidneys from chimpanzees or baboons were transplanted into patients[37-40] with graft survival being as long as six months. In some cases the xenotransplant served as a bridge, until a conventional donor organ could be found. However, all grafts failed in the end, either due to immunological rejection or to death of the patient from sepsis. Heart and liver xenotransplants, with primate donors, have also been carried out, again with no long term graft survival.[41-43]

More recently, the primate has been abandoned as a donor. This is because of public concern and they are relatively slow to breed and are often endangered species. In addition, the risk of zoonotic infection is unacceptably high. The favoured donor is, of course, the pig. This is because they are of an appropriate size. Porcine organs are in general physiologically compatible with humans (with the probable exception of livers). There is considerable experience in pig farming and husbandry. In addition transgenic technology is well set up for pigs, while the recent description of cloning of pigs[44] opens the possibility of rapid genetic manipulation, including gene deletion.

There are, however, fundamental incompatibilities between pig organs and human recipients. These can lead to violent and rapid rejection of pig organs transplanted into primates, termed hyperacute rejection.[45] This is similar to the rejection seen in allografts in the face of pre-existing antibody reactive with the donor tissue. In the case of xenotransplantation, the hyperacute rejection is mediated by the binding of pre-existing "natural" human antibodies that recognise a galactosyl α-1,3 galactosyl carbohydrate residue (α-galactosyl epitope) expressed on porcine cells.[46,47] Other epitopes may also be recognised, but these are less important.[48] The α-galactosyl residue is made by an α-1,3 galactosyl transferase enzyme that is present in all mammals except *homo sapiens* and Old World monkeys and apes.[46,47]

When preformed natural antibodies bind to the α-galactosyl epitope on porcine endothelial cells, they initiate the complement cascade. This can both kill endothelial cells and activate them. The activation of the endothelial cells results in rapid loss of expression of anti-coagulant molecules from the cell surface. In addition the cells retract, exposing pro-coagulant molecules such as Tissue Factor that are expressed below the endothelial cells. The result is rapid initiation of the blood clotting cascade, with thrombus formation and death of the tissue. This can occur within minutes of exposure of the organ to human blood.

There are several alternative strategies for overcoming hyperacute rejection. These include depleting natural antibody from the circulation,[49] or blocking their activity with soluble carbohydrates.[50] In addition genetic modification of the porcine cells to abolish expression of α-galactosyl can be performed, either by expression of alternative glycosylation enzymes[51,52] or by expression of intracellular antibodies that bind to and block the action of the α-1,3 galactosyl transferase.[53] However, the approach that has produced the most encouraging data has been to make pigs transgenic for human complement regulatory proteins, such as CD55 (Decay accelerating factor, DAF), CD46 (membrane co-factor, MCF) and CD59. Transplantation of hearts transgenic for DAF and CD59 into non-human primates survived 4–30 hours, compared to 60–90 minutes for controls.[54] In addition 4/9 DAF-transgenic kidneys grafted into bilaterally nephrectomised cynomologous monkeys survived for more than 50 days.[55] These data indicate that the problems of hyperacute rejection can be circumvented.

Once hyperacute rejection has been dealt with, the next barrier is delayed xenograft rejection, also known as acute vascular rejection (Fig. 3). The pathology of this rejection response includes platelet aggregation and fibrin deposition as well as a cellular infiltrate. The inflammatory cells are most commonly polymorphonuclear cells, monocytes and NK cells (i.e. from the innate branch of the immune system). Following is delayed xenograft rejection, it would be expected that the next stages would be acute T cell mediated rejection and chronic rejection. The mechanisms involved in these processes are likely to be similar to those seen in allografts.

We will not attempt in this short chapter to review all the strategies being developed to prevent xenograft rejection, several recent reviews cover this area,[56–60] as does other chapters in this book (Teranishi *et al.* and Tai and Platt). Instead we will concentrate on just one of the approaches being developed in Imperial College. If hyperacute graft rejection is prevented, many of the grafts fail showing signs of thrombosis and coagulation. The aetiology

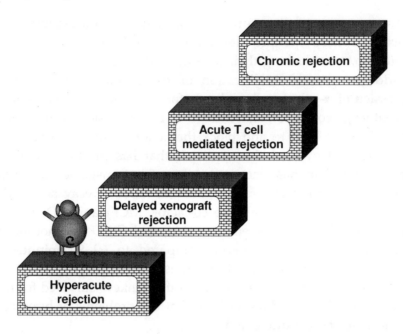

Fig. 3. Stages in xenograft rejection. The hurdles to long term xenograft survival are shown. Hyperacute rejection has to a great extent been overcome. Ahead lies the further obstacles of delayed xenograft rejection and acute (T-cell mediated) and chronic rejection.

of this is unknown. In order to prevent this, we have generated novel constructs in which either Tissue Factor Pathway Inhibitor (normally a soluble inhibitor of the coagulation pathway) or hirudin (used by leaches to prevent clotting) are made as fusion proteins with transmembrane regions.[61,62] When the genes encoding these constructs are transfected into endothelial cells, the anti-coagulant molecules are expressed on the cell surface, where they are active as shown both by the binding of the appropriate components of the blood clotting cascade and by their ability to reduce the clotting time.[61-63] Expression of these molecules can, therefore, increase the anti-coagulant properties of the endothelium. However, if these gene constructs were used to generate transgenic pigs, there is a potential problem that the blood vessels will constitutively express these molecules, rendering the animals susceptible to bleeding disorders. In order to overcome this the molecules were fused to the transmembrane regions of P selectin. This causes them to be packaged into the Weibel-Palade bodies within endothelial cells.[64] These vesicles contain a number of preformed proteins (including P-selectin) and, when the endothelial cell is activated, they rapidly fuse with the plasma membrane, leading to surface expression of their contents. When the Tissue Factor Pathway Inhibitor and hirudin contained the P-selectin tail, they were not expressed on the surface of the resting cells. However, upon activation they were rapidly exteriorised and could inhibit the clotting cascade.

The above example indicates one way in which transgenesis can be used to reduce rejection. The long term aim of preventing xenograft rejection must be to induce tolerance to the graft. This is one area were xenotransplantation has an enormous potential advantage over cadaveric allotransplantation. The transplant operation will in most cases be elective, and will be performed between a known donor-recipient combination, unlike the situation seen in allotransplantation when the operation is unscheduled. This gives the opportunity to prepare the recipient for the transplant, as well as using a donor that has been manipulated (most probably genetically), to increase the chances of tolerance induction.

WILL PIGS FLY?

The simple answer to the question as to whether pigs will fly is simply — we do not know. It is fairly certain that the current scientific challenges will be overcome. However, we do not know what awaits us over the horizon. In addition, at present we do not have enough data to satisfactorily assess the risk to the population due to the presence of infective agents in pigs.

However, if that seems somewhat pessimistic, then we should analyse where we have got to. Massive advances have been made in a short period, and the potential reward of xenotransplantation is enormous both for patients, researchers and the pharmaceutical industry. The risk of zoonosis has stalled any advance to clinical trials, but data on the infectivity of PERV is being rapidly collected.

We should also not be too upset if the pigs never quite manage to fly. The research effort of endeavouring to get them airborne has yielded a lot of important information about vascular biology and the immunology of rejection. It has produced a lot of therapeutic strategies that may have other applications. For example, the anti-thrombotic strategy outlined above could be useful in a variety of conditions, including prevention of restenosis following coronary bypass surgery. Similarly, the tolerance induction strategies will be applicable for autoimmune disease and for conventional transplantation.

ACKNOWLEDGEMENTS

The authors would like to thank the MRC for current support of their xenotransplantation programme, and gratefully acknowledge previous support by PPL. We would also like to thank Zhao An-zhu for secretarial assistance.

REFERENCES

1. UK Transplant, Yearly Transplant Statistics for the UK and Republic of Ireland, Bristol 2001.
2. Dark J. Cardiac Transplantation. In *Oxford Textbook of Medicine*, eds. D Weatherall, J Ledingham, D Warrell. Oxford University Press, Oxford 1999.
3. Hosenpud JD, Bennett LE, Keck BM, Fiol B, Boucek MM, Novick RJ. The Registry of the International Society for Heart and Lung Transplantation: Fifteenth official report — 1998. *J Heart Lung Transplant* 1998; **17**:656–668.

4. Woo KT. Social and cultural aspects of organ donation in Asia. *Ann Acad Med Singapore* 1992; **21**:421–427.

5. UKTSSA, Transplant activity report 1998, Bristol 1999.

6. Alexander JW, Zola JC. Expanding the donor pool: Use of marginal donors for solid organ transplantation. *Clin Transplant* 1996; **10**:1–19.

7. Milano A, Livi U, Casula R, Bortolotti U, Gambino A, Zenati M, Valente M, Angelini A, Thiene G, Casarotto D. Influence of marginal donors on early results after heart transplantation. *Transplant Proc* 1993; **25**:3158–3159.

8. Balupuri S, Buckley P, Snowden C, Mustafa M, Sen B, Griffiths P, Hannon M, Manas D, Kirby J, Talbot D. The trouble with kidneys derived from the non-heart-beating donor: A single center 10-year experience. *Transplantation* 2000; **69**:842–846.

9. Reich DJ, Munoz SJ, Rothstein KD, Nathan HM, Edwards JM, Hasz RD, Manzarbeitia CY. Controlled non-heart-beating donor liver transplantation: A successful single center experience, with topic update. *Transplantation* 2000; **70**:1159–1166.

10. Steen S, Sjoberg T, Pierre L, Liao Q, Eriksson L, Algotsson L. Transplantation of lungs from a non-heart-beating donor. *Lancet* 2001; **357**:825–829.

11. Abecassis M, et al. Consensus statement on the live organ donor. *J Am Med Assoc* 2000; **284**: 2919–2926.

12. Holland A, Johnson A. *Animal Biotechnology and Ethics.* Chapman & Hall, London 1998.

13. George AJT. Animal biotechnology in medicine. In *Animal Biotechnology and Ethics,* eds. A Holland, A Johnson. Chapman & Hall, London, 1998; 27–49.

14. Gundry SR. Is it time for clinical xenotransplantation (again)? *Xeno* 1994; **2**:60–61.

15. Nuffield Council on Bioethics, Animal-to-human transplants: The ethics of xenotransplantation, London 1996.

16. John Paul II. Special address by His Holiness John Paul II, Rome, August 29, 2000. *Transplant Proc* 2001; **33**:31–32.

17. Swindle MM. Defining appropriate health status and management programs for specific-pathogen-free swine for xenotransplantation. *Ann NY Acad Sci* 1998; **862**:111–120.

18. Armstrong JA, Porterfield JS, De Madrid AT. C-type virus particles in pig kidney cell lines. *J Gen Virol* 1971; **10**:195–198.

19. Benveniste RE, Todaro GJ. Homology between type-C viruses of various species as determined by molecular hybridization. *Proc Natl Acad Sci, USA,* 1973; **70**:3316–3320.

20. Breese SS, Jr. Virus-like particles occurring in cultures of stable pig kidney cell lines. Brief report. *Arch Gesamte Virusforsch* 1970; **30**:401–404.

21. Martin U, Kiessig V, Blusch JH, Haverich A, von der Helm K, Herden T, Steinhoff G. Expression of pig endogenous retrovirus by primary porcine endothelial cells and infection of human cells. *Lancet* 1998; **352**:692–694.

22. Patience C, Takeuchi Y, Weiss RA. Infection of human cells by an endogenous retrovirus of pigs. *Nat Med* 1997; **3**:282–286.

23. Wilson CA, Wong S, Muller J, Davidson CE, Rose TM, Burd P. Type C retrovirus released from porcine primary peripheral blood mononuclear cells infects human cells. *J Virol* 1998; **72**: 3082–3087.

24. Deng YM, Tuch BE, Rawlinson WD. Transmission of porcine endogenous retroviruses in severe combined immunodeficient mice xenotransplanted with fetal porcine pancreatic cells. *Transplantation* 2000; **70**:1010–1016.

25. van der Laan LJ, Lockey C, Griffeth BC, Frasier FS, Wilson CA, Onions DE, Hering BJ, Long Z, Otto E, Torbett BE, Salomon DR. Infection by porcine endogenous retrovirus after islet xenotransplantation in SCID mice. *Nature* 2000; **407**:90–94.

26. Dinsmore JH, Manhart C, Raineri R, Jacoby DB, Moore A. No evidence for infection of human cells with porcine endogenous retrovirus (PERV) after exposure to porcine fetal neuronal cells. *Transplantation* 2000; **70**:1382–1389.

27. Heneine W, Tibell A, Switzer WM, Sandstrom P, Rosales GV, Mathews A, Korsgren O, Chapman LE, Folks TM, Groth CG.

No evidence of infection with porcine endogenous retrovirus in recipients of porcine islet-cell xenografts. *Lancet* 1998; **352**: 695–699.

28. Paradis K, Langford G, Long Z, Heneine W, Sandstrom P, Switzer WM, Chapman LE, Lockey C, Onions D, Otto E. Search for cross-species transmission of porcine endogenous retrovirus in patients treated with living pig tissue. *Science* 1999; **285**: 1236–1241.

29. Patience C, Patton GS, Takeuchi Y, Weiss RA, McClure MO, Rydberg L, Breimer ME. No evidence of pig DNA or retroviral infection in patients with short-term extracorporeal connection to pig kidneys. *Lancet* 1998; **352**:699–701.

30. Pitkin Z, Mullon C. Evidence of absence of porcine endogenous retrovirus (PERV) infection in patients treated with a bioartificial liver support system. *Artif Organs* 1999; **23**:829–833.

31. Akiyoshi DE, Denaro M, Zhu H, Greenstein JL, Banerjee P, Fishman JA. Identification of a full-length cDNA for an endogenous retrovirus of miniature swine. *J Virol* 1998; **72**: 4503–4507.

32. Le Tissier P, Stoye JP, Takeuchi Y, Patience C, Weiss RA. Two sets of human-tropic pig retrovirus. *Nature* 1997; **389**:681–682.

33. Bösch S, Arnauld C, Jestin A. Study of full-length porcine endogenous retrovirus genomes with envelope gene polymorphism in a specific-pathogen-free Large White swine herd. *J Virol* 2000; **74**:8575–8581.

34. Qari SH, Magre S, Garcia-Lerma JG, Hussain AI, Takeuchi Y, Patience C, Weiss RA, Heneine W. Susceptibility of the porcine endogenous retrovirus to reverse transcriptase and protease inhibitors. *J Virol* 2001; **75**:1048–1053.

35. Tacke SJ, Kurth R, Denner J. Porcine endogenous retroviruses inhibit human immune cell function: Risk for xenotransplantation? *Virology* 2000; **268**:87–93.

36. UK Xenotransplantation Interim Regulatory Authority, Third Annual Report September 1999–November 2000, Department of Health: London 2000.

37. Hitchcock CR, Kiser JC, Telander RL, Seljeskog EL. Baboon renal grafts. *J Am Med Assoc* 1964; **189**:934–937.
38. Reemtsma K, McCracken BH, Schlegel JU, Pearl MA, DeWitt CW, Creech Jr, O. Reversal of early graft rejection after renal heterotransplantation in man. *J Am Med Assoc* 1964; **187**:691–696.
39. Reemtsma K, McCracken BH, Schlegel JU, Pearl MA, Pearce CW, DeWitt CW, Smith PE, Hewitt RL, Flinner RL, Creech Jr O. Renal heterotransplantation in man. *Ann Surg* 1964; **160**:384–410.
40. Starzl TE, Marchioro TL, Peters GN, Kirkpatrick CH, Wilson WEC, Porter KA, Rifkind D, Ogden DA, Hitchcock CR, Waddell WR. Renal heterotransplantation from baboon to man: Experience with 6 cases. *Transplantation* 1964; **2**:752–776.
41. Bailey LL, Nehlsen-Cannarella SL, Concepcion W, Jolley WB. Baboon-to-human cardiac xenotransplantation in a neonate. *J Am Med Assoc* 1985; **254**:3321–3329.
42. Barnard CN, Wolpowitz A, Losman JG. Heterotopic cardiac transplantation with a xenograft for assistance of the left heart in cardiogenic shock after cardiopulmonary bypass. *South African Medical Journal* 1977; **52**:1035–1038.
43. Starzl TE, Fung J, Tzakis A, Todo S, Demetris AJ, Marino IR, Doyle H, Zeevi A, Warty V, Michaels M, Kusne S, Rudert WA, Trucco M. Baboon-to-human liver transplantation. *Lancet* 1993; **341**:65–71.
44. Polejaeva IA, Chen SH, Vaught TD, Page RL, Mullins J, Ball S, Dai Y, Boone J, Walker S, Ayares DL, Colman A, Campbell KH. Cloned pigs produced by nuclear transfer from adult somatic cells. *Nature* 2000; **407**:86–90.
45. Calne RY. Organ transplantation between widely disparate species. *Transplant Proc* 1970; **2**:550–556.
46. Galili U. Interaction of the natural anti-Gal antibody with α-galactosyl epitopes: A major obstacle for xenotransplantation in humans. *Immunol Today* 1993; **14**:480–482.

47. Sandrin MS, Vaughan HA, Dabkowski PL, McKenzie IF. Anti-pig IgM antibodies in human serum react predominantly with Gal(α 1-3)Gal epitopes. *Proc Natl Acad Sci, USA,* 1993; **90**:11391–11395.

48. Zhu A. Binding of human natural antibodies to nonalphaGal xenoantigens on porcine erythrocytes. *Transplantation* 2000; **69**: 2422–2428.

49. Good AH, Cooper DK, Malcolm AJ, Ippolito RM, Koren E, Neethling FA, Ye Y, Zuhdi N, Lamontagne LR. Identification of carbohydrate structures that bind human antiporcine antibodies: Implications for discordant xenografting in humans. *Transplant Proc* 1992; **24**:559–562.

50. Ye Y, Neethling FA, Niekrasz M, Koren E, Richards SV, Martin M, Kosanke S, Oriol R, Cooper DK. Evidence that intravenously administered alpha-galactosyl carbohydrates reduce baboon serum cytotoxicity to pig kidney cells (PK15) and transplanted pig hearts. *Transplantation* 1994; **58**:330–337.

51. Sandrin MS, Fodor WL, Mouhtouris E, Osman N, Cohney S, Rollins SA, Guilmette ER, Setter E, Squinto SP, McKenzie IF. Enzymatic remodelling of the carbohydrate surface of a xenogenic cell substantially reduces human antibody binding and complement-mediated cytolysis. *Nat Med* 1995; **1**:1261–1267.

52. Sepp A, Skacel P, Lindstedt R, Lechler RI. Expression of alpha-1,3-galactose and other type 2 oligosaccharide structures in a porcine endothelial cell line transfected with human alpha-1,2-fucosyltransferase cDNA. *J Biol Chem* 1997; **272**:23104–23110.

53. Sepp A, Farrar T, Dorling A, Cairns T, George AJT, Lechler RI. Inhibition of expression of Galα1-3Gal epitope on porcine cells using an intracellular single-chain antibody directed against α1,3Galactosyltransferase. *J Immunol Meth* 1999; **231**:191–205.

54. McCurry KR, Kooyman DL, Alvarado CG, Cotterell AH, Martin MJ, Logan JS, Platt JL. Human complement regulatory proteins protect swine-to-primate cardiac xenografts from humoral injury. *Nature Medicine* 1995; **1**:423–427.

55. Cozzi E, Bhatti F, Schmoeckel M, Chavez G, Smith KG, Zaidi A, Bradley JR, Thiru S, Goddard M, Vial C, Ostlie D, Wallwork J, White DJ, Friend PJ. Long-term survival of non-human primates receiving life-supporting transgenic porcine kidney xenografts. *Transplantation* 2000; **70**:15–21.

56. Bach FH. Xenotransplantation: Problems and prospects. *Annu Rev Med* 1998; **49**:301–310.

57. Dorling A, Lechler RI. Glaxo/MRS Young Investigator Prize. Xenotransplantation: Immune barriers beyond hyperacute rejection. *Clin Sci Colch* 1997; **93**:493–505.

58. Dorling A, Riesbeck K, Warrens A, Lechler R. Clinical xenotransplantation of solid organs. *Lancet* 1997; **349**:867–871.

59. Platt JL, Nagayasu T. Current status of xenotransplantation. *Clin Exp Pharmacol Physiol* 1999; **26**:1026–1032.

60. Wekerle T, Sykes M. Mixed chimerism as an approach for the induction of transplantation tolerance. *Transplantation* 1999; **68**:459–467.

61. Reisbeck K, Chen D, Kemball-Cook G, McVey JH, George AJT, Tuddenham EGD, Dorling A, Lechler RI. Hirudin expressed in mammalian cells and tethered to the cell surface confers thrombin binding activity. *Circulation* 1998; **98**:2744–2752.

62. Riesbeck K, Dorling A, Kemball C-G, McVey JH, Jones M, Tuddenham EG, Lechler RI. Human tissue factor pathway inhibitor fused to CD4 binds both FXa and TF/FVIIa at the cell surface. *Thromb Haemost* 71997; **8**:1488–1494.

63. Chen D, Riesbeck K, Kemball C-G, McVey JH, Tuddenham EG, Lechler RI, Dorling A. Inhibition of tissue factor-dependent and -independent coagulation by cell surface expression of novel anticoagulant fusion proteins. *Transplantation* 1999; **67**:467–474.

64. Chen D, Riesbeck K, McVey JH, Kemball-Cook G, Tuddenham EG, Lechler RI, Dorling A. Regulated inhibition of coagulation by porcine endothelial cells expressing P-selectin-tagged hirudin and tissue factor pathway inhibitor fusion proteins. *Transplantation* 1999; **68**:832–839.

Chapter 10

Therapeutic Strategies for Xenotransplantation

Katsuhito Teranishi, Ian PJ Alwayn, Leo Bühler, David H Sachs & David KC Cooper

ABBREVIATIONS

ADCC = antibody-dependent cell-mediated cytotoxicity
AHXR = acute humoral xenograft rejection
AIA = anti-idiotypic antibodies
CVF = cobra venom factor
EIA = extracorporeal immunoadsorption
Gal = Galα1-3Gal
HAR = hyperacute rejection
IVIg = intravenous immunoglobulin
mAb = monoclonal antibody
sCR1 = soluble complement receptor 1

INTRODUCTION

Organ transplantation is an established and successful therapy for patients with end-stage organ failure. Its very success, however, has resulted in a critical shortage of appropriate human cadaveric donor organs. Xenotransplantation — the transplantation of organs/tissues between animals of different species — would be one solution to this organ shortage. For a number of reasons, the pig is viewed by many as the most likely source of donor organs.[1,2]

The Pig as a Source of Organs and Tissues for Humans (see also Chapters 9 and 12)

Many organ systems of pigs have been shown to be similar physiologically to their human counterparts, including the cardiovascular, renal, pulmonary, and digestive systems. Pigs can be typed for ABO blood type. All are A or H(O). For the purposes of organ transplantation, group H(O) swine have been selectively bred to ensure no immune response directed against AB antigens. Pigs have favourable breeding characteristics for the production of organ-source animals, and are one of the few large animal species in which it is possible to carry out a breeding program to establish genetic characteristics. Pigs have large litter sizes (5–10 offsprings), early sexual maturity (five months), short gestation time (114 days), and frequent estrus cycles (every three weeks). It is theoretically possible to incorporate any number of transgenes into a line of pigs to modify the animal to make it more appropriate as a donor species, e.g. by the introduction of a gene for a human complement-regulatory protein.

In the authors' laboratory, partially-inbred miniature swine have been produced by a selective breeding program over the past 25 years.[3] Animals homozygous for the major histocompatibility complex have been produced, and one inbred line has currently reached >70% coefficient of inbreeding. Miniature swine have a number of advantages as potential xenograft donors. They achieve adult weights of approximately 120–140 kg, making it possible to obtain a miniature swine of appropriate size as an organ donor for any potential human recipient, from a newborn baby to a large adult. In contrast, domestic swine reach mature weights of over 450 kg, clearly larger than necessary for organ donation, and difficult to handle as a laboratory animal.

If knowledge of the tissue type of the potential donor pig herd is available, genetic engineering of the recipient's tissues is also feasible. For example, it may be possible to induce tolerance to the products of some of the most important xenogeneic antigens by introducing the corresponding swine genes into the bone marrow stem cells of the recipient, utilizing gene therapy techniques. After a gene ubiquitous in the herd has been introduced into the potential recipient, a subsequent organ transplant can be carried out using an organ from any member of the herd. This procedure, which has been termed "molecular chimerism", has been shown to be effective for prolonging the survival of allotransplants.[4,5] Preliminary studies have been carried out in the pig-to-nonhuman primate model,[6] and are discussed below.

Concerns have been expressed that transplantation of porcine organs into humans may result in a public health risk.[7] This concern was triggered largely by data demonstrating the *in vitro* transmission of porcine endogenous retroviruses from pig cells to human cells.[8] It is feared that porcine endogenous retroviruses might be transferred to the human recipient with a transplanted organ, and then may be spread to other members of the community by contact with the recipient. To date, however, more than one hundred patients have been exposed to living porcine tissue in various forms (liver perfusion, splenic perfusion, islet xenotransplantation, etc.), with no definite evidence for cross-species infection with porcine endogenous retroviruses.[9] Thus, despite some concerns, the overall climate for pig-to-human transplantation remains favourable, and many scientists, including ourselves, urge that the field should proceed actively but with caution.[10]

Immunological response to a transplanted pig organs in a primate

The advantages of the pig as a source of organs for humans are to some extent offset, however, by their immunological disparity with primates. Xenografts between widely disparate species, such as the pig-to-primate, are generally rejected within minutes or hours by a humoral mechanism, known as hyperacute rejection (HAR).[11,12] This phenomenon results from the presence of antibodies in humans, apes and Old World monkeys that are directed against Galα1-3Gal (Gal) epitopes on the pig vascular endothelium,[13-17] which is the first structure with which the host's blood comes into contact. Anti-Gal antibodies consist of subclasses of IgG, IgM and IgA,[18,19] and make up >85% of circulating natural anti-pig antibody in humans. In baboons, commonly-used surrogate animals for humans in experimental studies, essentially all anti-pig antibodies are anti-Gal.

Binding of anti-Gal antibodies to the Gal epitopes on pig tissues results in activation of complement through the classical pathway, with rapid complement-mediated destruction of the pig organ.[20-22] The barrier to HAR is the first that needs to be overcome if xenotransplantation involving pig organs in humans is to be successful. HAR can be prevented by several approaches. These include depletion of anti-Gal antibody by extracorporeal immunoadsorption using plasmapheresis or immunoaffinity columns of synthetic αGal. Such depletion, coupled with standard pharmacologic immunosuppressive therapy, extends xenograft survival from minutes or hours to several days. Complement depletion or inhibition by, for example, cobra venom factor or soluble complement receptor 1 (sCR1), is an alternative approach that can extend xenograft survival successfully. The transplantation of organs from pigs transgenic for one or more human complement regulatory protein (e.g. human decay-accelerating factor) has to date proven the most successful approach, with life-supporting renal graft survival continuing for one to two months in pharmacologically-immunosuppressed monkeys.

However, even if HAR is successfully prevented, a delayed form of humoral rejection (variously known as acute vascular rejection, delayed xenograft rejection, or acute humoral xenograft rejection (AHXR)) develops after days or weeks.[23,24] This form of rejection is believed to be related to the continuing presence (or return) of anti-Gal antibodies, possibly caused by a different mechanism, such as antibody-dependent cell-mediated cytotoxicity (ADCC), rather than by complement activation, although this remains uncertain.

Even if AHXR could be prevented, there is evidence from *in vitro* studies that a strong acute cellular rejection will develop.[24-27] The *in vitro* human anti-pig cellular response is at least as strong as the corresponding allogeneic immune response. Both direct and indirect pathways of porcine endothelial cell antigen recognition by T cells take place. It is also likely that chronic rejection, e.g. graft atherosclerosis, will develop early in xenografted organs, but survival of xenografts has to date been too limited for meaningful observations to be made.

Efforts to overcome the barriers of xenotransplantation are being directed at both the recipient primate and the donor pig. Genetic modification of the donor pig is discussed elsewhere in this volume.

IMMUNE MODULATION OF THE RECIPIENT

Several approaches are being investigated:

i) Depletion of anti-Gal antibodies or inhibition of their binding to the antigens of the xenograft.
ii) Depletion or inhibition of complement.
iii) Depletion or suppression of function of B and T lymphocytes.
iv) Development of immunologic tolerance by creation of mixed hematopoietic cell chimerism, molecular chimerism, or the transplantation of donor thymic tissue.

All of the above approaches are currently under study in the pig-to-baboon model at our centre. Progress utilizing these approaches will be discussed in this chapter.

Depletion of Xenoreactive (Anti-Gal) Antibodies

Anti-Gal antibodies can be depleted by (i) blood or plasma perfusion through a donor-specific organ, (ii) plasma exchange, (iii) blood or plasma perfusion through non-specific antibody sorbents, (iv) blood or plasma perfusion through specific antibody sorbents.

Hemoperfusion through a pig organ

Initial experiments attempting to remove xenoreactive antibody in the 1960s and 1970s were carried out by perfusing the recipient's blood through donor-specific organs, such as a pig liver or kidney.[28,29] Anti-pig antibodies were adsorbed on to the vascular endothelium of the perfused organ and the blood was temporarily depleted of antibody. This approach provided the first prolongation of survival of a pig heart transplanted into a baboon.[21] The pig liver may possibly be a better organ for immunoadsorption than the kidney [21,30-33] and, more recently, Macchiarini et al. have demonstrated that the pig lung is also a good organ for this purpose.[34]

Plasma exchange

Plasma exchange, which removes the subject's plasma with replacement of volume by saline, crystalloid fluid or antibody-depleted plasma, is used in the treatment of certain diseases, such as myasthenia gravis, thrombotic microangiopathies, cryoglobulinemias, and severe liver dysfunction.[35-37] Because all circulating antibodies, including anti-Gal antibody, are removed, a temporary state of agammaglobulinemia is achieved. This state increases the risk of infection and bleeding by loss of gamma globulin and coagulation factors.

Accommodation

Using plasma exchange, circulating anti-A and anti-B blood group antibodies have been depleted from potential kidney transplant patients, enabling ABO-incompatible kidney allotransplantation to be performed successfully.[38-41] HAR is avoided by transplantation of the organ during the period of antibody depletion. Accommodation, a state where return of antibody directed against ABO antigens expressed on the graft vascular endothelium does not lead to rejection even in the presence of normal levels of complement, has commonly been documented in these cases.

In some rodent models of xenotransplantation, xenografts are not rejected by AHXR and continue to function in the presence of xenoreactive antibodies and complement.[42,43] When antibodies are removed by plasma exchange in the pig-to-baboon organ transplantation model, xenograft survival has been prolonged up to 22 days.[22] However, accommodation has not been documented and graft failure develops from AHXR. In our experiments, using a protocol aimed at inducing tolerance through mixed chimerism in pig-to-baboons (see below), although low levels of anti-Gal antibody and complement activity are observed for several days, accommodation is not achieved.

Potential mechanisms responsible for accommodation have been discussed by Bach *et al.*[44]

i) The antibodies that return to the circulation may be different in isotype, affinity and/or specificity from the original and cannot initiate rejection.
ii) The surface antigens on the graft endothelial cells may change during the absence of antibody, and recognition by the returning antibodies is prevented.

iii) Endothelial cells may adapt during the return of antibody, and react differently to these antibodies. During accommodation, the expression of "protective genes" has been demonstrated in vascularized xenografts, and these are believed to induce resistance of the vascular endothelium to antibody-mediated destruction.[42]

If accommodation could be achieved after discordant xenotransplantation in the pig-to-baboon model, maintenance immunosuppressive treatment might be required only to prevent the cellular response. However, to date no group has been able to achieve this goal.

Nonspecific antibody sorbents

Perfusion of plasma through a protein immunoaffinity column, such as staphylococcal protein A or protein G, has a similar effect as plasma exchange, i.e., all immunoglobulin is removed by non-specific binding of the Fc portion of antibodies to the protein, leading to a temporary state of hypo or agammaglobulinemia.[45] The depleted plasma is returned to the patient with other replacement fluids. Using this technique, some success has been achieved in depleting potential kidney transplant recipients of anti-HLA antibodies, thus allowing successful allotransplantation.[46–48] Xenoantibodies have also been depleted experimentally in non-human primates.[49]

Specific antibody sorbents

Specific antibody sorbents can remove anti-Gal antibodies alone from plasma by binding the antibody to specific natural or synthetic Gal oligosaccharide immunoaffinity columns.[14,33,50–55] As only antibodies of a single specificity are removed, hypo- or agammaglobulinemia does not occur. Ye *et al.*[55] demonstrated that extracorporeal immuno-adsorption (EIA) of baboon plasma through immunoaffinity columns of melibiose (a Gal sugar of low specificity) on four consecutive days could reduce serum cytotoxicity to pig kidney (PK15) cells by 80%. If highly-specific synthetic Gal oligosaccharides, such as a Gal trisaccharide (Galα1-3Galβ1-4Glc),[56,57] are used, a three-day course of EIA can deplete anti-Gal IgM and IgG by 99% and 97%, respectively.[58] However, continuing production of anti-Gal antibodies results in their steady return to baseline level within a few days,[50-52] and leads to AHXR.[33,53,54]

Although all of the above methods are successful in depleting anti-pig antibodies, return of antibody has not yet been successfully suppressed by pharmacologic immunosuppression or through other approaches.

Inhibition or Neutralization of Xenoreactive (Anti-Gal) Antibodies

Infusion of natural or synthetic gal oligosaccharides

An alternative approach to depletion of antibody is the continuous infusion of the specific Gal oligosaccharide into the blood.[14,16,55,59-61] The circulating anti-Gal antibodies bind to the oligosaccharides and are therefore theoretically no longer free to bind to a transplanted pig organ. Early studies involved the intravenous infusion of melibiose and/or arabinogalactan,[55] both of which have terminal non-reducing alpha-galactose but which are of low specificity. Their infusion at high dosage significantly decreased the serum cytotoxicity to PK15 cells,[56] but proved toxic to the baboon.[55] However, highly specific oligosaccharides, such as the Gal disaccharide (Galα1-3Gal) or trisaccharide (Galα1-3Galβ1-4Glc), proved nontoxic and decreased serum cytotoxicity further.[60,61] Rejection of transplanted pig organs, although delayed, was not totally prevented as binding of the antibody to the oligosaccharide was insufficiently strong to prevent some binding to the transplanted organ.

Intravenous infusion of anti-idiotypic antibodies (AIA)

The infusion of anti-idiotypic antibodies (AIAs) directed against idiotypes expressed on anti-Gal antibodies, is a similar approach that has been explored. Koren et al. reported that murine monoclonal antibodies (mAbs) directed against human anti-pig antibodies bound to human anti-Gal IgG and F(ab)^2 fragments, and reduced serum cytotoxicity to pig cells.[62] Some of these AIAs also bound to human peripheral B lymphocytes, suggesting that they might delete these cells and reduce anti-Gal antibody production. The combination of two or three AIAs was more effective than the infusion of one alone. When two AIAs were infused into baboons, serum cytotoxicity was reduced to approximately 10%.

Furthermore, the repeated administration of pig polyclonal anti-human AIAs (produced by immunizing a pig with human anti-Gal antibodies) to a baboon (following EIA of anti-Gal antibodies) was able to delay the return of anti-Gal and reduce the cytotoxicity of the serum to pig cells.[63] It appeared that a reduction in the number of B cells producing cytotoxic anti-Gal antibody may have been achieved.

Intravenous infusion of human immunoglobulin (IVIg)

The intravenous administration of high doses of human IVIg has been used in the treatment of a variety of autoimmune and inflammatory disorders, including idiopathic thrombocytopenic purpura (ITP), myasthenia gravis, and inflammatory myopathies.[64–67] IVIg has also been used to reduce anti-HLA antibodies in highly-sensitized patients both pre- and post-transplant.[68–72] Magee et al.[73] and Gautreau et al.[74] demonstrated that the intravenous infusion of purified human IgG into nonhuman primates delayed the HAR of transplanted porcine hearts. Several mechanisms of action of IVIg have been proposed, including the possible action of IVIg as a complement modulator by inhibiting C3 binding to target cell membranes, thereby suppressing complement-mediated graft rejection. However, the exact mechanism of action of IVIg remains unclear. Studies have revealed that a specialized intracellular Fc receptor is saturated by IVIg, allowing the degradation of IgG in proportion to its total plasma concentration.[75] Recent data indicated that recurrent IVIg therapy reduced HLA panel-reactive antibodies in patients awaiting organ transplantation, but did not reduce anti-Gal antibody levels.[76] The infusion of IVIg depleted of anti-Gal IgG would be more likely to be successful in achieving a decrease in anti-Gal antibody levels. In our laboratory, administration of IVIg to baboons undergoing a protocol aimed at inducing mixed chimerism (see below), resulted in a modest increase in the initial levels of chimerism observed.[77,78]

Depletion or Inhibition of Complement

Complement plays a major role in the HAR of porcine xenografts in primates, primarily through activation of the classical pathway, although the alternative pathway may possibly play a role. The binding of host xenoreactive antibodies to the donor organ endothelial cells activates complement, inducing HAR. Depletion or inhibition of complement therefore impairs the development of HAR and prolongs xenograft survival.

Depletion of complement

In the 1960s, cobra venom factor was used as a tool to investigate the complement pathway.[79] Purified cobra venom factor (CVF), which is a functional analogue of C3b, has been shown to activate complement by binding to mammalian factor B.[80,81] Because CVF is approximately five times more stable than C3bBb and is resistant to decay acceleration and proteolytic inactivation, its administration causes continuous activation of complement, resulting in its rapid depletion.[82] CVF therapy can, therefore, prolong discordant xenograft survival significantly.[28,83–85] However, in the presence of xenoreactive antibodies, even when complement is continuously depleted, histo-pathological features of AHXR begin to develop within several days, presumably by complement-independent mechanisms or from the effect of early complement fractions (e.g. C1, C4).[84–86] The addition of concomitant pharmacologic immunosuppressive therapy, to suppress function of B and T cells, delays graft rejection further (up to 25 days) but eventually proves inadequate and AHXR occurs.[85]

Inhibition of complement

Human complement receptor I (CR1, C3b/C4b receptor) is a single-chain, cell surface glycoprotein found on erythrocytes, most white blood cells, and some other cells.[87] It is also found circulating in a soluble form in plasma at low concentrations. The interaction of CR1 with C3b and C4b can regulate complement activation through its convertase decay-accelerating factor activity and its factor I cofactor activity. Weisman et al.[88] constructed a soluble form of complement receptor 1 (sCR1) which lacks the transmembrane and cytoplasmic protein domains. sCR1 has been demonstrated to be a potent and selective inhibitor of both the classical and alternative complement pathways. The administration of sCR1 has prolonged cardiac graft survival in the guinea pig-to-rat and in the pig-to-primate models. sCR1 therapy combined with pharmacologic immunosuppression has prolonged pig cardiac xenograft survival for up to six weeks.[89–93]

Pharmacologic inhibition of complement

FUT-175, which is a synthetic inhibitor of serine proteases, has been demonstrated to inhibit C3a and C5a anaphylatoxin generation. Kobayashi et al.[94] demonstrated that in high concentration, it inhibited complement-mediated hemolysis of sheep erythrocytes and marginally delayed pig organ graft rejection in baboons. However, it proved not only less effective than either CVF or sCR1 but also had an adverse effect by markedly prolonging prothrombin and partial thromboplastin times.

K76COOH, which is a monocarboxylic acid derivative of a fungal product, inhibits complement activity at the C5 stage. High doses of K76COOH are required to maintain a high serum level but even these levels proved inadequate to fully suppress complement activation.[94] It also failed to significantly prolong porcine organ survival after transplantation into a baboon.

Attempts to Suppress Anti-Gal Antibody Production

Antibodies, including anti-Gal antibodies, are produced by B lymphocytes and plasma cells. Selective suppression of the cells responsible for the production of anti-Gal antibodies is likely to contribute significantly to long-term xenograft survival. However, the technology to selectively deplete only anti-Gal-secretory cells is currently not available. We have therefore concentrated our efforts on the temporary suppression of **all** antibody production. We have investigated whole body irradiation, various pharmacologic agents, and certain anti-B cell or anti-plasma cell monoclonal antibodies and immunotoxins.

Whole body irradiation (WBI)

Irradiation which causes death of rapidly-dividing cells, such as malignant cells and hematopoietic cells, has been used as an element in conditioning regimens aimed at facilitating bone marrow transplantation. Of relevance to transplantation, it has been used to achieve mixed hematopoietic chimerism and the induction of immunologic tolerance.[95-98] Almost total, but temporary, B cell ablation in peripheral blood, bone marrow and lymph nodes is induced by whole body irradiation at the non-lethal dose of 300 cGy.[99] However, this dose induces only a small and transient reduction of anti-Gal antibody production.[54,99,100] Within 2–3 weeks, B cell numbers are recovering rapidly and anti-Gal antibody levels have returned to baseline.

Pharmacologic agents

Pharmacologic immunosuppressive therapy to suppress anti-Gal antibody production following antibody depletion by EIA was investigated extensively by Lambrigts et al.[52] After a course of three EIAs, without concomitant immunosuppressive therapy, anti-Gal antibody levels returned to pre-EIA levels within 48 hours. Even though the rate of return of anti-Gal antibodies after EIA could be reduced by several combinations of immunosuppressive drugs, it proved impossible to suppress antibody production completely. Antibody levels could be maintained at approximately 20–45% of baseline during the course of EIA, but rose to 50–70% of baseline when EIA was discontinued, despite continuous and intensive immunosuppressive drug therapy. Furthermore, this partial suppression of anti-Gal antibody production was associated with several unwanted side effects of the drug therapy, such as thrombocytopenia and/or leukopenia, often necessitating discontinuation of therapy or reduction in dosage. Further studies by Alwayn,[99] and Teranishi (unpublished) and their colleagues, using drugs as diverse as cladribine, methotrexate and zidovudine (AZT) have all been similarly disappointing.

In summary, pharmacologic agents, even in combination in relatively high dosage, have proved capable of only partially suppressing anti-Gal antibody production. We therefore turned our attention to other forms of therapy that might deplete antibody-producing cells.

Depletion of B cells/plasma cells

Several mAbs and immunotoxins directed against B cell- or plasma cell-specific surface markers have been developed for treatment of B cell or plasma cell neoplastic conditions.[101-105] We investigated whether these mAbs were effective in depleting normal (non-neoplastic) B cells and plasma cells in baboons.

Anti-CD20 monoclonal antibody

The chimeric murine-human anti-CD20 mAb, (IDEC-C2B8, Rituximab, Rituxan; IDEC Pharmaceuticals, San Diego, CA, and Genentech, Inc., San Francisco, CA) targets CD20 on the surface of B cells. It has significant activity in depleting malignant B cells in the blood of patients with non-Hodgkin's lymphoma.[101-105] In our experience in the baboon,[99,106,107] after four doses at weekly intervals, no B cells could be detected by flow cytometry in the blood or bone marrow, and there was an 80% depletion in the lymph nodes. With the addition of low-dose whole body irradiation (150 cGy) after the course of mAb treatment, B cells were depleted even in the lymph nodes. No B cells could be detected in blood, bone marrow or lymph nodes for >3 months. After a course of EIA carried out during this period of B cell depletion, anti-Gal IgG remained at approximately 50% of pretreatment level but anti-Gal IgM returned almost to baseline within two weeks.

This suggested to us that the major source of anti-Gal antibody is plasma cells, rather than B cells. This conclusion is supported by *in vitro* data from our centre that indicates that anti-Gal antibody-producing cells are enriched following depletion of CD20$^+$ cells (B cells) but are deleted following depletion of cells bearing the CD38 marker (which is predominantly found on plasma cells).[107] Depletion of plasma cells by anti-CD38-ricin immunotoxin *in vivo*, however, has not proved successful, almost certainly because of inadequate conjugation of ricin to the mAb (see below).

Anti-CD22 and anti-CD38 immunotoxins

B cell-restricted immunotoxins have been constructed by conjugating mAbs directed at B cell determinants (e.g. CD22) to toxins such as ricin A or saporin.[108–110] Such immunotoxins have been administered to selected patients with B cell non-Hodgkin's lymphoma.[111–113] CD22 is a surface molecule involved with the generation of mature B cells within the bone marrow, blood and marginal zones of lymphoid tissues. A murine anti-CD22 mAb conjugated to the ricin A chain (kindly provided by Dr. Ellen Vitetta) has been tested in healthy baboons.[107] At the dose given, though there was little depletion of B cells in the lymph nodes, a rapid B cell depletion in the blood and bone marrow was observed. The response was partly impaired by the development of anti-murine and anti-ricin antibodies. Its effect on anti-Gal antibody production (after EIA) was relatively modest and transient.

CD38 is a surface antigen present on plasma cells and certain other cells, such as activated T cells.[114–116] The deficiency of an antibody response to T cell-dependent and independent antigens in CD38–knockout mice suggests that cells bearing CD38 may play an important role in regulating the humoral immune response.[117] Although we have *in vitro* evidence that CD38+ cells are major secretors of anti-Gal antibody, our *in vivo* studies to delete these cells have been unsuccessful.

Prevention of the Induced Antibody Response

Induced high-affinity anti-Gal IgG, and possibly antibodies directed against new porcine (non-Gal) antigenic determinants, are considered to play a major role in AHXR. It is believed that these newly-synthesized antibodies are produced by a T cell-dependent mechanism. Levels of anti-Gal IgM and, particularly, IgG are markedly increased after transplantation of porcine cartilage, kidney, hearts, bone marrow cells, or pancreatic islets.[54,100,118,119] CD40 plays a major role in the interaction of T cells with many antigen-presenting cells, including B cells, endothelial cells and dendritic cells.[120–123] The co-stimulatory pathway of CD40 and the T cell ligand, CD40L (or CD154), is crucial for effective activation of T cells to antigen,[124] and plays an important role in the activation of resting B cells by helper T cells.[125] Blockade of the CD40 pathway alone or in combination with the CD28 pathway prolongs survival of skin and cardiac allografts in rodents and of renal allografts in primates.[126,127] Simultaneous blockade of the CD28 and CD40 pathways, combined with a non-myeloablative regimen, allows the induction of mixed chimerism and allogeneic pluripotent stem cell engraftment in a rodent model.[128] Furthermore, co-stimulatory blockade was shown to prolong rat-to-mouse cardiac and skin xenografts.[129]

In our laboratory, the intravenous infusion of large numbers of mobilized porcine leukocytes (which include approximately 2% of progenitor cells) into baboons led to sensitization to Gal, with a 100-fold increase in anti-Gal IgG, and the development of antibody to new porcine (non-Gal) determinants.[99,130] The addition to the immunosuppressive regimen of anti-CD40L mAb prevented sensitization to all pig antigens.[99,130] Although, after EIA, return of anti-Gal IgM and IgG to baseline levels was observed, there was no increase in either immunoglobulin above baseline. This suggested to us that blockade of the CD40-CD40L pathway prevented the T cell-dependent (induced) antibody response, but had no effect on the production of T cell-independent antibodies (so-called "natural" antibodies). We have subsequently confirmed that anti-CD40L mAb therapy prevents (or greatly reduces) the induced antibody response to a transplanted pig organ in baboons (Fig. 2).[131]

Anti-CD40L mAb therapy may therefore have considerable potential in facilitating the development of mixed chimerism after porcine hematopoietic cell transplantation and the induction of immunological tolerance to pig antigens. Even if tolerance is not induced, anti-CD40L mAb may prolong pig organ survival in

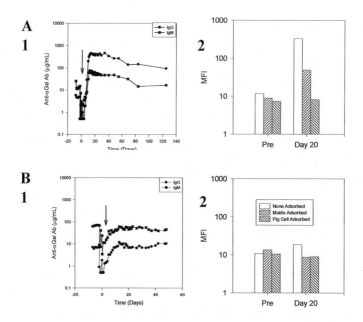

Fig. 1. Anti-Gal IgM and IgG and non-Gal-reactive antibody responses (as measured by median fluorescence intensity) following mobilized porcine leukocyte transplantation in representative baboons receiving an immunomodulatory regimen (aimed at inducing mixed chimerism and tolerance) without (A) and with (B) an anti-CD40L mAb. The arrows in (A1 and B1) indicate the first day of mobilized porcine leukocytes transplantation, which was administered after the third and final extracorporeal immunoadsorption (day 0). In (A2 and B2), column 1 represents the serum level of anti-pig antibody, column 2 represents the same serum after immunoadsorption of anti-Gal antibody, and column 3 represents this serum depleted of both Gal-reactive and nonGal-reactive antibodies. The difference between columns 1 and 2 therefore indicates the amount of anti-Gal antibody, and the difference between columns 2 and 3 indicates the amount of anti-nonGal antibody. A1; A rise in both anti-Gal IgM and IgG (to markedly higher levels than those measured before mobilized porcine leukocyte transplantation) occurred by day 10, indicating sensitization to the Gal determinants on the mobilized porcine leukocytes. A2; Antibody directed to both Gal and porcine nonGal determinants on the mobilized porcine leukocytes developed within 20 days. B1; No rise in Gal-reactive IgM or IgG above the levels measured before mobilized porcine leukocyte transplantation occurred, indicating that sensitization to Gal did not develop when anti-CD40L mAb was administered. B2; No antibody against porcine nonGal determinants developed.

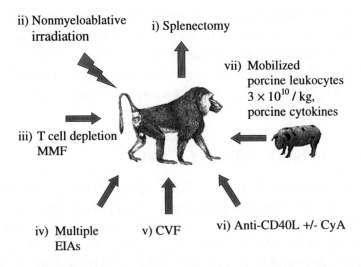

Fig. 2. Schematic representation of our immunomodulation regimen (aimed at obtaining mixed hematopoietic chimerism and inducing tolerance) in baboons. (i) A splenectomy is performed, followed by (ii) 300 cGy whole body (WBI) and thymic (TI) irradiation or induction therapy with cyclophosphamide (CPP). (iii) T cell depletion is with anti-thymocyte globulin. A continuous intravenous infusion of mycophenolate mofetil (MMF) is commenced and continued for 30 days. (iv) The baboon undergoes three extracorporeal immunoadsorptions (EIAs) using immunoaffinity columns of a synthetic Gal oligosaccharide. (v) Anti-complement therapy is initiated with cobra venom factor, dosed according to the results of the measurement of CH50, and continued for 30 days. (vi) Anti-CD40L mAb therapy (20 mg/kg) is initiated prior to the first infusion of porcine mobilized porcine leukocytes and continued on alternate days for 14 days (total eight doses). In some cases, cyclosporine (CyA) and/or macrophage blockade with medronate liposomes have been added to the protocol. (vii) Mobilized porcine leukocytes (to a total of 3×10^{10}/kg) are infused over three days and porcine hematopoietic growth factors (cytokines) are administered to help engraftment. A pig organ may also be transplanted at this time.

primates without the need for excessive dosages of pharmacologic immunosuppressive agents.

Our studies indicate, however, that AHXR can occur in the absence of induced antibody both to Gal and non-Gal porcine epitopes.[131]

Anti-Gal IgM alone may be sufficient to initiate either AHXR or endothelial cell activation sufficient to cause a state of disseminated intravascular coagulation, necessitating urgent graft removal.[132] Both AHXR and disseminated intravascular coagulation occurred in the absence of evidence of IgG or complement deposition on the grafted tissues.

Suppression or Depletion of T Cells

Many of the drugs that have been administered in an effort to suppress B cell function (discussed above) also suppress T cell function or deplete T cells. They may therefore have an effect in delaying both AHXR and acute cellular rejection. Because of early graft destruction from either HAR or AHXR, little is known about *in vivo* xenogeneic cellular immune responses. Most information regarding the human cellular response to pig cells has been obtained from *in vitro* models.[25–27,133–138] Both CD4+ and CD8+ T cells may recognize porcine xenoantigens directly without the presence of host antigen-presenting cells, although the indirect T cell pathway may be particularly important. In contrast to the allogenic response, NK cells play an important role in the rejection of xenografts.[27,137–141] The cellular response may be stronger than that to alloantigens.

Current methods at our centre to deplete or suppress T cells include the use of thymic irradiation, antithymocyte globulin, a novel anti-T cell mAb (LoCD2b, a rat anti-human mAb directed against CD2 markers), an anti-CD3 immunotoxin, anti-CD40L mAb, and several pharmacologic agents. Some of these will be briefly reviewed.

Thymic irradiation

T lymphocytes are more radioresistant than B lymphocytes. Furthermore, activated T lymphocytes are more radio-resistant than resting T lymphocytes. Approaches to deplete T cells have included total lymphoid irradiation and thymic irradiation. Total lymphoid irradiation has prolonged survival of kidney and liver allografts in the baboon.[142] Using a non-myeloablative whole body and thymic irradiation preparative regimen in miniature swine, stable mixed chimerism and donor-specific tolerance has been achieved.[143] Tolerance to renal allografts has also been achieved in cynomolgus monkeys with a protocol including thymic irradiation. We are exploring this approach in our pig-to-baboon xenotransplantation model aimed at achieving mixed hematopoietic cell chimerism and tolerance (see below).

Antithymocyte globulin (ATG)

The first clinical use of antilymphocyte sera in renal transplantation was reported in 1967.[144] Since then, antithymocyte globulin has been used for depleting mature T cells in allograft recipients. In our tolerance-inducing protocol, horse anti-human thymocyte globulin is administered to baboons at a dose of 50 mg/kg/day on days −3, −2, −1 before transplantation of an organ or hematopoietic cells on day 0.[54] Good, but not complete, T cell depletion is obtained but, in the absence of anti-CD40L mAb, this is probably insufficient to allow tolerance induction.

Anti-CD2 monoclonal antibody (Lo-CD2b)

CD2, which is known as lymphocyte function-associated antigen-2, binds to lymphocyte function-associated antigen 3, which is expressed on T and NK cells.[145,146] Lymphocyte function-associated antigen-2 may regulate the function of other cell surface molecules and signaling receptors, and affect CD4+ and CD8+ T cell-dependent responses, including cytotoxic lympholysis.[147] Its action is most effective at an early phase of the immune response. Anti-CD2 mAb has been shown to suppress immunity *in vitro* and *in vivo*. A single dose of anti-CD2 mAb administered at the time of transplantation was associated with indefinite graft survival in a rodent allograft model.[148] Donor-specific tolerance was induced with upregulation of expression of both Th 1 and Th 2 cytokine mRNA profiles. Chavin et al.[149] demonstrated prolongation of islet xenografts by anti-CD2 mAb in the rat-to-mouse model. In large animal models, its administration has been well-tolerated. A new anti-CD2 mAb (Lo-CD2b), produced from rat ascites,[150] has proved effective in reducing T cells in baboons in our laboratory and has been incorporated into a tolerance-inducing regimen (see below).

Anti-CD3 immunotoxin

FN18-CRM9, an anti-CD3 immunotoxin, has been developed to deplete T cells.[151] It is composed of FN18 (an anti-rhesus monkey CD3 mAb) and a mutant diphtheria toxin, CRM9,[152] and has the potential to deplete T cells to less than 1% of baseline level in both peripheral blood and lymph nodes.[153] Therapy with FN18-CRM9 has prolonged survival of renal allografts in MHC class I- and II-mismatched rhesus monkeys and induced tolerance in this model.[152,154–156] Although it results in the most efficient depletion of T cells of any agent we have tested to date, it is in limited supply and cannot be used in baboons that have significant levels of anti-diphtheria antibody.

Pharmacologic agents

Pharmacologic immunosuppressive agents have been widely tested in suppressing the cellular response to a xenograft. Indeed, extensive experience has been obtained at several centres, particularly by White and his colleagues in the UK.[157,158] More intensive pharmacologic immunosuppressive therapy is required in discordant xenotransplantation than in allotransplantation. This has resulted in an increased risk of infection and malignancy. Some agent, e.g. cyclophosphamide, mycophenolate mofetil, 15-deoxyspergualin, can suppress both T and B cells, and therefore may be particularly useful in xenotransplantation.

White *et al.* have demonstrated significant prolongation of organ graft survival when cyclophosphamide therapy was combined with cyclosporine and steroids in their hDAF transgenic pig-to-primate heart and kidney transplantation model.[159–161] Although the grafts were ultimately lost by AHXR, xenograft survival for up to three months was documented, the best results achieved to date in any preclinical xenotransplantation model. However, it is difficult to use cyclophosphamide long-term at the dosage required because it induces marked leukopenia and other side effects. White's group has had good results with an induction course of cyclophosphamide, followed by maintenance therapy with cyclosporine and corticosteroids in combination with a third agent, such as mycophenolate mofetil or rapamycin.[157,158,162] Xenograft survival, however, remains at a mean of approximately one month.

INDUCTION OF IMMUNOLOGICAL TOLERANCE

As current evidence suggests that suppression of both the humoral and cellular response to a discordant xenograft will require extremely intensive immunosuppressive therapy, with its associated increased risks, it seems likely that xenotransplantation will only prove successful if immunologic tolerance to the transplanted pig organ can be achieved. The induction of donor-specific immunologic tolerance allows a state of permanent specific unresponsiveness to donor antigen by the recipient without continuing immunosuppressive treatment, and prevents the development of acute and chronic rejection in allografts. Because the organ-source animal is available well before the transplant procedure (in contrast to cadaveric allotransplantation), the induction of tolerance may be facilitated in xenotransplantation. At our centre, tolerance in small and large animal allotransplantation models has been achieved, based on the induction of mixed hematopoietic cell chimerism and/or molecular chimerism.

Mixed Hematopoietic Cell Chimerism (B and T Cell Tolerance)

The aim of this approach is to deplete the host's immune response and, during this period of depletion, to engraft donor hematopoietic cells. Tolerance develops probably through the intrathymic clonal deletion of alloreactive cells. A conditioning regimen involving T cell depletion and bone marrow and thymic "space" creation is required before bone marrow transplantation. This is achieved by whole body and thymic irradiation with the administration of a polyclonal antibody directed against T cells. High-dose cyclophosphamide has been used successfully in some models to replace whole body irradiation. Non-human primate renal allografts have survived without maintenance immunosuppressive therapy for >five years.[98] In concordant xenogeneic models, the same approach has induced tolerance in rodents and specific hyporesponsiveness in nonhuman primates (baboon-to-cynomolgus monkey).[163]

This successful method for inducing allogeneic transplantation tolerance has been adapted to the discordant pig-to-nonhuman primate model.[164,165] For xenotransplantation, it is necessary to achieve B cell tolerance in addition to T cell tolerance. Sykes and her colleagues[166,167] demonstrated B cell tolerance following bone marrow transplantation from Gal-positive wild-type mice into irradiated Gal-knockout mice, which produce anti-Gal antibodies. The presence of Gal-positive bone marrow cells in Gal-knockout mice suppressed anti-Gal antibody production. Combination therapy, including temporary (i) depletion or inhibition of anti-Gal antibody, (ii) depletion or inhibition of complement, (iii) suppression or depletion of B and T cell function, and (iv) pig hematopoietic cell engraftment may all be required to achieve tolerance in the pig-to-primate model.

In attempts to achieve this state, we have included EIA to remove xenoreactive antibodies from the recipient baboon,[51,54,168] successfully preventing HAR. Initially, relatively small numbers ($5-30 \times 10^8$/kg) of porcine bone marrow cells were infused.[54,100] Although pig cells were detected by polymerase chain reaction, they could not be detected by flow cytometry, suggesting that the number of pig cells infused was insufficient. As the infused porcine cells could not be detected after a few hours, we hypothesised that they are rapidly phagocytosed by host macrophages. More recently, large numbers ($2-4 \times 10^{10}$ cells/kg) of hematopoietic growth factor-mobilized leukocytes, including progenitor cells, obtained from MHC-inbred miniature swine, have been infused into preconditioned baboons.[130]

The basic conditioning regimen that we are currently testing (Fig. 1) consists of induction therapy with splenectomy, whole body (300 cGy) and thymic (700 cGy) irradiation, multiple EIAs, and horse anti-human thymocyte globulin.[130] Porcine mobilized leukocytes (including progenitor cells) are infused on days 0, 1 and 2. Porcine hematopoietic growth factors are administered,[169-171] and have been demonstrated to increase the chimerism obtained.[172] Additional maintenance treatment with CVF, anti-CD40L mAb, mycophenolate mofetil, and porcine growth factors is given. With this regimen, porcine cells are always detected by flow cytometry until days 4-6, consistently by polymerase chain reaction (PCR) through day 33, and intermittently by PCR <140 days.[173,174] Occasional engraftment between days 10-20 has been detected by return of chimerism detectable by flow cytometry. However, return of xenoreactive antibodies following EIA has to date prevented sustained mixed chimerism and the induction of tolerance. We hypothesise that, if xenoreactive antibody can be fully depleted for a sufficient period of time to allow pig cell engraftment, the induction of tolerance through mixed hematopoietic cell chimerism will be possible.

As the infused porcine cells are rapidly phagocytized by the recipient's macrophages, a diphosphonate (medronate) encapsulated in liposomes has been administered to deplete the macrophages.[77,78] After intravenous infusion, the liposomes are phagocytized by the macrophages, which are killed by the release of the medronate. Much higher initial levels of chimerism are obtained following medronate liposome therapy. However, although it prolongs survival of porcine cells, this therapy abrogates the beneficial effect of anti-CD40L mAb in preventing an induced antibody response.[175] We presume that antigen presentation by macrophages is important if the induced antibody response is to be successfully prevented by CD40L blockade.

Baboons receiving infusions of pig leukocytes have all developed a thrombotic microangiopathy similar to the thrombotic thrombocytopenic purpura sometimes seen after clinical bone marrow allotransplantation. The condition we have seen consists of marked thrombocytopenia, schistocytosis, elevation of lactic dehydrogenase, mild purpura, and occasional minor neurologic and renal dysfunctions.[132,176,177] The platelet count falls dramatically (to 20,000–50,000/mm^3) after the first infusion of pig leukocytes, and is reduced to <10,000 by subsequent leukocyte infusions. Following induction therapy with whole body irradiation, the platelet count does not recover for about 14 days, at which time bone marrow recovery is taking place. When cyclophosphamide induction therapy is used, recovery of platelet count occurs within hours of leukocyte infusion. Prophylactic therapy with prostacyclin, corticosteroids, and low-dose heparin only partially reduces the extent and duration of the thrombocytopenia.

Molecular Chimerism

An alternative approach to induce tolerance, termed molecular chimerism, utilizes a gene therapy approach.[178] This approach is being investigated in attempts to induce both B and T cell tolerance.

B cell tolerance

The basis of this approach is that expression of the Gal epitope in the recipient baboon's native hematopoietic cells should result in suppression of anti-Gal antibody production. The gene for the enzyme α1,3galactosyltransferase, which leads to the production of the Gal epitope, has been introduced *in vitro* into the bone marrow of Gal-knockout mice.[179] The autologous gene-transduced bone marrow cells were then infused into the mouse after a myeloablative regimen that allowed bone marrow engraftment. Because the autologous bone marrow cells transduced with α1,3galactosyltransferase led to expression of Gal in the Gal-knockout mice, anti-Gal antibody production was suppressed.

We are currently attempting to duplicate this result in the baboon. Transient expression of Gal, demonstrated by flow cytometry, has been achieved rarely following the infusion of transduced autologous bone marrow cells. However, Gal expression was eventually lost and anti-Gal antibody production was not suppressed. Further studies are underway to improve the efficiency of transduction of bone marrow cells in the baboon.

T cell tolerance

If the anti-Gal antibody problem can be resolved by the induction of B cell tolerance by the above approach, then additional gene therapy could be undertaken to suppress the T cell response. Following successful studies in a pig renal allograft model,[5] we have pursued this approach in the pig-to-baboon model.[6,180] A gene encoding a swine leukocyte class II antigen was transduced into the recipient baboon bone marrow cells. After the infusion of the autologous transduced cells into the baboon, the presence of the donor-specific class II antigen induced what we believe was T cell tolerance to transplanted pig kidneys. Although the kidneys were rejected after 8–22 days by a humoral response, in one case long-term follow-up demonstrated no induced Gal or non-Gal IgG in the SLA class II-transduced baboon (in the absence of anti-CD40L mAb therapy). This strongly suggested that the presence of the porcine class II gene in the baboon bone marrow cells prevented a T cell-dependent response to the porcine xenograft.

Thymic Transplantation

Donor-specific tolerance has been induced in T/NK cell-depleted and thymectomized mice by the transplantation of fetal pig thymic and liver tissue under the kidney capsule. Donor-matched pig skin grafts survived permanently, whereas allogeneic mouse skin grafts were rapidly rejected.[181,182] Non-tolerant control mice rejected pig skin within 26 days. *In vitro* unresponsiveness has been demonstrated in thymectomized and T cell-depleted baboons after fetal porcine thymic tissue grafting, but these thymic grafts have eventually been rejected, presumably by a humoral mechanism following continuing production of anti-Gal antibody.[183]

Autologous thymic grafting under the renal capsule results in the preparation of a thymokidney which, within weeks, becomes vascularized and functions normally.[184] When a composite thymo-kidney was transplanted across allogeneic barrier in a miniature swine model, T cell tolerance was inducible in a T cell-depleted and thymectomized recipient.[185,186] Transplantation of a vascularized porcine thymokidney might induce T-cell tolerance in the recipient primate. Thus, this strategy would be a potential method for the induction of tolerance in xenotransplantation if anti-Gal antibody depletion could be sustained. Studies in the pig-to-baboon model are currently underway at our centre.

COMMENT

It will almost certainly be necessary to combine several therapeutic techniques and/or agents if xenotransplantation is to be successful, as has proved to be the case with allotransplantation today. However, although the barriers are great, xenotransplantation offers us the first opportunity for modifying the donor rather than just the recipient. This opens up new possibilities, such as genetic engineering and cloning (discussed elsewhere in this volume). The breeding of a pig with a vascular endothelial structure against which humans have no preformed antibodies would prove a major advance. There are likely to be many further steps before we achieve our ultimate goal of the induction of tolerance in the human recipient to the transplanted pig organ or tissues.

Considerable effort is being directed at several centers to overcoming the barriers that remain. Slow but steady progress has been made and continues to be made. Although we are far from a solution to all of the problems associated with xenografting, none of the obstacles appears to be insurmountable. The potential benefits of an unlimited source of porcine organs and cells are so immense that the stimulus to persist in working towards this goal is great. Xenotransplantation has the potential to radically change transplantation medicine as we know it today. It is hoped that further progress will enable discordant xenotransplantation become a clinical reality in the near future.

REFERENCES

1. Cooper DKC, Ye Y, Rolf LL Jr, Zuhdi N. The pig as potential organ donor for man. In *Xenotransplantation: The Transplantation of Organs and Tissues Between Species*, eds. DKC Cooper, E Kemp, K Reemtsma and DJG White, Springer, Heidelberg, 1991; pp. 481–500.
2. Sachs DH. The pig as a potential xenograft donor. *Vet Immunol Immunopathol* 1994; **43**:185–191.
3. Sachs DH. MHC homozygous miniature swine. In *Swine as Models in Biomedical Research*, eds. MM Swindle, DC Moody

and LD Phillips, Iowa State University Press, Ames, Iowa, 1992; 3–15.

4. Sykes M, Sachs DH, Nienhuis AW, Pearson DA, Moulton AD, Bodine DM. Specific prolongation of skin graft survival following retroviral transduction of bone marrow with an allogeneic major histocompatibility complex gene. *Transplantation* 1993; **55**:197–202.

5. Emery DW, Sablinski T, Shimada H, Germana S, Gianello P, Foley A, Shulman S, Arn S, Fishman J, Lorf T, Nickeleit V, Colvin RB, Sachs DH, LeGuern C. Expression of an allogeneic MHC DRB transgene, through retroviral transduction of bone marrow, induces specific reduction of alloreactivity. *Transplantation* 1997; **64**:1414–1423.

6. Ierino FL, Gojo S, Banerjee PT, Giovino M, Xu Y, Gere J, Kaynor C, Awwad M, Monroy R, Rembert J, Hatch T, Foley A, Kozlowski T, Yamada K, Neethling FA, Fishman J, Bailin M, Spitzer TR, Cooper DKC, Cosimi AB, LeGuern C, Sachs DH. Transfer of swine major histocompatibility complex class II genes into autologous bone marrow cells of baboons for the induction of tolerance across xenogeneic barriers. *Transplantation* 1999; **67**:1119–1128.

7. Bach FH, Fishman JA, Daniels N, Proimos J, Anderson B, Carpenter CB, Forrow L, Robson SC, Fineberg HV. 1998. Uncertainty in xenotransplantation: Individual benefit versus collective risk. *Nat Med* 1998; **4**:141–144.

8. Patience C, Takeuchi Y, Weiss RA. Infection of human cells by an endogenous retrovirus of pigs. *Nat Med* 1997; **3**:282–286.

9. Paradis K, Langford G, Long Z, Heneine W, Sandstrom P, Switzer WM, Chapman LE, Lockey C, Onions D, Otto E. Search for cross-species transmission of porcine endogenous retrovirus in patients treated with living pig tissue. *Science* 1999; **285**:1236–1241.

10. Sachs DH, Colvin RB, Cosimi AB, Russell PS, Sykes M, McGregor CG, Platt JL. Xenotransplantation — caution, but no moratorium. *Nat Med* 1998; **4**:372–373.

11. Auchincloss H Jr, Sachs DH. Xenogeneic transplantation. *Annu Rev Immunol* 1998; **16**:433–470.

12. Sablinski T, Cooper DKC, Sachs DH. Xenotransplantation. In *Therapeutic Immunology*, eds. Austen KF, Burakoff SJ, Rosen FS, Strom TB, Blackwell Science, Cambridge, MA, 2001, pp. 535–549.

13. Galili U, Shohet SB, Kobrin E, Stults CL, Macher BA. Man, apes, and old world monkeys differ from other mammals in the expression of alpha-galactosyl epitopes on nucleated cells. *J Biol Chem* 1988; **263**:17755–17762.

14. Good AH, Cooper DKC, Malcolm AJ, Ippolito RM, Koren E, Neethling FA, Ye Y. Zuhdi N, Lamontagne LR. Identification of carbohydrate structures that bind human antiporcine antibodies: Implications for discordant xenografting in humans. *Transplant Proc* 1992; **24**:559–562.

15. Cooper DKC. Depletion of natural antibodies in non-human primates — A step towards successful discordant xenografting in humans. *Clin Transpl* 1992; **6**:178–183.

16. Cooper DKC, Koren E, Oriol R. Oligosaccharides and discordant xenotransplantation. *Immunol Rev* 1994; **141**:31–58.

17. Oriol R, Ye Y, Koren E, Cooper DKC. Carbohydrate antigens of pig tissues reacting with human natural antibodies as potential targets for hyperacute vascular rejection in pig-to-man organ xenotransplantation. *Transplantation* 1993; **56**:1433–1442.

18. Koren E, Neethling FA, Richards S, Koscec M, Ye Y, Zuhdi N, Cooper DKC. Binding and specificity of major immunoglobulin classes of preformed human anti-pig heart antibodies. *Transpl Int* 1993; **6**:351–353.

19. Kujundzic M, Koren E, Neethling FA, Milotic F, Kosec M, Kujundzic T, Martin M, Cooper DKC. Variability of anti-alphaGal antibodies in human serum and their relation to serum cytotoxicity against pig cells. *Xenotransplantation* 1994; **1**:58–65.

20. Perper RJ, Najarian JS. Experimental renal heterotransplantation. I. In widely divergent species. *Transplantation* 1966; 4:377–388.

21. Cooper DKC, Human PA, Lexer G, Rose AG, Rees J, Keraan M, Du Toit E. Effects of cyclosporine and antibody adsorption on pig cardiac xenograft survival in the baboon. *J Heart Transplant* 1988; 7:238–246.

22. Alexandre GPJ, Gianello P, Latinne D, Carlier M, Dewaele A, Van Obbergh L, Moriau M, Marbaix E, Lambotte JL, Lambotte L, Squifflet JP. Plasmapheresis and splenectomy in experimental renal xenotransplantation. In *Xenograft 25*, ed. MA Hardy, Excerpta Medica, New York, 1989; pp. 259–266.

23. Platt JL, Lin SS, McGregor CG. Acute vascular rejection. *Xenotransplantation* 1998; 5:169–175.

24. Bach FH, Robson SC, Winkler H, Ferran C, Stuhlmeier KM, Wrighton CJ, Hancock WW. Barriers to xenotransplantation. *Nat Med* 1995; 1:869–873.

25. Yamada K, Sachs DH, DerSimonian H. Human anti-porcine xenogeneic T-cell response. Evidence for allelic specificity of MLR and for both direct and indirect pathways of recognition. *J Immunol* 1995; 155:5249–5256.

26. Rose ML, Yacoub MH. Heart transplantation: Cellular and humoral immunity. *Springer Semin Immunopathol* 1989; 11:423–438.

27. Yamada K, Seebach JD, DerSimonian H, Sachs DH. Human anti-pig T-cell mediated cytotoxicity. *Xenotransplantation* 1996; 3:179–187.

28. Moberg AW, Shons AR, Gewurz H, Mozes M, Najarian JS. Prolongation of renal xenografts by the simultaneous sequestration of preformed antibody, inhibition of complement, coagulation and antibody synthesis. *Transplant Proc* 1971; 3:538–541.

29. Slapak M, Greenbaum M, Bardawil W, Saravis C, Joison J, McDermott WVJ. Effect of heparin, arvin, liver perfusion, and

heterologous antiplatelet serum on rejection of pig kidney by dog. *Transplant Proc* 1971; **3**:558–561.

30. Azimzadeh A, Meyer C, Watier H, Beller JP, Chenard-Neu MP, Kieny R, Boudjema K, Jaeck D, Cinqualbre J, Wolf P. Removal of primate xenoreactive natural antibodies by extracorporeal perfusion of pig kidneys and livers. *Transpl Immunol* 1998; **6**:13–22.

31. Latinne D, Gianello P, Smith CV, Nickeleit V, Kawai T, Beadle M, Haug C, Sykes M, Lebowitz E, Bazin H, Colvin R, Cosimi AB, Sachs DH. Xenotransplantation from pig to cynomolgus monkey: Approach toward tolerance induction. *Transplant Proc* 1993; **25**:336–338.

32. Tanaka M, Latinne D, Gianello P, Sablinski T, Lorf T, Bailin M, Nickeleit V, Colvin R, Lebowitz E, Sykes M, Cosimi AB, Sachs DH. Xenotransplantation from pig to cynomolgus monkey: The potential for overcoming xenograft rejection through induction of chimerism. *Transplant Proc* 1994; **26**: 1326–1327.

33. Sablinski T, Latinne D, Gianello P, Bailin M, Bergen K, Colvin RB, Foley A, Hong H-Z, Lorf T, Meehan S, Monroy R, Powelson JA, Sykes M, Tanaka M, Cosimi AB, Sachs DH. Xenotransplantation of pig kidneys to nonhuman primates: I. Development of the model. *Xenotransplantation* 1995; **2**: 264–270.

34. Macchiarini P, Oriol R, Azimzadeh A, de Montpreville V, Rieben R, Bovin N, Mazmanian M, Dartevelle P. Evidence of human non-alpha-galactosyl antibodies involved in the hyperacute rejection of pig lungs and their removal by pig organ perfusion. *J Thorac Cardiovasc Surg* 1998; **116**:831–843.

35. Siami GA, Siami FS. Plasmapheresis and paraproteinemia: Cryoprotein-induced diseases, monoclonal gammopathy, Waldenstrom's macroglobulinemia, hyperviscosity syndrome, multiple myeloma, light chain disease, and amyloidosis. *Ther Apher* 1999; **3**:8–19.

36. Mahalati K, Dawson RB, Collins JO, Mayer RF. Predictable recovery from myasthenia gravis crisis with plasma exchange: Thirty-six cases and review of current management. *J Clin Apheresis* 1999; **14**:1–8.

37. Bosch T, Buhmann R, Lennertz A, Samtleben W, Kolb HJ. Therapeutic plasma exchange in patients suffering from thrombotic microangiopathy after allogeneic bone marrow transplantation. *Ther Apher* 1999; **3**:252–256.

38. Chopek MW, Simmons RL, Platt JL. ABO-incompatible kidney transplantation: initial immunopathologic evaluation. *Transplant Proc* 1987; **19**:4553–4557.

39. Alexandre GPJ, Squifflet JP, De Bruyere M, Latinne D, Reding R, Gianello P, Carlier M, Pirson Y. Present experiences in a series of 26 ABO-incompatible living donor renal allografts. *Transplant Proc* 1987; **19**:4538.

40. Slapak M, Digard N, Ahmed M, Shell T, Thompson F. Renal transplantation across the ABO barrier — a 9-year experience. *Transplant Proc* 1990; **22**:1425–1428.

41. Takahashi K, Yagisawa T, Sonda K, Kawaguchi H, Yamaguchi Y, Toma H, Agishi T, Ota K. ABO-incompatible kidney transplantation in a single-center trial. *Transplant Proc* 1993; **25**:271–273.

42. Bach FH, Ferran C, Hechenleitner P, Mark W, Koyamada N, Miyatake T, Winkler H, Badrichani A, Candinas D, Hancock WW. Accommodation of vascularized xenografts: Expression of "protective genes" by donor endothelial cells in a host Th2 cytokine environment. *Nat Med* 1997; **3**:196–204.

43. Soares MP, Lin Y, Sato K, Stuhlmeier KM, Bach FH. Accommodation. *Immunol Today* 1999; **20**:434–437.

44. Bach FH, Platt JL, Cooper DKC. Accommodation — The role of natural antibody and complement in discordant xenograft rejection. In *Xenotransplantation: The Transplantation of Organs and Tissues Between Species*, eds. DKC Cooper, E Kemp, K Reemtsma and DJG White, Springer, Heidelberg, 1991, pp. 81–99.

45. Benny WB, Sutton DM, Oger J, Bril V, McAteer MJ, Rock G. Clinical evaluation of a staphylococcal protein A immunoadsorption system in the treatment of myasthenia gravis patients. *Transfusion* 1999; **39**:682–687.

46. Palmer A, Welsh K, Gjorstrup P, Taube D, Bewick M, Thick M. Removal of anti-HLA antibodies by extracorporeal immunoadsorption to enable renal transplantation. *Lancet* 1989; **1**:10–12.

47. Pretagostini R, Berloco P, Poli L, Cinti P, Di Nicuolo A, De Simone P, Colonnello M, Salerno A, Alfani D, Cortesini R. Immunoadsorption with protein A in humoral rejection of kidney transplants. *ASAIO J* 1996; **42**:M645–M648.

48. Hickstein H, Korten G, Bast R, Barz D, Templin R, Schneidewind JM, Kittner C, Nizze H, Schmidt R. Protein A immunoadsorption (i.a.) in renal transplantation patients with vascular rejection. *Transfus Sci Suppl* 1998; **19**:53–57.

49. Leventhal JR, John R, Fryer JP, Witson JC, Derlich JM, Remiszewski J, Dalmasso AP, Matas AJ, Bolman RM. Removal of baboon and human antiporcine IgG and IgM natural antibodies by immunoadsorption. Results of *in vitro* and *in vivo* studies. *Transplantation* 1995; **59**:294–300.

50. Taniguchi S, Neethling FA, Korchagina EY, Bovin N, Ye Y, Kobayashi T, Niekrasz M, Li S, Koren E, Oriol R, Cooper DKC. *In vivo* immunoadsorption of antipig antibodies in baboons using a specific Galα1-3Gal column. *Transplantation* 1996; **62**:1379–1384.

51. Kozlowski T, Ierino FL, Lambrigts D, Foley A, Andrews D, Awwad M, Monroy R, Cosimi AB, Cooper DKC, Sachs DH. Depletion of anti-Gal(alpha)1-3Gal antibody in baboons by specific alpha-Gal immunoaffinity columns. *Xenotransplantation* 1998; **5**:122–131.

52. Lambrigts D, Van Calster P, Xu Y, Awwad M, Neethling FA, Kozlowski T, Foley A, Watts A, Chae SJ, Fishman J, Thall AD, White-Scharf ME, Sachs DH, Cooper DKC. Pharmacologic

immunosuppressive therapy and extracorporeal immuno-adsorption in the suppression of anti-alphaGal antibody in the baboon. *Xenotransplantation* 1998; **5**:274–283.

53. Xu Y, Lorf T, Sablinski T, Gianello P, Bailin M, Monroy R, Kozlowski T, Awwad M, Cooper DKC, Sachs DH. Removal of anti-porcine natural antibodies from human and nonhuman primate plasma *in vitro* and *in vivo* by a Galα1-3Galβ1-4βGlc-X immunoaffinity column. *Transplantation* 1998; **65**: 172–179.

54. Kozlowski T, Shimizu A, Lambrigts D, Yamada K, Fuchimoto Y, Glaser R, Monroy R, Xu Y, Awwad M, Colvin RB, Cosimi AB, Robson SC, Fishman J, Spitzer TR, Cooper DKC, Sachs DH. Porcine kidney and heart transplantation in baboons under-going a tolerance induction regimen and antibody adsorption. *Transplantation* 1999; **67**:18–30.

55. Ye Y, Neethling FA, Niekrasz M, Koren E, Richards SV, Martin M, Kosanke S, Oriol R, Cooper DKC. Evidence that intravenously administered alpha-galactosyl carbohydrates reduce baboon serum cytotoxicity to pig kidney cells (PK15) and transplanted pig hearts. *Transplantation* 1994; **58**:330–337.

56. Neethling FA, Cooper DKC. Serum cytotoxicity to pig cells and anti-alphaGal antibody level and specificity in humans and baboons. *Transplantation* 1999; **67**:658–665.

57. Neethling FA, Joziasse D, Bovin N, Cooper DKC, Oriol R. The reducing end of alpha Gal oligosaccharides contributes to their efficiency in blocking natural antibodies of human and baboon sera. *Transpl Int* 1996; **9**:98–101.

58. Watts A, Foley A, Awwad M, Treter S, Oravec G, Buhler L, Alwayn IPJ, Kozlowski T, Lambrigts D, Gojo S, Basker M, White-Scharf ME, Andrews D, Sachs DH, Cooper DKC. Plasma perfusion by apheresis through a Gal immunoaffinity column successfully depletes anti-Gal antibody: Experience with 320 aphereses in baboons. *Xenotransplantation* 2000; **7**:181–185.

59. Cooper DKC, Good AH, Koren E, Oriol R, Malcolm AJ, Ippolito RM, Neethling FA, Ye Y, Romano E, Zuhdi N. Identification of alpha-galactosyl and other carbohydrate epitopes that are bound by human anti-pig antibodies: Relevance to discordant xenografting in man. *Transpl Immunol* 1993; **1**:198–205.

60. Simon PM, Neethling FA, Taniguchi S, Goode PL, Zopf D, Hancock WW, Cooper DKC. Intravenous infusion of Galα1-3Gal oligosaccharides in baboons delays hyperacute rejection of porcine heart xenografts. *Transplantation* 1998; **65**:346–353.

61. Romano E, Neethling FA, Nilsson K, Kosanke S, Shimizu A, Magnusson S, Svensson L, Samuelsson B, Cooper DKC. Intravenous synthetic alphaGal saccharides delay hyperacute rejection following pig-to-baboon heart transplantation. *Xenotransplantation* 1999; **6**:36–42.

62. Koren E, Milotic F, Neethling FA, Koscec M, Fei D, Kobayashi T, Taniguchi S, Cooper DKC. Monoclonal antiidiotypic antibodies neutralize cytotoxic effects of anti-alphaGal antibodies. *Transplantation* 1996; **62**:837–843.

63. Buhler L, Treter S, McMorrow I, Neethling FA, Alwayn I, Awwad M, Thall A, LeGuern C, Sachs DH, Cooper DKC. Injection of porcine anti-idiotypic antibodies to primate anti-Gal antibodies leads to active inhibition of serum cytotoxicity in a baboon. *Transplant Proc* 2000; **32**:1102.

64. Wordell CJ. Use of intravenous immune globulin therapy: An overview. *DICP* 1991; **25**:805–817.

65. Kaveri SV, Dietrich G, Hurez V, Kazatchkine MD. Intravenous immunoglobulins (IVIg) in the treatment of autoimmune diseases. *Clin Exp Immunol* 1991; **86**:192–198.

66. Dwyer JM. Manipulating the immune system with immune globulin. *N Engl J Med* 1992; **326**:107–116.

67. Smiley JD, Talbert MG. Southwestern Internal Medicine Conference: High-dose intravenous gamma globulin therapy: How does it work? *Am J Med Sci* 1995; **309**:295–303.

68. Glotz D, Haymann JP, Sansonetti N, Francois A, Menoyo-Calonge V, Bariety J, Druet P. Suppression of HLA-specific alloantibodies by high-dose intravenous immunoglobulins (IVIg). A potential tool for transplantation of immunized patients. *Transplantation* 1993; **56**:335–337.

69. Tyan DB, Li VA, Czer L, Trento A, Jordan SC. Intravenous immunoglobulin suppression of HLA alloantibody in highly sensitized transplant candidates and transplantation with a histoincompatible organ. *Transplantation* 1994; **57**:553–562.

70. McIntyre JA, Higgins N, Britton R, Faucett S, Johnson S, Beckman D, Hormuth D, Fehrenbacher J, Halbrook H. Utilization of intravenous immunoglobulin to ameliorate alloantibodies in a highly sensitized patient with a cardiac assist device awaiting heart transplantation. Fluorescence-activated cell sorter analysis. *Transplantation* 1996; **62**:691–693.

71. De Marco T, Damon LE, Colombe B, Keith F, Chatterjee K, Garovoy MR. Successful immunomodulation with intravenous gamma globulin and cyclophosphamide in an alloimmunized heart transplant recipient. *J Heart Lung Transplant* 1997; **16**:360–365.

72. Jordan SC, Tyan D, Czer L, Toyoda M. Immunomodulatory actions of intravenous immunoglobulin (IVIG): Potential applications in solid organ transplant recipients. *Pediatr Transplant* 1998; **2**:92–105.

73. Magee JC, Collins BH, Harland RC, Lindman BJ, Bollinger RR, Frank MM, Platt JL. Immunoglobulin prevents complement-mediated hyperacute rejection in swine-to-primate xeno-transplantation. *J Clin Invest* 1995; **96**:2404–2412.

74. Gautreau C, Kojima T, Woimant G, Cardoso J, Devillier P, Houssin D. Use of intravenous immunoglobulin to delay xenogeneic hyperacute rejection. An *in vivo* and *in vitro* evaluation. *Transplantation* 1995; **60**:903–907.

75. Yu Z, Lennon VA. Mechanism of intravenous immune globulin therapy in antibody-mediated autoimmune diseases. *N Engl J Med* 1999; **340**:227–228.

76. Buhler L, Pidwell D, Dowling RD, Newman D, Awwad M, Cooper DKC. Different responses of human anti-HLA and anti-alphagal antibody to long-term intravenous immunoglobulin therapy. *Xenotransplantation* 1999; **6**:181–186.

77. Alwayn IPJ, Basker M, Buhler L, Rieben R, Harper D, Appel JZ, Awwad M, Down J, White-Scharf M, Sachs DH, Thall A, Cooper DKC. Porcine hematopoietic progenitor cells are rapidly cleared by the phagocytic reticuloendothelial system in baboons. *Transplantation* 2000; **69**:S420 (Abstract).

78. Basker M, Alwayn IPJ, Buhler L, Harper D, Abraham S, Kruger Gray H, De Angelish Awwad M, Down J, Rieben R, White-Scharf ME, Sachs DH, Thall A, Cooper DKC. Clearance of mobilized porcine peripheral blood progenitor cells is delayed by depletion of the phagocytic reticuloendothelial system in baboons. *Transplantation* 2001; **72**:1278–1285.

79. Muller-Eberhard HJ, Nilsson UR, Dalmasso AP, Polley MJ, Calcott MA. A molecular concept of immune cytolysis. *Arch Pathol* 1966; **82**:205–217.

80. Alper CA, Balavitch D. Cobra venom factor: Evidence for its being altered cobra C3 (the third component of complement). *Science* 1976; **191**:1275–1276.

81. Vogel CW, Smith CA, Muller-Eberhard HJ. Cobra venom factor: Structural homology with the third component of human complement. *J Immunol* 1984; **133**:3235–3241.

82. Lachmann PJ, Halbwachs L. The influence of C3b inactivator (KAF) concentration on the ability of serum to support complement activation. *Clin Exp Immunol* 1975; **21**:109–114.

83. Scheringa M, Schraa EO, Bouwman E, van Dijk H, Melief MJ, Ijzermans JN, Marquet RL. Prolongation of survival of guinea pig heart grafts in cobra venom factor-treated rats by splenectomy. No additional effect of cyclosporine. *Transplantation* 1995; **60**:1350–1353.

84. Leventhal JR, Dalmasso AP, Cromwell JW, Platt JL, Manivel CJ, Bolman RM, Matas AJ. Prolongation of cardiac

xenograft survival by depletion of complement. *Transplantation* 1993; **55**:857–865.

85. Kobayashi T, Taniguchi S, Neethling FA, Rose AG, Hancock WW, Ye Y, Niekrasz M, Kosanke S, Wright LJ, White DJ, Cooper DKC. Delayed xenograft rejection of pig-to-baboon cardiac transplants after cobra venom factor therapy. *Transplantation* 1997; **64**:1255–1261.

86. Adachi H, Rosengard BR, Hutchins GM, Hall TS, Baumgartner WA, Borkon AM, Reitz BA. Effects of cyclosporine, aspirin and cobra venom factor on discordant cardiac xenograft survival in rats. *Transplant Proc* 1987; **19**:1145–1148.

87. Marsh HC, Ryan US. Therapeutic effect of soluble complement receptor type 1 in xenotransplantation. In *Xenotransplantation*. 2nd ed., eds. DKC Cooper, E Kemp, JL Platt and DJG White, Springer, Heidelberg, 1997; pp. 437–455.

88. Weisman HF, Bartow T, Leppo MK, Marsh HCJ, Carson GR, Concino MF, Boyle MP, Roux KH, Weisfeldt ML, Fearon DT. Soluble human complement receptor type 1: *In vivo* inhibitor of complement suppressing post-ischemic myocardial inflammation and necrosis. *Science* 1990; **249**:146–151.

89. Davis EA, Pruitt SK, Greene PS, Ibrahim S, Lam TT, Levin JL, Baldwin III WM, Sanfilippo F. Inhibition of complement, evoked antibody, and cellular response prevents rejection of pig-to-primate cardiac xenografts. *Transplantation* 1996; **62**:1018–1023.

90. Pruitt SK, Baldwin WM, Marsh Jr HC, Lin SS, Yeh CG, Bollinger RR. The effect of soluble complement receptor type 1 on hyperacute xenograft rejection. *Transplantation* 1991; **52**: 868–873.

91. Pruitt, SK, Kirk AD, Bollinger RR, Marsh HCJ, Collins BH, Levin JL, Mault JR, Heinle JS, Ibrahim S, Rudolph AR. The effect of soluble complement receptor type 1 on hyperacute rejection of porcine xenografts. *Transplantation* 1994; **57**: 363–370.

92. Xia W, Fearon DT, Kirkman RL. Effect of repetitive doses of soluble human complement receptor type 1 on survival of discordant cardiac xenografts. *Transplant Proc* 1993; **25**: 410–411.

93. Pruitt SK, Bollinger RR, Collins BH, Marsh HC Jr, Levin JL, Rudolph AR, Baldwin III WM, Sanfilippo F. Effect of continuous complement inhibition using soluble complement receptor type 1 on survival of pig-to-primate cardiac xenografts. *Transplantation* 1997; **63**:900–902.

94. Kobayashi T, Neethling FA, Taniguchi S, Ye Y, Niekrasz M, Koren E, Hancock WW, Takagi H, Cooper DKC. Investigation of the anti-complement agents, FUT-175 and K76COOH, in discordant xenotransplantation. *Xenotransplantation* 1996: **3**: 237–245.

95. Lindsley K, Deeg HJ. Total body irradiation for marrow or stem-cell transplantation. *Cancer Invest* 1998; **16**:424–425.

96. Sykes M, Sachs DH. Xenogeneic tolerance through hematopoietic cell and thymic transplantation. In *Xenotransplantation: The Transplantation of Organs and Tissues Between Species*. 2nd ed., eds. DKC Cooper, E Kemp, JL Platt and DJG White. Springer, Heidelberg, 1997; pp. 496–518.

97. Kimikawa M, Kawai T, Sachs DH, Colvin RB, Bartholomew A, Cosimi AB. Mixed chimerism and transplantation tolerance induced by a nonlethal preparative regimen in cynomolgus monkeys. *Transplant Proc* 1997; **29**:1218.

98. Kawai T, Poncelet A, Sachs DH, Mauiyyedi S, Boskovic S, Wee SL, Ko DS, Bartholomew A, Kimikawa M, Hong HZ, Abrahamian G, Colvin RB, Cosimi AB. Long-term outcome and alloantibody production in a non-myeloablative regimen for induction of renal allograft tolerance. *Transplantation* 1999; **68**:1767–1775.

99. Alwayn IP, Basker M, Buhler L, Cooper DKC. The problem of anti-pig antibodies in pig-to-primate xenografting: Current and novel methods of depletion and/or suppression of production of anti-pig antibodies. *Xenotransplantation* 1999; **6**:157–168.

100. Kozlowski T, Monroy R, Xu Y, Glaser R, Awwad M, Cooper DKC, Sachs DH. Anti-Galα1-3Gal antibody response to porcine bone marrow in unmodified baboons and baboons conditioned for tolerance induction. *Transplantation* 1998; **66**:176–182.

101. Reff ME, Carner K, Chambers KS, Chinn PC, Leonard JE, Raab R, Newman RA, Hanna N, Anderson DR. Depletion of B cells *in vivo* by a chimeric mouse human monoclonal antibody to CD20. *Blood* 1994; **83**:435–445.

102. Maloney DG, Grillo-Lopez AJ, Bodkin DJ, White CA, Liles TM, Royston I, Varns C, Rosenberg J, Levy R. IDEC-C2B8: Results of a phase I multiple-dose trial in patients with relapsed non-Hodgkin's lymphoma. *J Clin Oncol* 1997; **15**:3266–3274.

103. Coiffier B, Haioun C, Ketterer N, Engert A, Tilly H, Ma D, Johnson P, Lister A, Feuring-Buske M, Radford JA, Capdeville R, Diehl V, Reyes F. Rituximab (anti-CD20 monoclonal antibody) for the treatment of patients with relapsing or refractory aggressive lymphoma: A multicenter phase II study. *Blood* 1998; **92**:1927–1932.

104. McLaughlin P, Grillo-Lopez AJ, Link BK, Levy R, Czuczman MS, Williams ME, Heyman MR, Bence-Bruckler I, White CA, Cabanillas F, Jain V, Ho AD, Lister J, Wey K, Shen D, Dallaire BK. Rituximab chimeric anti-CD20 monoclonal antibody therapy for relapsed indolent lymphoma: Half of patients respond to a four-dose treatment program. *J Clin Oncol* 1998; **16**: 2825–2833.

105. Tobinai K, Kobayashi Y, Narabayashi M, Ogura M, Kagami Y, Morishima Y, Ohtsu T, Igarashi T, Sasaki Y, Kinoshita T, Murate T. Feasibility and pharmacokinetic study of a chimeric anti-CD20 monoclonal antibody (IDEC-C2B8, rituximab) in relapsed B-cell lymphoma. The IDEC-C2B8 Study Group. *Ann Oncol* 1998; **9**:527–534.

106. Basker M, Alwayn IPJ, Treter S, Harper D, Buhler L, Andrews D, Thall A, Lambrigts D, Awwad M, White-Scharf M, Sachs DH,

and Cooper DKC. 2000. Effects of B cell/plasma cell depletion or suppression on anti-Gal antibody in the baboon. *Transplant Proc* 2000; **32**:1069.

107. Alwayn IPJ, Xu Y, Basker M, Wu C, Buhler L, Lambrigts D, Treter S, Harper D, Kitamura H, Vitetta E, Awwad M, White-Scharf ME, Sachs DH, Thall A, Cooper DKC. Effects of specific anti-B and/or anti-plasma cell immunotherapy on antibody production in baboons: Depletion of CD20- and CD22-positive B cells does not result in significantly decreased production of anti-αGal antibody. *Xenotransplantation* 2001; **8**:157–171.

108. Bregni M, Siena S, Formosa A, Lappi DA, Martineau D, Malavasi F, Dorken B, Bonadonna G, Gianni AM. B-cell restricted saporin immunotoxins: Activity against B-cell lines and chronic lymphocytic leukemia cells. *Blood* 1989; **73**:753–762.

109. Sausville EA, Headlee D, Stetler-Stevenson M, Jaffe ES, Solomon D, Figg WD, Herdt J, Kopp WC, Rager H, Steinberg SM. Continuous infusion of the anti-CD22 immunotoxin IgG-RFB4-SMPT-dgA in patients with B-cell lymphoma: A phase I study. *Blood* 1995; **85**:3457–3465.

110. Bolognesi A, Tazzari PL, Olivieri F, Polito L, Lemoli R, Terenzi A, Pasqualucci L, Falini B, Stirpe F. Evaluation of immunotoxins containing single-chain ribosome-inactivating proteins and an anti-CD22 monoclonal antibody (OM124): *In vitro* and *in vivo* studies. *Br J Haematol* 1998; **101**:179–188.

111. Vitetta ES, Stone M, Amlot P, Fay J, May R, Till M, Newman J, Clark P, Collins R, Cunningham D. Phase I immunotoxin trial in patients with B-cell lymphoma. *Cancer Res* 1991; **51**: 4052–4058.

112. Vose JM. Antibody-targeted therapy for low-grade lymphoma. *Semin Hematol* 1999; **36**:15–20.

113. Tedder TF, Tuscano J, Sato S, Kehrl JH. CD22, a B lymphocyte-specific adhesion molecule that regulates antigen receptor signaling. *Annu Rev Immunol* 1997; **15**:481–504.

114. Leo R, Boeker M, Peest D, Hein R, Bartl R, Gessner JE, Selbach J, Wacker G, Deicher H. Multiparameter analyses of normal and malignant human plasma cells: $CD38^{++}$, $CD56^+$, $CD54^+$, cIg^+ is the common phenotype of myeloma cells. *Ann Hematol* 1992; **64**:132–139.

115. Harada H, Kawano MM, Huang N, Harada Y, Iwato K, Tanabe O, Tanaka H, Sakai A, Asaoku H, Kuramoto A. Phenotypic difference of normal plasma cells from mature myeloma cells. *Blood* 1993; **81**:2658–2663.

116. Hata H, Matsuzaki H, Matsuno F, Sonoki T, Takemoto S, Kuribayashi N, Nagasaki A, Takatsuki K. Establishment of a monoclonal antibody to plasma cells: A comparison with CD38 and PCA-1. *Clin Exp Immunol* 1994; **96**:370–375.

117. Cockayne DA, Muchamuel T, Grimaldi JC, Muller-Steffner H, Randall TD, Lund FE, Murray R, Schuber F, Howard MC. Mice deficient for the ecto-nicotinamide adenine dinucleotide glycohydrolase CD38 exhibit altered humoral immune responses. *Blood* 1998; **92**:1324–1333.

118. Galili U, LaTemple DC, Walgenbach AW, Stone KR. Porcine and bovine cartilage transplants in cynomolgus monkey: II. Changes in anti-Gal response during chronic rejection. *Transplantation* 1997; **63**:646–651.

119. Groth CG, Tibell A, Wennberg L, Korsgren O. Xenoislet transplantation: Experimental and clinical aspects. *J Mol Med* 1999; **77**:153–154.

120. Roy M, Aruffo A, Ledbetter J, Linsley P, Kehry M, Noelle R. Studies on the interdependence of gp39 and B7 expression and function during antigen-specific immune responses. *Eur J Immunol* 1995; **25**:596–603.

121. Karmann K, Hughes CC, Schechner J, Fanslow WC, Pober JS. CD40 on human endothelial cells: Inducibility by cytokines and functional regulation of adhesion molecule expression. *Proc Natl Acad Sci, USA* 1995; **92**:4342–4346.

122. Hollenbaugh D, Mischel-Petty N, Edwards CP, Simon JC, Denfeld RW, Kiener PA, Aruffo A. Expression of functional CD40 by vascular endothelial cells. *J Exp Med* 1995; **182**:33–40.

123. Caux C, Massacrier C, Vanbervliet B, Dubois B, Van Kooten C, Durand I, Banchereau J. Activation of human dendritic cells through CD40 cross-linking. *J Exp Med* 1994; **180**:1263–1272.

124. Armitage RJ, Fanslow WC, Strockbine L, Sato TA, Clifford KN, Macduff BM, Anderson DM, Gimpel SD, Davis-Smith T, Maliszewski CR. Molecular and biological characterization of a murine ligand for CD40. *Nature* 1992; **357**:80–82.

125. Noelle RJ, Roy M, Shepherd DM, Stamenkovic I, Ledbetter JA, Aruffo A. A 39-kDa protein on activated helper T cells binds CD40 and transduces the signal for cognate activation of B cells. *Proc Natl Acad Sci, USA* 1992; **89**:6550–6554.

126. Larsen CP, Elwood ET, Alexander DZ, Ritchie SC, Hendrix R, Tucker-Burden C, Cho HR, Aruffo A, Hollenbaugh D, Linsley PS, Winn KJ, Pearson TC. Long-term acceptance of skin and cardiac allografts after blocking CD40 and CD28 pathways. *Nature* 1996; **381**:434–438.

127. Kirk AD, Harlan DM, Armstrong NN, Davis TA, Dong Y, Gray GS, Hong X, Thomas D, Fechner Jr JH, Knechtle SJ. CTLA4-Ig and anti-CD40 ligand prevent renal allograft rejection in primates. *Proc Natl Acad Sci, USA* 1997; **94**: 8789–8794.

128. Wekerle T, Sayegh MH, Hill J, Zhao Y, Chandraker A, Swenson KG, Zhao GL, Sykes M. Extrathymic T cell deletion and allogeneic stem cell engraftment induced with costimulatory blockade is followed by central T cell tolerance. *J Exp Med* 1998; **187**:2037–2044.

129. Elwood ET, Larsen CP, Cho HR, Corbascio M, Ritchie SC, Alexander DZ, Tucker-Burden C, Linsley PS, Aruffo A, Hollenbaugh D, Winn KJ, Pearson TC. Prolonged acceptance of concordant and discordant xenografts with combined

CD40 and CD28 pathway blockade. *Transplantation* 1998; **65**: 1422–1428.

130. Buhler L, Awwad M, Basker M, Gojo S, Watts A, Treter S, Nash K, Oravec G, Chang Q, Thall A, Down JD, Sykes M, Andrews D, Sackstein R, White-Scharf ME, Sachs DH, Cooper DKC. High-dose porcine hematopoietic cell transplantation combined with CD40 ligand blockade in baboons prevents an induced anti-pig humoral response. *Transplantation* 2000; **69**:2296–2304.

131. Buhler L, Yamada K, Kitamura H, Alwayn I, Basker M, Barth RN, Appel III JZ, Andrews D, Colvin RB, White-Scharf ME, Sachs DH, Robson RC, Awwad M, Cooper DKC. Miniature swine and hDAF pig kidney transplantation in baboons: Anti-Gal IgM alone can initiate acute vascular rejection and disseminated intravascular coagulation. *Transplantation*. In press.

132. Buhler L, Basker M, Alwayn IPJ, Goepfert C, Kitamura H, Kawai T, Gojo S, Kozlowski T, Ierino FL, Awwad M, Sachs DH, Sackstein R, Robson SC, Cooper DKC. Coagulation and thrombotic disorders associated with pig organ and hematopoietic cell transplantation in nonhuman primates. *Transplantation* 2000; **70**:1323–1331.

133. Moses RD, Auchincloss H Jr. Mechanism of cellular xenograft rejection. In *Xenotransplantation: The Transplantation of Organs and Tissues Between Species*. 2nd ed. eds. Cooper DKC, Kemp E, Platt JL, White DJG. Springer, Heidelberg, 1997; pp. 140–174.

134. Murray AG, Khodadoust MM, Pober JS, Bothwell AL. Porcine aortic endothelial cells activate human T cells: Direct presentation of MHC antigens and costimulation by ligands for human CD2 and CD28. *Immunity* 1994; **1**:57–63.

135. Dorling A, Lombardi G, Binns R, Lechler RI. Detection of primary direct and indirect human anti-porcine T cell responses using a porcine dendritic cell population. *Eur J Immunol* 1996; **26**:1378–1387.

136. Dorling A, Binns R, Lechler RI. Cellular xenoresponses: Observation of significant primary indirect human T cell anti-pig xenoresponses using co-stimulator-deficient or SLA class II-negative porcine stimulators. *Xenotransplantation* 1996; 3:112–119.

137. Chan DV, Auchincloss H Jr. Human anti-pig cell-mediated cytotoxicity *in vitro* involves non-T as well as T cell components. *Xenotransplantation* 1996; 3:158–165.

138. Lin Y, Vandeputte M, Waer M. Natural killer cell- and macrophage-mediated rejection of concordant xenografts in the absence of T and B cell responses. *J Immunol* 1997; **158**: 5658–5667.

139. Seebach JD, Yamada K, McMorrow IM, Sachs DH, DerSimonian H. Xenogeneic human anti-pig cytotoxicity mediated by activated natural killer cells. *Xenotransplantation* 1996; 3:188–197.

140. Seebach JD, Comrack C, Germana S, LeGuern C, Sachs DH, DerSimonian H. HLA-Cw3 expression on porcine endothelial cells protects against xenogeneic cytotoxicity mediated by a subset of human NK cells. *J Immunol* 1997; **159**:3655–3661.

141. Seebach JD, Waneck GL. Natural killer cells in xenotransplantation. *Xenotransplantation* 1997; **4**:201–211.

142. Myburgh JA, Smit JA, Stark JH, Browde S. Total lymphoid irradiation in kidney and liver transplantation in the baboon: Prolonged graft survival and alterations in T cell subsets with low cumulative dose regimens. *J Immunol* 1984; **132**:1019–1025.

143. Huang CA, Fuchimoto Y, Scheier-Dolberg R, Murphy MC, Neville Jr DM, Sachs DH. Stable mixed chimerism and tolerance using a nonmyeloablative preparative regimen in a large-animal model. *J Clin Invest* 2000; **105**:173–181.

144. Starzl TE, Marchioro TL, Hutchinson DE, Porter KA, Cerilli GJ, Brettschneider L. The clinical use of antilymphocyte globulin in renal homotransplantation. *Transplantation* 1967; **5**:Suppl. 1000–1005.

145. Moingeon P, Chang HC, Sayre PH, Clayton LK, Alcover A, Gardner P, Reinherz EL. The structural biology of CD2. *Immunol Rev* 1989; **111**:111–144.

146. Bierer BE, Burakoff SJ. T-lymphocyte activation: The biology and function of CD2 and CD4. *Immunol Rev* 1989; **111**:267–294.

147. Bromberg JS, Chavin KD, Altevogt P, Kyewski BA, Guckel B, Naji A, Barker CF. Anti-CD2 monoclonal antibodies alter cell-mediated immunity *in vivo*. *Transplantation* 1991; **51**:219–225.

148. Krieger NR, Most D, Bromberg JS, Holm B, Huie P, Sibley RK, Dafoe DC, Alfrey EJ. Coexistence of Th1- and Th2-type cytokine profiles in anti-CD2 monoclonal antibody-induced tolerance. *Transplantation* 1996; **62**:1285–1292.

149. Chavin KD, Lau HT, Bromberg JS. Prolongation of allograft and xenograft survival in mice by anti-CD2 monoclonal antibodies. *Transplantation* 1992; **54**:286–291.

150. Bombil F, Kints JP, Havaux X, Scheiff JM, Bazin H, Latinne D. A rat monoclonal anti-(human CD2) and L-leucine methyl ester impacts on human/SCID mouse graft and B lymphoproliferative syndrome. *Cancer Immunol Immunother* 1995; **40**:383–389.

151. Neville DM, Scharff J Jr, Srinivasachar K. *In vivo* T-cell ablation by a holo-immunotoxin directed at human CD3. *Proc Natl Acad Sci, USA* 1992; **89**:2585–2589.

152. Knechtle SJ, Vargo D, Fechner J, Zhai Y, Wang J, Hanaway MJ, Scharff J, Hu HZ, Knapp L, Watkins D, Neville Jr DM. FN18-CRM9 immunotoxin promotes tolerance in primate renal allografts. *Transplantation* 1997; **63**:1–6.

153. Neville DMJ, Scharff J, Hu HZ, Rigaut K, Shiloach J, Slingerland W, Jonker M. A new reagent for the induction of T-cell depletion, anti-CD3-CRM9. *J Immunother Emphasis Tumor Immunol* 1996; **19**:85–92.

154. Fechner JHJ, Vargo DJ, Geissler EK, Graeb C, Wang J, Hanaway MJ, Watkins DI, Piekarczyk M, Neville DMJ, Knechtle SJ. Split tolerance induced by immunotoxin in a

rhesus kidney allograft model. *Transplantation* 1997; **63**: 1339–1345.

155. Contreras JL, Wang PX, Eckhoff DE, Lobashevsky AL, Asiedu C, Frenette L, Robbin ML, Hubbard WJ, Cartner S, Nadler S, Cook WJ, Sharff J, Shiloach J, Thomas FT, Neville Jr DM, Thomas JM. Peritransplant tolerance induction with anti-CD3-immunotoxin: A matter of proinflammatory cytokine control. *Transplantation* 1998; **65**:1159–1169.

156. Armstrong N, Buckley P, Oberley T, Fechner Jr J, Dong YC, Hong XN, Kirk A, Neville Jr D, Knechtle S. Analysis of primate renal allografts after T-cell depletion with anti-CD3-CRM9. *Transplantation* 1998; **66**:5–13.

157. Schmoeckel M, Bhatti FN, Zaidi A, Cozzi E, Waterworth PD, Tolan MJ, Goddard M, Warner RG, Langford GA, Dunning JJ, Wallwork J, White DJ. Orthotopic heart transplantation in a transgenic pig-to-primate model. *Transplantation* 1998; **65**: 1570–1577.

158. Zaidi A, Schmoeckel M, Bhatti F, Waterworth P, Tolan M, Cozzi E, Chavez G, Langford G, Thiru S, Wallwork J, White D, Friend P. Life-supporting pig-to-primate renal xenotransplantation using genetically modified donors. *Transplantation* 1998; **65**:1584–1590.

159. White DJG. hDAF transgenic pig organs: Are they concordant for human transplantation? *Xeno* 1996; **4**:50–54.

160. Bhatti FN, Zaidi A, Schmoeckel M, Cozzi E, Chavez G, Wallwork J, White DJ, Friend PJ. Survival of life-supporting HDAF transgenic kidneys in primates is enhanced by splenectomy. *Transplant Proc* 1998 **30**:2467.

161. Schmoeckel M, Bhatti FN, Zaidi A, Cozzi E, Chavez G, Wallwork J, White DJ, Friend PJ. Splenectomy improves survival of HDAF transgenic pig kidneys in primates. *Transplant Proc* 1999; **31**:961.

162. Vial CM, Ostlie DJ, Bhatti FN, Cozzi E, Goddard M, Chavez GP, Wallwork J, White DJ, Dunning JJ. Life supporting

function for over one month of a transgenic porcine heart in a baboon. *J Heart Lung Transplant* 2000; **19**:224–229.

163. Ko DS, Bartholomew A, Poncelet AJ, Sachs DH, Huang C, Le Guern A, Abraham KI, Colvin RB, Boskovic S, Hong HZ, Wee SL, Winn HJ, Cosimi AB. Demonstration of multilineage chimerism in a nonhuman primate concordant xenograft model. *Xenotransplantation* 1998; **5**:298–304.

164. Sachs DH, Sablinski T. Tolerance across discordant xenogeneic barriers. *Xenotransplantation* 1995; **2**:234–239.

165. Sablinski T, Gianello PR, Bailin M, Bergen KS, Emery DW, Fishman JA, Foley A, Hatch T, Hawley RJ, Kozlowski T, Lorf T, Meehan S, Monroy R, Powelson JA, Colvin RB, Cosimi AB, Sachs DH. Pig to monkey bone marrow and kidney xenotransplantation. *Surgery* 1997; **121**:381–391.

166. Yang YG, DeGoma E, Ohdan H, Bracy JL, Xu YX, Iacomini J, Thall AD, Sykes M. Tolerization of anti-Galα1-3Gal natural antibody-forming B cells by induction of mixed chimerism. *J Exp Med* 1998; **187**:1335–1342.

167. Ohdan H, Yang YG, Shimizu A, Swenson KG, Sykes M. Mixed chimerism induced without lethal conditioning prevents T cell- and anti-Gal alpha 1,3Gal-mediated graft rejection. *J Clin Invest* 1999; **104**:281–290.

168. Buhler L, Friedman T, Iacomini J, Cooper DKC. Xenotransplantation — state of the art — update 1999. *Front Biosci* 1999; **4**:D416–D432.

169. Hawley RJ, Abraham S, Akiyoshi DE, Arduini R, Denaro M, Dickerson M, Meshulam DH, Monroy RL, Schacter BZ, Rosa MD. Xenogeneic bone marrow transplantation: I. Cloning, expression, and species specificity of porcine IL-3 and granulocyte-macrophage colony-stimulating factor. *Xenotransplantation* 1997; **4**:103–111.

170. Giovino MA, Hawley RJ, Dickerson WM, Glaser R, Meshulam D, Arduini R, Rosa MD, Monroy RL. Xenogeneic bone marrow transplantation: II. Porcine-specific growth factors

enhance porcine bone marrow engraftment in an *in vitro* primate microenvironment. *Xenotransplantation* 1997; **4**:112–1119.

171. Kozlowski T, Monroy R, Giovino M, Hawley RJ, Glaser R, Li Z, Meshulam DH, Spitzer TR, Cooper DKC, Sachs DH. Effect of pig-specific cytokines on mobilization of hematopoietic progenitor cells in pigs and on pig bone marrow engraftment in baboons. *Xenotransplantation* 1999; **6**:17–27.

172. Sablinski T, Emery DW, Monroy R, Hawley RJ, Xu Y, Gianello P, Lorf T, Kozlowski T, Bailin M, Cooper DKC, Cosimi AB, Sachs DH. Long-term discordant xenogeneic (porcine-to-primate) bone marrow engraftment in a monkey treated with porcine-specific growth factors. *Transplantation* 1999; **67**:972–977.

173. Buhler L, Awwad M, Treter S, Basker M, Ericson T, Lachance A, Oldmixon B, Kurilla-Mahon B, Gojo S, Huang C, Thall A, Down JD, White-Scharf ME, Sachs DH, Cooper DKC. Induction of mixed hematopoietic chimerism in the pig-to-baboon model. *Transplant Proc*, 2000; **32**:1101.

174. Buhler L, Awwad M, Treter S, Chang Q, Basker M, Alwayn IPJ, Teranishi K, Ericsson T, Moran K, Harper D, Kurilla-Mahon B, Huang C, Sackstein R, Sykes M, White-Scharf ME, Sachs DH, Down JD, Cooper DKC. Pig hematopoietic cell chimerism in baboons conditioned with a nonmyeloablative regimen and CD154 blockade. *Transplantation*. In Press.

175. Buhler L, Basker M, Alwayn I, Awwad M, Appel J, Thall A, White-Scharf M, Sachs DH, Cooper DKC. CD40L blockade requires host macrophages to induce humoral unresponsiveness to pig hematopoietic cells in baboons. *Transplantation* 2000; **69**:S415. (Abstract)

176. Buhler L, Goepfert C, Basker M, Gojo S, Guckelberger O, Nash K, Chang Q, Watts A, Treter S, Awwad M, Shimizu A, Andrews D, White-Scharf M, Sachs DH, Robson SC, Cooper DKC, Sackstein R. A thrombotic disorder complicating xenogeneic peripheral blood stem cell transplantation: A

primate model of thrombotic thrombocytopenic purpura (TTP). *Blood* 2000; **92**:706a. (Abstract)

177. Buhler L, Goepfert C, Kitamura H, Basker M, Gojo S, Alwayn IPJ, Chang Q, Down JD, Tsai H, Wise R, Sachs DH, Cooper DKC, Robson SC, Sackstein R. Porcine hematopoietic cell xenotransplantation in nonhuman primates is complicated by thrombotic microangiopathy. *Bone Marrow Transplant.* 2001; **27**:1227–1236.

178. Gojo S, Cooper DKC, Iacomini J, Le Guern C. Gene therapy and transplantation. *Transplantation* 2000; **69**:1995–1999.

179. Bracy JL, Sachs DH, Iacomini J. Inhibition of xenoreactive natural antibody production by retroviral gene therapy. *Science* 1998; **281**:1845–1847.

180. Gojo S, Shimizu A, Ierino FL, Banerjee PT, Cooper DKC, LeGuern C, Sachs DH. Xenogeneic and allogeneic skin grafting after retrovirus-mediated SLA class II DR gene transfer in baboons. *Transplant Proc* 2000; **32**:289–290.

181. Zhao Y, Fishman JA, Sergio JI, Oliveros JL, Pearson DA, Szot GL, Wilkinson RA, Arn JS, Sachs DH, Sykes M. Immune restoration by fetal pig thymus grafts in T cell-depleted, thymectomized mice. *J Immunol* 1997; **158**:1641–1649.

182. Zhao Y, Swenson K, Sergio JJ, Arn JS, Sachs DH, Sykes M. Skin graft tolerance across a discordant xenogeneic barrier. *Nat Med* 1996; **2**:1211–1216.

183. Wu A, Esnaola NF, Yamada K, Awwad M, Shimizu A, Huang C, Wain J, Zhao Y, Neville DMJ, Cooper DKC, Sykes M, Sachs DH. Xenogeneic thymic transplantation in a pig-to-nonhuman primate model. *Transplant Proc* 1999; **31**:957.

184. Yamada K, Shimizu A, Ierino FL, Utsugi R, Barth R, Esnaola N, Colvin RB, Sachs DH. Thymic transplantation in miniature swine. I. Development and function of the "thymokidney". *Transplantation* 1999; **68**:1684–1692.

185. Yamada K, Shimizu A, Ierino FL, Gargollo P, Barth R, Colvin RB, Sachs DH. Allogeneic thymo-kidney transplants

induce stable tolerance in miniature swine. *Transplant Proc* 1999; **31**:1199–1200.

186. Yamada K, Shimizu A, Utsugi R, Ierino FL, Gargollo P, Haller GW, Colvin RB, Sachs DH. Thymic transplantation in miniature swine. II. Induction of tolerance by transplantation of composite thymokidneys to thymectomized recipients. *J Immunol* 2000; **164**:3079–3086.

Chapter 11

Encapsulation as a Strategy for Cell Xenotransplantation

BL Schneider & P Aebischer

INTRODUCTION

Cell transplantation represents a promising tool for *in vivo* delivery of therapeutic proteins. It may provide an alternative to the treatment of numerous diseases necessitating repetitive administrations of recombinant proteins. Patient compliance and long-term complications related to poorly regulated administration, such as in diabetes, highlight the need to investigate cell-based therapies. Nevertheless, the outcome of these techniques for clinical purposes depends on future strategies to improve

i) the survival of transplanted cells
ii) the predictability and regulation of secreted protein amounts
iii) the safety of the procedure.

Since the mid-1970s, the immunoisolation of various cell types for transplantation purposes has been described. This concept regroups various techniques that allow the physical separation of grafted cells from the host immune system. Immunoisolation is achieved by enclosing the transplanted cells within a synthetic biocompatible membrane. The presence of pores in the membrane permits the outward diffusion of therapeutic substances and the inward diffusion of oxygen, nutrients as well as trophic factors essential for the metabolism of the implanted tissue. Ideally, the pore size should be tight enough to prevent the diffusion of antibodies and complement components having a higher molecular weight. By segregating the transplanted tissue, encapsulation avoids any cell-to-cell contact with the host immune cells, and thus blocks some essential pathways belonging to the recognition as well as effector functions of cellular immune components. Moreover, host immune cells have access only to the antigens diffusing outward from the device. The capsule may thus recreate an artificial immunoprivileged site permitting the transplantation of cells across barriers due to immune incompatibilities.

The source of tissue for transplantation purposes appears as an important matter subject to debate. Human cells offer the advantage of their close immunocompatibility, thus limiting the risk of rejection. On the other hand, practical and ethical obstacles relative to human sources generally limit the available amount of tissue. The use of xenogeneic cells circumvents this problem, giving access to a wide panel of tissue sources with different phenotypes. They include

   i) primary cells.
  ii) common cell lines that can be genetically modified.
 iii) immortalised cell lines derived from transgenic animals. Nevertheless, the testing of animal cells for the presence of pathogens has become a major concern since some risk of zoonoses, due to contamination over species barriers, has been highlighted.

It has been hypothesised that encapsulation may procure sufficient immunoprotection to reasonably envisage the use of xenogeneic cells for transplantation without the need of permanent immuno-suppressive drug therapy. However, successful transplantation of encapsulated xenogeneic cells often correlates with immunoprivileged conditions, and remains seldomly reported in scientific literature. Recent enlightenment in cell transplantation immunology may benefit the use of this technique for challenging purposes such as the treatment of autoimmune diabetes.

PRINCIPLE AND STRUCTURE OF THE DEVICE

Device and Membrane Design

To date, different implant designs have been investigated for the immunoisolation of transplanted cells. We distinguish between intra- and extravascular devices, depending on whether or not they are in contact with the vascular system.

Intravascular devices have been developed to support the survival of large amounts of cells drawing oxygen and nutrients from the blood. Transplanted cells deliver the therapeutic compound into the blood so as to compensate for a deficient endocrine function. Intravascular systems have been mainly investigated in the design of a bioartificial pancreas intended for the treatment of diabetic patients.[1,2] Vascular grafts divert the blood flow into a tubular porous membrane surrounded by a compartment containing the secretory cells. If needed, cells can be replaced using ports which give access to the chamber. The diffusion of molecules through the permeable membrane allows both the survival of cells and the release of the secreted compound into the blood. However, this type of devices necessitates invasive vascular surgery often associated with blood clotting. Intra-tubing clotting represents the major limitation to long-term clinical applications; thus, research now focuses on designs requiring a minimal surgical procedure.

Among extravascular devices, nomenclature currently distinguishes between micro- and macrocapsules. The former includes a set of techniques that allow the confinement of secretory cells within microporous beads having an external diameter of less than 1000 micrometres.[3] These devices are particularly suited for the immunoisolation of islets of Langerhans shaped as small spherical clusters. A system formed by the ionic cross-linking of a negatively charged polyelectrolyte, such as alginate, entraps the cells within a mechanically stable shell, which can then be coated with a layer of positively charged polyelectrolytes, such as polylysine, forming a semipermeable membrane.[4,5] Some *in vivo* studies have reported the long-term survival of xenogeneic islets in the peritoneal cavity using uncoated alginate microspheres.[6]

Microcapsules have two main advantages: they are easily produced and have a maximal area interfacing the secretory cells and their environment, so as to allow fast responses to external stimuli such as variations in blood glucose concentration. Nevertheless, the reliance on ionic bonds for shell cohesion leads to a chemical instability and a mechanical fragility. In addition, the difficulty to find and retrieve microcapsules as well as the total implant volume needed for therapeutic amounts of cells, such as in diabetes, represent major obstacles to their clinical application. More recent studies have investigated a conformal coating of cell clusters minimising the distance between transplanted cells and the surrounding tissue.[7,8] The reduction of the dead space allows the decrease of the total volume of the implant. Problems related to biocompatibility and difficulty to ensure an even and complete coating still hamper this approach.

Macroencapsulation currently appears as the most advanced format aiming at immunoisolation of various cell types for clinical purposes (Fig. 1). In macroencapsulation, preformed hollow fibres are loaded with the cell quantity needed for treatment prior to be sealed with glue. Membranes are engineered from thermoplastic polymers, such as polyethersulfone (PES) or acrylonitrile-vinylchloride copolymer (PAN–PVC), which have a foam-like structure. They are mechanically and chemically significantly more stable than microcapsules and can be shaped or assembled in self-supporting designs combined with internal reinforcements, such

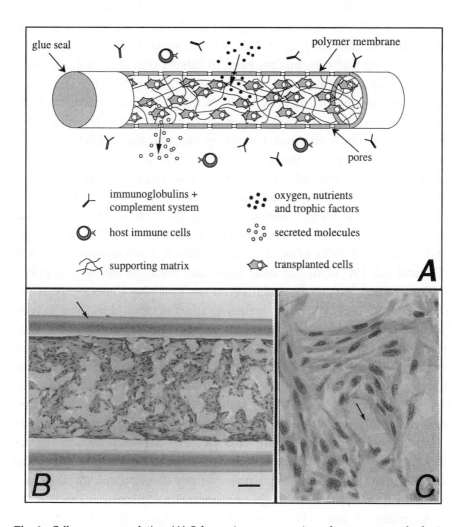

Fig. 1. *Cell macroencapsulation.* **(A)** Schematic representation of a macrocapsule device. **(B)** Micrograph showing a longitudinal section of a macrocapsule containing mouse C_2C_{12} myoblasts maintained during 4 weeks *in vitro*. Note the polymer membrane (arrow) and the homogenous distribution of the cells throughout the matrix. Scale bar: 100 μm. **(C)** 10X magnification of (B): the arrow shows the polyvinylalcohol polymer (PVA) matrix supporting viable C_2C_{12} myoblasts.

as titanium coils, if needed. To facilitate both localisation and retrieval, a silicone tether containing an X-ray opaque plug can be integrated to the implant. The structure and chemistry of the membrane aim at minimising protein adsorption and cell adhesion on the external surface to improve its biocompatibility. The outer skin must be smooth to provoke a minimal fibrotic reaction of the host tissue. Some researchers have also shaped the external surface of the device in a manner favouring neovascularization, so as to improve the access to oxygen and nutrients.[9] Pore diameter can be tuned in order to exclude cell-to-cell contacts and limit the diffusion of molecules belonging to the humoral immune system. Membranes with a molecular weight cut-off ranging from 50 to 280 kDa enable the metabolic supply of transplanted cells. On the other hand, IgG (150 kDa), IgM (900 kDa) and complement components (C1q, 410 kDa and C3, 190 kDa) can only penetrate through the largest pores.

The device has to be engineered so that the distance between encapsulated cells and the host vascular system does not exceed 500 μm. This distance seems to be critical in order to guarantee both nutrient supply and sufficient oxygen tension permitting the *in vivo* survival of non-vascularized tissues. Also, the choice of adequate macrocapsule geometry, either sealed tubes or disc-shaped flat sheets, depends on required internal volume and mechanical resistance. Five centimetre-long hollow tubular fibres can accommodate up to 5 million cells. If higher amounts of cells are required, discs are more appropriate: with a diameter of 5 cm, they have capacity for 50 to 100 million cells, while maintaining mechanical stability and facilitating surgical handling.

As most of the cells intended for encapsulation do not settle under large aggregates, they need a mechanical support to grow and survive. Selection of the appropriate matrix depends essentially on cell type. Commonly, loaded cells are mixed with collagen or polysaccharide-based hydrogels (alginate and agarose) prior to encapsulation so as to create an artificial intercellular semisolid matrix favouring the three-dimensional structuring of the tissue.

More recently, a solid polymer sponge made of polyvinylalcohol polymer (PVA) has been inserted within the membrane to partition the inner space in a reticular manner[10] (Fig. 1). This matrix avoids the formation of large cell clusters, thereby diminishing the necrotic core in the centre zone of the capsule. Additionally, collagen-coated PVA may provide a mechanical support encouraging some cell types, such as myoblasts, to differentiate into a post-mitotic state avoiding cell overgrowth in the device.

The difficulty to support the viability of large cell amounts appears as the main limitation to the clinical application of macrocapsules. As more than 100 million cells may be required to treat diabetes, up scaling of devices for human therapy challenges biomaterial engineers. Nevertheless, macroencapsulation offers many advantages: the opportunity to check membrane integrity before implantation, a convenient surgical handling as well as retrievability of the implant. Thus, this approach appears as the most reliable and practical modality in immunoisolation.

Encapsulated Cells

Primary cells versus cell lines

The main questions to be addressed when considering the selection of tissue supply are the following: (1) the needed amount of cells, according to the expected secretion rate, and (2) the safety in terms of presence of pathogens.

Xenogeneic sources provide an answer to the first question as they circumvent many practical and ethical issues relative to human cells. Actually, the debate focuses on the selection of primary cells as opposed to cell lines having an unlimited expansion. Primary cells have to be considered for applications requiring a precise control of the delivery of biotherapeutics upon physiological parameters. In order to replace efficiently the endocrine pancreatic function, grafted cells must tightly regulate their insulin secretion in response to blood glucose concentration. Until now, most approaches toward a cell-based therapy of diabetes relied on primary islets retaining their original function. The use of xenogeneic islets appears attractive since foreign insulin, such as porcine insulin, is bioactive in humans and does not induce a neutralising antibody response.

In some specific cases, primary cells are the only current alternative, as cell lines with similar characteristics simply do not exist. For example, a phase II clinical trial has been conducted for the treatment of chronic pain using bovine adrenal chromaffin cells which secrete a cocktail of analgesic substances.

Yet, it should be kept in mind that isolation of primary tissue has to circumvent major hurdles to get into routine clinical applications. Up-scaling and quality control of animal sources require burdensome procedures so as to guarantee the absence of pathogens as well as the purity of the tissue. With the threat of cross-species viral contaminations, the safety issue presents a major difficulty for the outcome of primary cell xenotransplantation. In contrast, the use of constantly dividing cells largely facilitates their testing before implantation in human beings.

For some diseases, highly specialised cells are needed to replace the deficient secretory function. Recently, the *in vitro* setting of stem cells from various tissues may provide a promising source of multi-potent cells. Their ability to divide answers the question of required cell amount. Using the right factors, they can generate progenitor cells, which can be guided on the route of differentiation to obtain the original phenotype required for tissue replacement. Nevertheless, the use of stem cells remains a futuristic strategy, still limited by their metabolic properties and the poor knowledge of their biology.

In recent years, some investigators have started to encapsulate cell lines bearing the advantage of unlimited availability, with a result of prolonged cell survival.[11–14] Remarkably, dividing cells do not die of massive metabolic failure due to overgrowth within the capsule inner space. On the contrary, they seem to attain a state of dynamic equilibrium where lost cells are constantly replaced by newly generated ones, thus ensuring the survival of the grafted tissue for a few months. When the growth rate is too fast, a necrotic core appears in the middle of the device. More recently, the C_2C_{12} mouse myoblasts have been successfully employed in the rat central nervous system (CNS) and the mouse subcutaneous tissue.[15,16] The goal was to take advantage of their capacity to differentiate into post-mitotic myotubes upon exposure to a low-serum containing medium. This permits, prior to implantation, to stop or at least reduce the division rate of encapsulated myoblasts so as to avoid accumulation of cell debris, while maintaining a constant secretion of the therapeutic compound.

Genetic engineering

Owing to recent advances in genetic engineering, cultured dividing cell lines become an attractive source of tissue. Different approaches can be investigated to generate cell lines suitable for encapsulation:

(1) Existent cell lines can be modified to achieve constant or even regulated expression of therapeutic molecules. Some trials have already demonstrated the feasibility of using xenogeneic cell lines to deliver *in situ* bioactive human factors: hamster fibroblasts and mouse C_2C_{12} myoblasts genetically engineered to constitutively secrete elevated amounts of ciliary neurotrophic factor have already been transplanted in patients suffering from amyotrophic lateral sclerosis[14] and remained functional up to 15 months. A similar phase I clinical trial is currently being performed in patients suffering from Huntington's disease.

In parallel to the gene of therapeutic interest, cell lines can be modified to express resistance or immunomodulating factors, which may prolong the survival of encapsulated cells after implantation. It has been demonstrated that anti-apoptotic genes, such as bcl-2, improve cell resistance to metabolic and immunological stress.[17] As for the host immune reaction, it could be artificially tempered by secreting an appropriate immunomodulating agent.

Beside problems relative to implant survival, it should be kept in mind that some targeted diseases require a tight regulation of the delivered compound to avoid side effects. Therefore, the development of *in vivo* inducible promoters remains a key issue for the outcome of cell-based therapies. Specific systems have been considered to control gene expression by either systemic administration of an exogenous molecule or physiological parameters. For example, a chimeric transactivating or transrepressing transcription factor has been engineered to bind tetracycline, a small antibiotic molecule.[18] The ability of this factor to either activate or repress gene transcription depends on whether tetracycline is present or not. Therefore, antibiotic administration may regulate the expression of a therapeutic proteins by transplanted cells.[19,20] Another study suggests that physiological parameters may directly control the release of the therapeutic compound[21]: in a delivery model using encapsulated myoblasts, hypoxia has increased the expression of erythropoietin driven by the mouse PGK-1 promoter. For certain diseases such as diabetes, current tools in genetic engineering may not be sufficient to restore both specificity and kinetics of regulation. In these cases, there is an obvious need for cells which originally possess the appropriate molecular machinery.

(2) Thus, primary cells can be conditionally immortalised from transgenic animals which overexpress an oncogene under control of an inducible promoter. For instance, a murine cell line has been derived from pancreatic islets expressing the large T antigen.[22] These cells have the ability to divide in a tetracycline-dependent manner, while their primary function is preserved.[23]

Safety

The use of xenogeneic cell lines raises two safety concerns: first, their tumorigenicity and second, the risk of zoonosis due to the presence of endogenous retroviruses.

Most of the cultured dividing cells can form tumours when implanted in immunocompatible hosts. In the unlikely event of capsule breakage, cells may escape from the device and start growing in an uncontrolled manner. Yet, different concepts and techniques ensure an adequate oncosafety in the transplantation of encapsulated xenogeneic cells:

(1) non-encapsulated cells facing the host immune system should be rapidly killed because of their animal origin, either by the host cellular immune response or by naturally preformed antibodies activating the complement system. Even in the immunoprivileged central nervous system, naked xenogeneic cells are rejected.[24,25]

(2) Highly tumorigenic cell lines can be genetically modified for the expression of molecules inducing their rejection through a cell contact-mediated process, which does not interfere with cell survival within the capsule. Expression of the IgG Fc portion in the reversed orientation on target cell surface has been demonstrated to mimick IgG opsonization and activate macrophages.[26] It is of importance that this molecule does not fix complement, which may affect the viability of grafted cell in the capsule compartment. Similarly, the constitutive over-expression of B7.1 promotes the stimulation of host T cells by the grafted cells themselves. If the transplanted cells present potentially antigenic peptides, B7.1 expression will exacerbate their rejection by the host CD8+ T cells.[27–29] Such strategies may also find application in encapsulated allotransplantation, as immortalised human cells may escape the control of the human immune system.

(3) In addition, incorporation of suicide genes such as the herpes simplex virus thymidine kinase gene allows, if needed, cell elimination upon oral administration of a small molecule such as ganciclovir.[24,30]

Recently, concerns have been raised about transplantation of xenogeneic organs or tissues containing endogenous retroviruses. Some of these pathogens, such as porcine endogenous retroviruses (PERV), could lead to the emergence of new diseases, as they are able to infect human cells.[31] Still, systematic searches for PERV infections in 160 patients treated with living pig tissue[32] and in 10 diabetic patients having received porcine foetal islets[33] did not evidence any cross-species transmission of PERV. In the particular case of encapsulated xenogeneic cell lines, some facts may also temper the risk of zoonosis: the use of dividing cells allows the constitution and cryopreservation of cell banks from clonal origin. Approved laboratories can test and certify these stocks free of eco- xeno- and amphotropic retroviruses. Moreover, the pore size of the permeable membrane may be selected so as to prevent the release of viral particles.[34,35] The conjunction of xenogeneic cell lines with encapsulation techniques presents biosafety features justifying their use in wide-base clinical trials.

Consequently, cell lines of xenogeneic origin offer many practical advantages, including an unlimited availability and a large flexibility to genetic modifications. In addition, safety considerations do not presently raise major obstacles to their encapsulated transplantation in humans. They may thereby appear as a promising tool for future large-scale clinical applications of cell-based therapies. Nevertheless, this depends on the technique's ability to prevent the immune-mediated rejection of xenogeneic cells.

THE IMMUNOLOGICAL CONCEPT

The encapsulating membrane provides a physical barrier excluding any cell-to-cell contact. This partition precludes most of the recognition and effector mechanisms mediating the specific destruction of the graft. The trafficking of antibodies and complement is controlled by the pore size which can be selected so as to prevent their diffusion into the graft. Therefore, researchers have hypothesised that immunoisolation may permit the allo- or even xenotransplantation of cells in the absence of host immunosuppression.

Recent advances in immunology have enlightened the protecting role of the encapsulating membrane. But they have also suggested that certain immune mechanisms may actually bypass this barrier and lead to a rejection process.

Mechanisms of Antigen Recognition

Basically, antigenic peptides originating from transplanted cells can induce an immune response following two distinct pathways. In both cases, activation of the immune system is subsequent to the interaction between T cells and antigen presenting cells (APC). The latter are able to process antigens and possess the costimulating machinery which fulfil the two-signal need for T cell activation and proliferation. The first antigen-specific signal results from the binding of the major histocompatibility complex (MHC), presenting the antigenic peptide, to the T cell receptor. Costimulatory molecules, such as B7 and ICAM-1, are highly expressed on APCs and provide the second non-specific signal. Depending on whether APCs are graft or host cells, the antigen presentation pathway is described as direct or indirect, respectively.

Direct antigen presentation

According to the present knowledge, direct antigen presentation appears especially relevant to the transplantation of allogeneic cells. In this case, grafted cells express MHC class I and/or II, which trigger host CD8+ and/or CD4+ T cells, respectively. In fact, the graft-versus-host polymorphism of MHC molecules induces vigorous immune responses even if the MHC groove binds peptides with few antigenic differences.[36] In addition, the ligand-to-receptor compatibility between donor and host molecules allows allogeneic grafted cells to function as APCs. In contrast, if the donor cells have a xenogeneic origin, the species-specificity of some ligand-receptor interactions can prevent the binding of accessory (CD4, CD8) and costimulatory (CD40) molecules, or even cytokines (IFNγ).[37] This compromises the ability of donor cells to activate host immune cells. Therefore, with increasing phylogenetic distance between host and graft tissues, the efficiency of direct presentation diminishes. As compared to allografts, the recognition of xenogeneic cells should be biased toward the indirect pathway.

Obviously, encapsulation completely averts the direct pathway of antigen presentation by preventing cell-to-cell contacts between host and graft cells. This may by itself explain that encapsulation has provided an efficient immunoprotection to allogeneic cells in numerous studies. Yet, the match between allotransplantation and direct antigen presentation has to be reconsidered since peptides derived from major histocompatibility complexes can indirectly participate to the host sensitization.[38,39]

Indirect antigen presentation

Immunoisolation has also been intended to improve the survival of xenotransplanted cells. However, does it prevent all graft antigens of being recognized by the host?

For the above-mentioned reasons, xenogeneic tissue should not be able to directly present antigens to the host immune system. Therefore, it has been postulated that xenoantigens may be preferentially prone to indirect presentation. We should also keep in mind that interspecies differences dramatically increase the antigenic feature of transplanted cells, as much quantitatively as qualitatively. For instance, differences in peptidic sequences as well as protein glycosylation may provoke or enhance immune responses against foreign molecules. In addition, the polymorphism is not limited to MHC molecules, but also affects secreted factors which can diffuse out of the device and come into contact with host APCs.

Thus, encapsulation of xenogeneic cells within a permeable membrane will not completely preclude the outward diffusion of small xenoantigens. In addition, the device is placed into an inflammatory context, due to the surgical trauma, which may favour the activation of APCs. These specialised cells, including dendritic cells and macrophages, can capture the released antigens and process them for binding to MHC class II molecules. APCs migrate to host lymphoid organs, where they interact with a wide population of naïve lymphocytes. Interestingly, researchers have recently demonstrated that peptides derived from extracellular proteins are not class II-restricted and can be presented on MHC class I molecules through a process termed "cross-priming"[40–42] suggesting that indirect presentation may also activate CD8+ T cells. Still, past experiments, using encapsulated COS cells implanted in mice, evidenced that depletion of CD8+ T cells did not improve implant survival, whereas CD4+ T cells were necessary for xenograft rejection.[43] The importance of CD4+ T cells has also been stressed in the NOD mouse model: a treatment with anti-CD4 monoclonal antibodies significantly prolonged the survival of rat and canine islets contained in polylysine/alginate microcapsules.[44] The pivotal role of helper T cells argues in favour of indirect recognition, by CD4+ T cells, of antigens bound to MHC class II molecules. Nevertheless, in the case of encapsulated transplants, the effector mechanisms subsequent to indirect presentation of xenoantigens have not yet been understood and will be discussed further on in this chapter.

Whereas the encapsulating membrane does not screen all xenoantigens released from the cells, it can still provide a partial immune "invisibility". The pore size does not allow apoptotic bodies or membrane fragments deriving from dying cells to pass through the microporous membrane. Therefore, macrophages or dendritic cells can not phagocytose these cellular debris so as to process and present donor antigens.[41] In addition, the absence of host cell trafficking inside the device may prevent APCs to pick up antigens bound to the surface of grafted cells, including MHC complexes. Thus, encapsulation should restrict both the quantity and variety of donor antigens presented through the indirect pathway.

Implantation site is likely to affect indirect presentation pathway: this depends on the presence of highly specialised APCs, such as dendritic cells, and the efficiency of the lymphatic drainage. With regard to these points, the subcutaneous site may raise major obstacles to the transplantation of encapsulated xenografts. In contrast, the lack of direct lymphatic drainage, the paucity of efficient APCs and the presence of the blood-brain barrier confer an immune privilege to the CNS. Remarkably, xenogeneic cells contained in small hollow fibres survived in the CNS without additional immunoprotection, holding therefore promises for the delivery of bioactive molecules in the CNS.

Effector Mechanisms

Subsequently to antigen recognition, specific T cells are activated and targeted back to the transplantation site by the local up-regulation of adhesion molecules, activated in response to chemokine expression. Activated helper T cells will provoke a strong inflammatory response which can destroy the graft through still poorly understood processes. Clarifying this mechanism may help to design strategies improving transplant immunoprotection.

Antibodies and complement

Rejection of xenogeneic organs has given rise to numerous studies which have clarified mechanisms potentially applicable to cell transplantation. Until recently, investigators have assumed that the main obstacle to xenotransplantation would be the hyperacute rejection, due to naturally preformed antibodies present in discordant donor/recipient combinations.[45,46] These bind xenoantigens expressed on the surface of donor endothelial cells lining graft blood vessels, resulting in a massive activation of the recipient complement cascade. Antibodies and complement activate endothelial cells inducing loss of vascular integrity, oedema, haemorrhage and thrombosis, leading to organ destruction within a few minutes. In concordant xeno-transplantation, the neoformation of antibodies causes the acute vascular rejection of the organ, a delayed process targeting the donor vascular system independently of complement activation.[45,46] In contrast to organs, cellular transplants are not vascularized. This particular situation precludes a hyperacute rejection, since complement activation can not induce a massive metabolical failure by destroying the vascular system.

In cell encapsulation, commonly used microporous membranes can reduce the diffusion of antibodies and complement. Still, even membranes with a small exclusion limit have a gaussian distribution of pore sizes that may permit the entry of large humoral immune components. Therefore, it has been suggested that antibodies and complement forming membrane attack complexes may destroy encapsulated xenografts by disrupting the cell osmotic balance. During organ rejection, the humoral system is particularly effective on endothelial cells, as their two-dimensional arrangement favours the binding of antibodies and complement components. Yet, in encapsulated cellular transplants, the dense three-dimensional arrangement of cells impedes both penetration of antibodies in avascular tissues and formation of complement complexes. Thus, necrosis remains limited to the marginal zone of encapsulated tissue. A study has stressed the secondary role of complement in rejecting immunoisolated transplants.[47] However, according to device geometry, cell density and implantation site, investigating the antibody/complement-mediated cell death may still have some rationale, as in the case of small microencapsulated islets.

Cellular immune system

With the setting of strategies inactivating hyperacute rejection, the survival of xenotransplanted organs remained limited by a strong cellular reaction. This process, quite resistant to most of immuno-suppressive protocols, is likely to result from antigen presentation through the indirect route. By analogy, encapsulated xenotransplants should undergo an inflammatory response, very similar to delayed-type hypersensitivity, due to the release of xenoantigens. In addition, a non-specific reaction, subsequent to the surgical trauma, can aggravate this reaction. Whatever the specificity of the response, numerous cell-mediated mechanisms may affect the survival of transplanted cells in spite of membrane protection. Helper CD4+ T cells seem to play a central role[43] by controlling the recruitment and activation of many other cell types, including T cells, macrophages and granulocytes.

A few years ago, it was assumed that isolating the graft from host cells should obviate most of the effector arms belonging to the cellular immune system. Actually, only CD8+ killer T cells and natural killer cells must be brought in contact with target cells to recognise and destroy foreign tissues. It is of particular interest that CD8+ T cells belong to the main effectors in allograft rejection; the fact that they cannot penetrate the capsule also accounts for the successful immunoprotection of allogeneic cells. But other mechanisms are susceptible of damaging the implant through the immunoisolating membrane. For example, macrophages may secrete deleterious molecules that can diffuse into immunoisolating chambers.

Two main schemes have been postulated to explain the destruction of cells behind the membrane.

(1) First, many cells participating to the inflammatory reaction release small soluble factors which can rapidly diffuse into the capsule. Secretion of cytokines contributes to the cross-activation of helper T cells and accessory cells including macrophages and

granulocytes. Yet, cytokines such as TNFα, IFNγ and IL-1 can also trigger apoptotic pathways in transplanted cells, provided the latter express specific receptors. In addition, activated macrophages and neutrophils produce reactive oxygen species, free radicals and nitric oxide. Thus, a cocktail of short-lived deleterious molecules may penetrate the membrane and kill the implant. Still, it remains difficult to point the culprit out, as several compounds probably act in cooperation to create a microenvironment noxious to transplanted cells.

(2) A second immune-dependent mechanism may affect the survival of encapsulated cells. The inflammatory reaction is known to stimulate the formation of fibrotic tissue surrounding the implant.[48] The thickness of this layer depends both on membrane biocompatibility and immunogenicity of transplanted cells. As this fibrotic shell is poorly vascularized, it may progressively starve the implant. Encapsulated cells may finally die of metabolic failure consequent upon nutrient and oxygen deprivation.

These two effector mechanisms share one *sine qua non* condition: an inflammatory reaction can be induced provided the capsule is fixed in the host tissue. Therefore, the implantation site may play a critical role. For example, when placed either in the peritoneal cavity or in the cerebrospinal fluid, the capsule does not necessarily adhere to the host tissue. Since the device is soaked in liquid, cells contributing to the inflammatory response do not accumulate on the membrane. This particular situation may account for the long-term survival observed in transplanting encapsulated islets into the peritoneal cavity. As for macrocapsules containing hamster or murine cells, they showed a minimal fibrotic reaction when retrieved from the intrathecal space of amyotrophic lateral sclerosis or cancer pain patients.[14,49]

Transplantation of xenogeneic cells mobilises various immune mechanisms. Interposing a porous membrane between host and graft cells does not fend off all effector arms belonging to the immune system. In fact, the complexity of involved mechanisms does not allow to draw a general scheme applying to immunoisolation since important parameters, including cell immunogenicity and implantation site, have to be taken into account to evaluate each case. Therefore, the transplantation of encapsulated cells could take advantage of particular immunoprivileged conditions. In the same way, long-term survival may be achieved using mild immuno-modulating techniques, which, in comparison, would not permit the transplantation of non-encapsulated xenogeneic cells.

Immunomodulation

The success obtained in the CNS gives hope for the outcome of encapsulated xenotransplantation. A number of pre-clinical and clinical trials have demonstrated the long-term survival of immunoisolated cells in many donor-to-host species combinations including mouse, bovine and hamster to rat,[15,50,51] hamster and bovine to sheep,[24,52] hamster and rat to primate,[53,54] and finally bovine and hamster to human.[14,49] The immune response against encapsulated xenografts is presumably tempered by the relative absence of APCs and direct lymphatic drainage as well as the non-inflammatory state of the CNS, associated with local production of immunomodulating factors such as TGFβ.

Outside the CNS, implantation of encapsulated cells or tissues has resulted in more variable results within the peritoneal cavity, including the prolonged acceptance of primary islets of Langerhans and the rejection of the NIT-1 islet β-cell line, primary rat and canine islets, COS cells and foetal lung tissue.[43,44,47,48,55] As for the subcutaneous site, only one study has reported the survival of primary islets isolated from rat donors, encapsulated and then implanted under the skin of mice recipients.[56] It is of importance that these results should be evaluated considering donor-to-host phylogenetic distance and overall implant immunogenicity. With respect to the latter point, primary islets are likely to be considered as a poorly immunogenic tissue. Yet, discrepancies in cell survival mainly reflect the immunological status of the different implantation sites. Nevertheless, transplantation of encapsulated xenogeneic cells appears feasible, provided it takes advantage of immunoprotecting conditions.

In fact, new approaches combining encapsulation with an active immunomodulation may benefit cell xenotransplantation outside the CNS. Immunomodulation may help to transiently repress acute immune responses, permanently inhibit the host immune system, or even render the host tolerant of the implant. In the optic of clinical applications, we have to prevent side-effects associated with prolonged administration of pan-immunosuppressive drugs. Different protocols are being investigated:

(1) Transient treatment and/or low doses of immunosuppressive drugs, such as FK506 or cyclosporin, may provide sufficient protection. A recent study has shown that a short-course treatment using FK506 in rat recipients permits the long-term survival of encapsulated C_2C_{12} mouse myoblasts, whereas the same immunosuppression does not prevent rejection of naked cells.[57] In addition, daily injections of FK506 in rats for a 4-week period following implantation are sufficient to achieve long-term tolerance of capsules containing C_2C_{12} cells.[58]

(2) Recently, new light has been shed on the ligand-receptor interactions governing the costimulation of T cells. This has allowed to design molecules blocking the activation of two major mechanisms, namely the CD28-B7 and CD40-CD154 pathways. Taking advantage of the strong affinity of CTLA4 for B7, the chimeric fusion protein CTLA4-Ig inhibits the activation of the CD28 receptor on T cells, and thus significantly protects cell or organ transplants[59,60] from T cell-mediated rejection. In a model of naked human islet transplantation, CTLA4-Ig administration has rendered mice tolerant of the implant.[61] Furthermore, a study has reported the prolonged survival of microencapsulated porcine islets in NOD mice following

repetitive injection of the fusion protein.[62] Similarly, the binding of CD154 to the CD40 receptor plays a major role in the cross-activation of T cells, dendritic cells, macrophages or even B cells. Therefore, injection of antibodies blocking CD154 have allowed the long-term survival of both cell[63] or organ[64] allografts. In xenotransplantation, CD154 blockade, combined with either inhibition of the CD28 pathway[65] or transfusion of donor spleen cells,[66] has resulted in prolonged acceptance.

To date, there is still a need to elucidate how costimulation blockade improves transplant survival and induces tolerance. Nevertheless, previous studies have evidenced the lack of secondary effects associated with this approach,[67] which may temper T cell-mediated responses against encapsulated xenotransplants.

(3) Finally, the device itself may deliver a local immunomodulation in order to avoid systemic immunosuppression. For instance, encapsulated cells can be engineered so as to secrete CTLA-Ig[68–71] which will diffuse in the implant environment as long as membrane pores are large enough. However, in the case of compounds acting on T cell costimulation, they probably have to reach lymphoid organs where antigen presentation takes place.

Still, clarifying recognition and effector mechanisms in the rejection process should permit the setting of new immunoprotecting techniques relevant to the clinical application of encapsulated xenogeneic cells. In addition, combining encapsulation with immunomodulation may steer the host immune system towards permanent tolerance of cell xenotransplants.

CONCLUSION

In comparison to transplantation of naked cells, encapsulation undoubtedly improves the reliability of therapies based on cell transfer. Preformed macrocapsules especially meet the requirements of clinical applications, since they guarantee both integrity and retrievability of the device. The use of cell lines, in terms of tissue sources, appears as the most advanced modality to access large amounts of cells, which can be tested for the presence of pathogens. Xenogeneic cell lines can be modified to secrete human therapeutic proteins in a regulated manner with the setting of sophisticated techniques in genetic engineering and tissue culturing. Therefore, it will certainly become feasible to adapt these cells for the treatment of numerous diseases. In addition, confining cells in a device with selectable diffusion properties circumvents many problems relative to safety of cell xenotransplantation. Thus, encapsulation may offer a realistic opportunity to bring animal tissue to wide-base clinical application.

The question of the immune response to encapsulated xenografts should however be carefully addressed. Numerous examples have demonstrated the long-term survival of immunoisolated allografts. Yet, we have to re-examine the enthusiasm of initial reports suggesting that encapsulation may obviate the risk of rejection following transplantation of xenogeneic cells. Some studies have recently evidenced an immune rejection of encapsulated xenografts. Concomitantly, the importance of the indirect pathway in presenting shed antigens and inducing vigorous immune responses has been highlighted. Considering the release of antigens through the membrane, the notion of "immunoisolation" has now to be revised: encapsulation does not preclude all interactions with the host immune system. It provides an artificial immune privilege mainly associated with the absence of host-to-graft cell contacts. But it still has to be clarified which effector arms remain involved in the rejection process. This may help to find specific immunomodulating strategies which do not induce undesirable side-effects, while permitting long-term graft acceptance.

These concerns contrast with the excellent viability of encapsulated cells transplanted in the CNS. Some reports have also pointed out the successful implantation of encapsulated xenogeneic islet cells in diabetic animal models. In these cases, the selection of immuno-privileged implantation sites or the use of poorly immunogenic cells can explain the long-term survival of the grafts. By taking advantage of these particular situations, encapsulation can, by itself, allow clinical applications employing xenogeneic cells.

REFERENCES

1. Sullivan SJ, Maki T, Borland KM, Mahoney MD, Solomon BA, Müller TE, Monaco AP, Chick WL. Biohybrid artificial pancreas: Long-term implantation studies in diabetic, pancreatectomized dogs. *Science* 1991; **252**:718–721.
2. Maki T, Otsu I, O'Neil JJ, Dunleavy K, Mullon CJ, Solomon BA, Monaco AP. Treatment of diabetes by xenogeneic islets without immunosuppression. Use of a vascularized bioartificial pancreas. *Diabetes* 1996; **45**:342–347.
3. Clayton HA, James RF, London NJ. Islet microencapsulation: A review. *Acta Diabetologica* 1993; **30**:181–189.
4. Lim F, Sun AM. Microencapsulated islets as bioartificial endocrine pancreas. *Science* 1980; **210**:908–910.
5. Sun Y, Ma X, Zhou D, Vacek I, Sun AM. Normalization of diabetes in spontaneously diabetic cynomologus monkeys by xenografts of microencapsulated porcine islets without immunosuppression. *Journal of Clinical Investigation* 1996; **98**: 1417–1422.
6. Lanza RP, Kuhtreiber WM, Ecker D, Staruk JE, Chick WL. Xenotransplantation of porcine and bovine islets without immunosuppression using uncoated alginate microspheres. *Transplantation* 1995; **59**:1377–1384.
7. Hill RS, Cruise GM, Hager SR, Lamberti FV, Yu X, Garufis CL, Yu Y, Mundwiler KE, Cole JF, Hubbell JA, Hegre OD, Scharp DW. Immunoisolation of adult porcine islets for the treatment of diabetes mellitus. The use of photopolymerizable polyethylene

glycol in the conformal coating of mass-isolated porcine islets. *Ann NY Acad Sci* 1997; **831**:332–343.

8. Sawhney AS, Pathak CP, Hubbell JA. Modification of Islet of Langerhans Surfaces with Immunoprotective Poly(ethylene glycol) Coatings via Interfacial Photopolymerization. *Biotech Bioeng* 1993; **44**:383–386.

9. Brauker JH, Carr-Brendel VE, Martinson LA, Crudele J, Johnston WD, Johnson RC. Neovascularization of synthetic membranes directed by membrane microarchitecture. *J Bio Mat Res* 1995; **29**:1517–1524.

10. Li RH, White M, Williams S, Hazlett T. Poly(vinyl alcohol) synthetic polymer foams as scaffolds for cell encapsulation. *J Bio Sci* 1998; **9**:239–258.

11. Jaeger CB, Greene LA, Tresco PA, Winn SR, Aebischer P. Polymer encapsulated dopaminergic cell lines as "alternative neural grafts". *Prog Brain Res* 1990; **82**:41–46.

12. Tresco PA, Winn SR, Tan S, Jaeger CB, Greene LA, Aebischer P. Polymer-encapsulated PC12 cells: Long-term survival and associated reduction in lesion-induced rotational behavior. *Cell Transplant* 1992; **1**:255–264.

13. Winn SR, Lindner MD, Lee A, Haggett G, Francis JM, Emerich DF. Polymer-encapsulated genetically modified cells continue to secrete human nerve growth factor for over one year in rat ventricles: Behavioral and anatomical consequences. *Exp Neurol* 1996; **140**:126–138.

14. Aebischer P, Schluep M, Deglon N, Baetge EE, Joseph JM, Hirt L, Heyd B, Goddard M, Hammang JP, Zurn AD, Kato AC, Regli F. Intrathecal delivery of CNTF using encapsulated genetically modified xenogeneic cells in amyotrophic lateral sclerosis patients. *Nat Med* 1996; **2**:696–699.

15. Deglon N, Heyd B, Tan SA, Joseph JM, Zurn AD, Aebischer P. Central nervous system delivery of recombinant ciliary neurotrophic factor by polymer encapsulated differentiated C2C12 myoblasts. *Hum Gen Ther* 1996; **7**:2135–2146.

16. Regulier E, Schneider BL, Deglon N, Beuzard Y, Aebischer P. Continuous delivery of human and mouse erythropoietin in mice by genetically engineered polymer encapsulated myoblasts. *Gen Ther* 1998; 5:1014–1022.

17. Dupraz P, Rinsch C, Pralong WF, Rolland E, Zufferey R, Thorens B. Lentivirus-mediated Bcl-2 expression in beta TC-tet cells improves resistance to hypoxia and cytokine-induced apoptosis while preserving *in vitro* and *in vivo* control of insulin secretion. *Gen Ther* 1999; 6:1160–1169.

18. Gossen M, Freundlieb S, Bender G, Muller G, Hillen W, Bujard H. Transcriptional activation by tetracyclines in mammalian cells. *Science* 1995; 268:1766–1769.

19. Serguera C, Bohl D, Rolland E, Prevost P, Heard JM. Control of erythropoietin secretion by doxycycline or mifepristone in mice bearing polymer-encapsulated engineered cells. *Hum Gen Ther* 1999; 10:375–383.

20. Bohl D, Naffakh N, Heard JM. Long-term control of erythropoietin secretion by doxycycline in mice transplanted with engineered primary myoblasts. *Nat Med* 1997; 3:299–305.

21. Rinsch C, Regulier E, Deglon N, Dalle B, Beuzard Y, Aebischer P. A gene therapy approach to regulated delivery of erythropoietin as a function of oxygen tension. *Hum Gen Ther* 1997; 8: 1881–1889.

22. Efrat S, Fusco-DeMane D, Lemberg H, al Emran O, Wang X. Conditional transformation of a pancreatic beta-cell line derived from transgenic mice expressing a tetracycline-regulated oncogene. *Proc Natl Acad Sci, USA* 1995; 92:3576–3580.

23. Fleischer N, Chen C, Surana M, Leiser M, Rossetti L, Pralong W, Efrat S. Functional analysis of a conditionally transformed pancreatic beta-cell line. *Diabetes* 1998; 47:1419–1425.

24. Aebischer P, Pochon NA, Heyd B, Deglon N, Joseph JM, Baetge EE, Hammang JP, Goddard M, Lysaght M, Kaplan F, Kato AC, Schluep M, Hirt L, Regli F, Porchet F, De Tribolet N. Gene therapy for amyotrophic lateral sclerosis (ALS) using a polymer

encapsulated xenogenic cell line engineered to secrete hCNTF. *Hum Gen Ther* 1996; **7**:851–860.

25. Ono T, Date I, Imaoka T, Shingo T, Furuta T, Asari S, Ohmoto T. Evaluation of intracerebral grafting of dopamine-secreting PC12 cells into allogeneic and xenogeneic brain. *Cell Transplant* 1997; **6**:511–513.

26. Stabila PF, Wong SC, Kaplan FA, Tao W. Cell surface expression of a human IgG Fc chimera activates macrophages through Fc receptors. *Nat Biotech* 1998; **16**:1357–1360.

27. Chen L, Ashe S, Brady WA, Hellstrom I, Hellstrom KE, Ledbetter JA, McGowan P, Linsley PS. Costimulation of antitumor immunity by the B7 counterreceptor for the T lymphocyte molecules CD28 and CTLA-4. *Cell* 1992; **71**:1093–1102.

28. Townsend SE, Allison JP. Tumor rejection after direct costimulation of CD8+ T cells by B7-transfected melanoma cells. *Science* 1993; **259**:368–370.

29. Jung D, Jaeger E, Cayeux S, Blankenstein T, Hilmes C, Karbach J, Moebius U, Knuth A, Huber C, Seliger B. Strong immunogenic potential of a B7 retroviral expression vector: Generation of HLA-B7-restricted CTL response against selectable marker genes. *Hum Gen Ther* 1998; **9**:53–62.

30. Mullen CA. Metabolic suicide genes in gene therapy. *Pharmac Ther* 1994; **63**:199–207.

31. Patience C, Takeuchi Y, Weiss RA. Infection of human cells by an endogenous retrovirus of pigs. *Nat Med* 1997; **3**:282–286.

32. Paradis K, Langford G, Long ZF, Heneine W, Sandstrom P, Switzer WM, Chapman LE, Lockey C, Onions D, Otto E. Search for cross-species transmission of porcine endogenous retrovirus in patients treated with living pig tissue. *Science* 1999; **285**: 1236–1241.

33. Heneine W, Tibell A, Switzer WM, Sandstrom P, Rosales GV, Mathews A, Korsgren O, Chapman LE, Folks TM, Groth CG. No evidence of infection with porcine endogenous retrovirus in recipients of porcine islet-cell xenografts. *Lancet* 1998; **352**: 695–699.

34. Aebischer P, Hottinger AF, Deglon N. Cellular xenotransplantation. *Nat Med* 1999; **5**:852.

35. Nyberg SL, Hibbs JR, Hardin JA, Germer JJ, Persing DH. Transfer of porcine endogenous retrovirus across hollow fiber membranes — Significance to a bioartificial liver. *Transplantation* 1999; **67**:1251–1255.

36. Sherman LA, Chattopadhyay S. The molecular basis of allorecognition. *Ann Rev Immunol* 1993; **11**:385–402.

37. Moses RD, Winn HJ, Auchincloss H, Jr. Evidence that multiple defects in cell-surface molecule interactions across species differences are responsible for diminished xenogeneic T cell responses. *Transplantation* 1992; **53**:203–209.

38. Gould DS, Auchincloss H. Direct and indirect recognition: The role of MHC antigens in graft rejection. *Immunol Today* 1999; **20**:77–82.

39. Auchincloss H, Jr, Sultan H. Antigen processing and presentation in transplantation. *Curr Opin Immunol* 1996; **8**:681–687.

40. Carbone FR, Bevan MJ. Class I-restricted processing and presentation of exogenous cell-associated antigen *in vivo*. *J Exp Med* 1990; **171**:377–387.

41. Albert ML, Sauter B, Bhardwaj N. Dendritic cells acquire antigen from apoptotic cells and induce class I-restricted CTLs. *Nature* 1998; **392**:86–89.

42. Rock KL, Gamble S, Rothstein L. Presentation of exogenous antigen with class I major histocompatibility complex molecules. *Science* 1990; **249**:918–921.

43. Loudovaris T, Mandel TE, Charlton B. CD4+ T cell mediated destruction of xenografts within cell-impermeable membranes in the absence of CD8+ T cells and B cells. *Transplantation* 1996; **61**:1678–1684.

44. Weber CJ, Zabinski S, Koschitzky T, Wicker L, Rajotte RD, D'Agati V, Peterson L, Norton J, Reemtsma K. The role of CD4+ helper T cells in the destruction of microencapsulated islet xenografts in nod mice. *Transplantation* 1990; **49**:396–404.

45. Auchincloss H, Jr, Sachs DH. Xenogeneic transplantation. *Ann Rev Immunol* 1998; **16**:433–470.
46. Parker W, Saadi S, Lin SS, Holzknecht ZE, Bustos M. Transplantation of discordant xenografts: A challenge revisited. *Immunol Today* 1996; **17**:373–378.
47. Loudovaris T, Charlton B, Mandel T. The role of T cells in the destruction of xenografts within cell-impermeable membranes. *Transplant Proc* 1992; **24**:2938.
48. Brauker J, Martinson LA, Young SK, Johnson RC. Local inflammatory response around diffusion chambers containing xenografts. Nonspecific destruction of tissues and decreased local vascularization. *Transplantation* 1996; **61**:1671–1677.
49. Buchser E, Goddard M, Heyd B, Joseph JM, Favre J, de Tribolet N, Lysaght M, Aebischer P. Immunoisolated xenogenic chromaffin cell therapy for chronic pain. Initial clinical experience. *Anesthesiology* 1996; **85**:1005–1012.
50. Decosterd I, Buchser E, Gilliard N, Saydoff J, Zurn AD, Aebischer P. Intrathecal implants of bovine chromaffin cells alleviate mechanical allodynia in a rat model of neuropathic pain. *Pain* 1998; **76**:159–166.
51. Sautter J, Tseng JL, Braguglia D, Aebischer P, Spenger C, Seiler RW, Widmer HR, Zurn AD. Implants of polymer-encapsulated genetically modified cells releasing glial cell line-derived neurotrophic factor improve survival, growth, and function of fetal dopaminergic grafts. *Exp Neurol* 1998; **149**:230–236.
52. Joseph JM, Goddard MB, Mills J, Padrun V, Zurn A, Zielinski B, Favre J, Gardaz JP, Mosimann F, Sagen J, Christenson L, Aebischer P. Transplantation of encapsulated bovine chromaffin cells in the sheep subarachnoid space: A preclinical study for the treatment of cancer pain. *Cell Transplant* 1994; **3**:355–364.
53. Emerich DF, Winn SR, Hantraye PM, Peschanski M, Chen EY, Chu Y, McDermott P, Baetge EE, Kordower JH. Protective effect of encapsulated cells producing neurotrophic factor CNTF in a monkey model of Huntington's disease. *Nature* 1997; **386**: 395–399.

54. Aebischer P, Goddard M, Signore AP, Timpson RL. Functional recovery in hemiparkinsonian primates transplanted with polymer-encapsulated PC12 cells. *Exp Neurol* 1994; **126**:151–158.
55. Loudovaris T, Charlton B, Hodgson RJ, Mandel TE. Destruction of xenografts but not allografts within cell impermeable membranes. *Transplant Proc* 1992; **24**:2291–2292.
56. Lacy PE, Hegre OD, Gerasimidi-Vazeou A, Gentile FT, Dionne KE. Maintenance of normoglycemia in diabetic mice by subcutaneous xenografts of encapsulated islets. *Science* 1991; **254**:1782–1784.
57. Peduto G, Rinsch C, Schneider BL, Rolland E, Aebischer P. Long-term host unresponsiveness to encapsulated xenogeneic myoblasts following transient immunosuppression. *Transplantation* 2000; **70**:78–85.
58. Rinsch C, Peduto G, Schneider BL, Aebischer P. Inducing host acceptance to encapsulated xenogeneic myoblasts. *Transplantation* 2001; **71**:345–351.
59. Levisetti MG, Padrid PA, Szot GL, Mittal N, Meehan SM, Wardrip CL, Gray GS, Bruce DS, Thistlethwaite JR, Jr, Bluestone JA. Immunosuppressive effects of human CTLA4Ig in a non-human primate model of allogeneic pancreatic islet transplantation. *J Immunol* 1997; **159**:5187–5191.
60. Kirk AD, Harlan DM, Armstrong NN, Davis TA, Dong Y, Gray GS, Hong X, Thomas D, Fechner JH, Jr, Knechtle SJ. CTLA4-Ig and anti-CD40 ligand prevent renal allograft rejection in primates. *Proc Natl Acad Sci, USA* 1997; **94**:8789–8794.
61. Lenschow DJ, Zeng Y, Thistlethwaite JR, Montag A, Brady W, Gibson MG, Linsley PS, Bluestone JA. Long-term survival of xenogeneic pancreatic islet grafts induced by CTLA4Ig. *Science* 1992; **257**:789–792.
62. Weber CJ, Hagler MK, Chryssochoos JT, Kapp JA, Korbutt GS, Rajotte RV, Linsley PS. CTLA4-Ig prolongs survival of microencapsulated neonatal porcine islet xenografts in diabetic NOD mice. *Cell Transplant* 1997; **6**:505–508.

63. Kenyon NS, Fernandez LA, Lehmann R, Masetti M, Ranuncoli A, Chatzipetrou M, Iaria G, Han D, Wagner JL, Ruiz P, Berho M, Inverardi L, Alejandro R, Mintz DH, Kirk AD, Harlan DM, Burkly LC, Ricordi C. Long-term survival and function of intrahepatic islet allografts in baboons treated with humanized anti-CD154. *Diabetes* 1999; **48**:1473–1481.

64. Kirk AD, Burkly LC, Batty DS, Baumgartner RE, Berning JD, Buchanan K, Fechner JH, Jr, Germond RL, Kampen RL, Patterson NB, Swanson SJ, Tadaki DK, TenHoor CN, White L, Knechtle SJ, Harlan DM. Treatment with humanized monoclonal antibody against CD154 prevents acute renal allograft rejection in nonhuman primates. *Nat Med* 1999; **5**:686–693.

65. Elwood ET, Larsen CP, Cho HR, Corbascio M, Ritchie SC, Alexander DZ, Tucker-Burden C, Linsley PS, Aruffo A, Hollenbaugh D, Winn KJ, Pearson TC. Prolonged acceptance of concordant and discordant xenografts with combined CD40 and CD28 pathway blockade. *Transplantation* 1998; **65**:1422–1428.

66. Gordon EJ, Markees TG, Phillips NE, Noelle RJ, Shultz LD, Mordes JP, Rossini AA, Greiner DL. Prolonged survival of rat islet and skin xenografts in mice treated with donor splenocytes and anti-CD154 monoclonal antibody. *Diabetes* 1998; **47**: 1199–1206.

67. Abrams JR, Lebwohl MG, Guzzo CA, Jegasothy BV, Goldfarb MT, Goffe BS, Menter A, Lowe NJ, Krueger G, Brown MJ, Weiner RS, Birkhofer MJ, Warner GL, Berry KK, Linsley PS, Krueger JG, Ochs HD, Kelley SL, Kang S. CTLA4Ig-mediated blockade of T-cell costimulation in patients with psoriasis vulgaris. *J Clin Invest* 1999; **103**:1243–1252.

68. Lew AM, Brady JL, Silva A, Coligan JE, Georgiou HM. Secretion of CTLA4Ig by an SV40 T antigen-transformed islet cell line inhibits graft rejection against the neoantigen. *Transplantation* 1996; **62**:83–89.

69. Chahine AA, Yu M, McKernan MM, Stoeckert C, Lau HT. Immunomodulation of pancreatic islet allografts in mice

with CTLA4Ig secreting muscle cells. *Transplantation* 1995; **59**: 1313–1318.

70. Feng S, Quickel RR, Hollister-Lock J, McLeod M, Bonner-Weir S, Mulligan RC, Weir GC. Prolonged xenograft survival of islets infected with small doses of adenovirus expressing CTLA4Ig. *Transplantation* 1999; **67**:1607–1613.

71. Gainer AL, Suarez-Pinzon WL, Min WP, Swiston JR, Hancock-Friesen C, Korbutt GS, Rajotte RV, Warnock GL, Elliott JF. Improved survival of biolistically transfected mouse islet allografts expressing CTLA4-Ig or soluble Fas ligand. *Transplantation* 1998; **66**:194–199.

Chapter 12

The Immunologic Barriers of Xenotransplantation and Application of Genetic Engineering

Joo Ho Tai & Jeffrey L Platt

Recent progress in overcoming the immunologic hurdles to transplanting organs and tissues between species has raised the prospect that interspecies transplantation or xenotransplantation may find application for the treatment of human disease. Enthusiasm for xenotransplantation stems from at least three factors. First, there is a severe and urgent shortage of human organs and tissues available for transplantation. This shortage limits the clinical application of organ transplantation to as little as 5–15% of the number of transplants that would be performed, were the supply unlimited.[1] Second, it is possible that animal tissues and organs might be less susceptible to the recurrence of disease compared to allotransplants.[2] Third, a xenograft might be used as a vehicle for introducing a novel gene or a biochemical process that could be of therapeutic value for the transplant recipient.

If interest in xenotransplantation is substantial, the hurdles are equally so. For the past three decades, the first and pre-eminent obstacle to transplanting organs and tissues between species has been the immune reaction of the host against the graft. Recent success in dealing with facets of this immune reaction has prompted investigators to begin focusing on other questions. A summary of some of the therapeutic approaches can be found in Chap. 10 by Teranishi *et al*. A second, and still theoretical, hurdle stems from the possibility that beyond the immune barrier, there might be physiologic limitations to the function or survival of a xenograft and the possibility that a xenotransplant might engender medical complications for the xenogeneic host. A third, and recently the most visible, obstacle is the possibility that a xenograft might act as a vector for transferring infectious agent from the donor to the host and that from the host, such an agents might spread to other members of society. Concerns regarding infection stemming from xenotransplantation are discussed in Chap. 9, by George and Lechler. This communication will consider the current state of efforts to overcome the various hurdles to xenotransplantation and will evaluate how genetic engineering in particular might be applied to this end.

THE PIG AS A SOURCE OF TISSUES AND ORGANS FOR CLINICAL XENOTRANSPLANTATION

Although it might be intuitive that the best source of xenogeneic tissues for clinical transplantation is the non-human primate, it is the pig that is the focus of most efforts in this field. The reasons for favouring the pig as a xenotransplant source include the availability of pigs in large numbers, the ease with which the pig can be bred, the limited risk of zoonotic disease engendered by the use of pigs, and the possibility of introducing new genes into the germline of the pig.

Genetic engineering of pigs using transgenic techniques has certain advantages over conventional gene therapy (Table 1). First, the genetic material is injected directly into the porcine germ cell, obviating the need for a vehicle, which may vary in reliability of gene delivery. Second, the genetic material introduced into the germline can be expressed constitutively in all cells, especially in

Table 1. Genetic engineering in xenotransplantation: Conventional gene therapy versus transgenic therapy versus cloning.

	Conventional Gene Therapy	Conventional Transgenic Techniques	Cloning
Delivery	Vector or vehicle required	Injection of genetic material directly into pro-nuclei of fertilized egg	Transfection of cultured somatic cells
Expression	Dependent on ability of each cell to take up genetic material. Requires treatment for every transplant or recipient. May require repeated treatment	Genetic material introduced into the germline, leading to expression in a line of animals. One manipulation	Genetic material introduced into the germline, leading to expression in a line of animals. One manipulation
Immunogenicity	Delivery vehicle or transgene may be immunogenic	The transgene may elicit immune response	The transgene may elicit immune response
Target of genetic manipulation	The recipient and the graft may be transduced	Genetic manipulation of the donor only	Genetic manipulation of the donor only
Genetic manipulation	Gene addition. Dominant negative	Gene addition. Dominant negative	Gene addition. Dominant negative. Gene knock out

stem cells, and passed on to subsequent generations. Third, the genetic material inserted in the porcine genome does not engender an immune response, in contrast to some vectors used in conventional gene therapy. Fourth, with the use of transgenic techniques, only the donor is manipulated; in conventional gene therapy, both the donor and the recipient may be affected.

The possibility of cloning pigs[3-5] through nuclear transfer offers yet additional opportunities for adding and "knocking out" genes. Besides the advantage of gene knock outs, nuclear transfer can be done with cultured somatic cells obviating the need for embryonic stem cells.

THE BIOLOGIC AND IMMUNOLOGIC RESPONSES TO XENOTRANSPLANTATION

The biologic and immunologic responses to xenotransplantation are dictated in part by the way in which the graft receives its vascular supply (Fig. 1). Isolated cells such as hepatocytes and "free" tissues such as pancreatic islets and skin derive their vascular supply through the in-growth of host blood vessels. The process of neovascularization, as such, might be impaired in a xenograft by incompatibility of donor growth factors with the host microvasculature. To the extent that neovascularization or graft function depends upon hormones and cytokines of host origin, the function of the xenograft might also be impaired. As the host microcirculation is established, however, a xenogeneic tissue may be relatively protected from attack by host immune elements. Whole-organ grafts provide their own microcirculation and growth factors, and, as a result, incompatibility between the donor and the recipient is less likely to have an impact on cellular function. On the other hand, because the circulation is of donor origin, the immune, inflammatory and coagulation systems of the recipient can act directly on donor cells, sometimes with dramatic and devastating consequences. Bearing in mind the distinction between free tissue

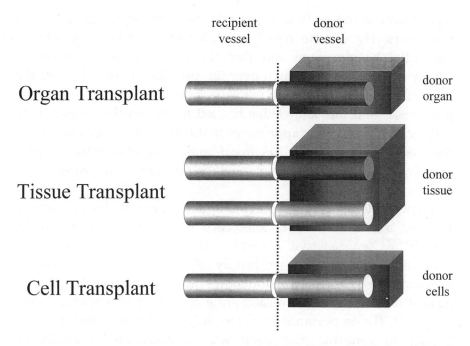

Fig. 1. Mechanisms of xenograft vascularization. Organ xenografts receive recipient blood exclusively through the donor blood vessels (bottom). Free tissue xenografts (e.g. pancreatic islets and skin) are vascularized partly by the in-growth of recipient blood vessels and partly by spontaneous anastomosis of donor and recipient capillaries (middle). Cellular xenografts (e.g. hepatocytes and bone marrow cells) are vascularized by the in-growth of recipient blood vessels (top).

grafts and organ grafts, this review will consider for the most part organ grafts because of the overwhelming importance of the organ donor shortage.

HYPERACUTE REJECTION

An organ transplanted from a pig into a primate is subject to hyperacute rejection. Hyperacute rejection begins immediately upon reperfusion of the graft and destroys the graft within minutes to a few hours. Hyperacute rejection is characterized histologically by interstitial haemorrhage and thrombosis, the thrombi consisting mainly of platelets.[6] Research over the past decade has clarified the molecular basis for the hyperacute rejection of pig organs by primates,[7,8] and this knowledge has led to the development of new and incisive therapeutic approaches to this problem. Once considered the most daunting hurdle to clinical application of xenotransplantation, hyperacute rejection can now be prevented in nearly every case.

Hyperacute rejection of porcine organ xenografts by primates is initiated by the binding of xenoreactive natural antibodies to the graft.[6,9–11] Xenoreactive natural antibodies are present in the circulation without a known history of sensitization.[12] Contrary to expectations, xenoreactive antibodies are predominantly directed against only one antigen, a saccharide consisting of terminal Galα1-3Gal.[13–16] The importance of Galα1-3Gal as the primary antigenic barrier to xenotransplantation was demonstrated recently by experiments in which anti-Galα1-3Gal antibodies were specifically depleted from baboons using immunoaffinity columns before transplantation of pig organs.[17] Antibody binding to the newly transplanted organs was largely curtailed, and hyperacute rejection did not occur.

Although the identification of the relevant antigen for pig-to-primate xenotransplantation allows specific depletion of the offending antibodies, more enduring and less intrusive forms of therapy would be preferred. One approach to overcoming the antibody-antigen reaction is to develop lines of pigs with low levels of antigen expression.[18] The most obvious approach to developing such lines of xenograft donors would be to genetically target or "knock out" the enzyme α1,3-galactosyltransferase, which catalyzes the synthesis of Galα1-3Gal. Embryonic stem cells were used to knock this gene out in mice[19] and it is hoped that nuclear transfer techniques with porcine somatic cells will accomplish the same result in the future.

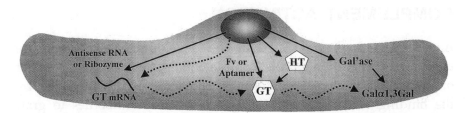

Fig. 2. Genetic approaches to decreasing expression of Galα1-3Gal. **Four approaches** are shown for reducing the level of Galα1-3Gal, the predominant antigen recognized by xenoreactive natural antibodies in porcine xenografts. 1) Antisense RNA, ribozyme, or other molecules disrupt or inhibit mRNA synthesis for α1,3-galactosyltransferase (GT), the enzyme that catalyzes the synthesis of Galα1-3Gal. 2) Fv or aptomer expressed in the cell binds to and inhibits the function of α1,3-galactosyltransferase. 3) A glycosyltransferase, such as α1,2-fucosyltransferase (HT), might generate a non-antigenic sugar and in doing so, might compete with α1,3-galactosyltransferase (GT). 4) A glycosidase, α-galactosidase (Gal'ase), might cleave the antigenic sugar. (Reprinted and adapted from *Nature* 1998: 392(Suppl) 11–17, with permission.)

Various other approaches have been proposed to alter the level of antigen expression through genetic means. Some of these approaches are summarized and illustrated in Fig. 2. Among the various ways in which antigen expression might be modulated, one approach, used by Sandrin and associates[20] and Sharma and colleagues,[21] involves the expression of a glycosyltransferase, which would catalyze the formation of cell-surface carbohydrates with a non-antigenic sugar instead of Galα1-3Gal. Transgenic pigs expressing the H-transferase express H antigen at the terminus of some sugar chains instead of Galα1-3Gal.[21] Another genetic approach to modify expression of Galα1-3Gal was proposed by Osman *et al.*[22] Expression of α-galactosidase which cleaves α-galactosyl residues[15] in conjunction with other galactosyltransferases significantly reduces expression of Galα1-3Gal.[22] The main limitation of genetic approaches, other than gene knock-out, is that residual Galα1-3Gal may be sufficient to allow rejection reactions to occur.[23]

COMPLEMENT ACTIVATION

A second and essential step in the development of hyperacute rejection is activation of the complement system of the recipient on donor blood vessels.[10] Complement activation is triggered by the binding of complement-fixing xenoreactive antibodies to graft endothelium and, to a smaller extent, perhaps, by reperfusion injury. Regardless of the mechanism leading to complement activation, a xenograft is extraordinarily sensitive to complement-mediated injury because of multiple defects in the regulation of complement (Fig. 3).[24–26] Under normal circumstances, the complement cascade is regulated or inhibited by various proteins in the plasma and on

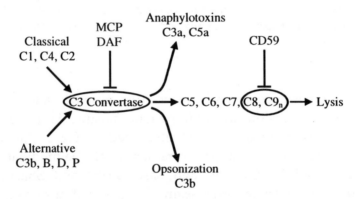

Fig. 3. Regulation of the complement system. The complement cascade, which can be activated via the classical or alternative pathway, is regulated under normal circumstances by various proteins in the plasma and on the cell surface. Three of the cell-surface complement regulatory proteins are shown here. Decay-accelerating factor (DAF) and membrane cofactor protein (MCP) regulates complement activation by dissociating or promoting the degradation of C3 convertase. CD59, also known as protectin, prevents the functions of terminal complement complexes by inhibiting C8 and C9. An organ graft transplanted into a xenogeneic recipient is especially sensitive to complement-mediated injury because decay-accelerating factor, membrane cofactor protein, and CD59 expressed on the xenograft endothelium cannot effectively regulate the complement system of the recipient.

the surface of cells. These proteins protect normal cells from suffering inadvertent injury during the activation of complement (some of the reactions of the complement cascade might occur on cells residing in the vicinity of infectious organisms). The proteins that regulate the complement cascade function in a species-restricted fashion; that is, complement regulatory proteins inhibit homologous complement far more effectively than heterologous complement.[25,27] Accordingly, the complement regulatory proteins expressed in a xenograft are ineffective at controlling the complement cascade of the recipient, and the graft is subject to severe complement-mediated injury.[24]

To address this problem, we and others have proposed that lines of animals might be developed that are transgenic for human complement regulatory proteins and that are able to control activation of complement in the xenograft.[25,28,29] We, for example, have focused on developing animals transgenic for human decay-accelerating factor, which regulates complement at the level of C3; and CD59, which regulates complement at the level of C8 and C9 (Fig. 3).[29] Others have focused on developing pigs transgenic for decay-accelerating factor.[30] Recent studies have demonstrated that the expression of even low levels of human decay-accelerating factor and CD59 in porcine-to-primate xenografts is sufficient to allow a xenograft to avoid hyperacute rejection.[31,32] These results, and the dramatic prolongation of xenograft survival achieved by expressing higher levels of human decay-accelerating factor in the pig,[33] underscore the importance of complement regulation as a determinant of xenograft outcomes.

ACUTE VASCULAR REJECTION

If hyperacute rejection of a xenograft is averted, a xenograft is subjected to the development of acute vascular rejection, so named because of its resemblance to acute vascular rejection of allografts.[34,35] Acute vascular rejection (sometimes called delayed xenograft rejection) may begin within 24 hours of reperfusion and leads to graft destruction over the following days and weeks.[34,36,37] Although the factors important in the pathogenesis of acute vascular rejection are incompletely understood, there is growing evidence that acute vascular rejection is triggered at least in part by the binding of xenoreactive antibodies to the graft. The importance of xenoreactive antibodies in triggering acute vascular rejection is suggested by three lines of evidence:

 i) Antidonor antibodies are present in the circulation of recipients whose grafts are subjected to acute vascular rejection.[10,34,38,39]

 ii) Depletion of antidonor antibodies delays or prevents acute vascular rejection.[40]

iii) Administration of antidonor antibodies leads to the development of acute vascular rejection.[41] Recent studies in our laboratory suggest that the antibodies that provoke acute vascular rejection include a large fraction of antibodies directed against Galα1-3Gal.[42,43] Regardless of which elements of the immune system trigger acute vascular rejection, it is commonly thought that this type of rejection and especially the intravascular coagulation characteristically associated with it are caused by the activation of endothelial cells in the transplant.[34,44,45] Activated endothelial cells express procoagulant molecules such as tissue factor and proinflammatory molecules such as E-selectin and cytokines.[25] The pathogenesis of acute vascular rejection is summarized in Fig. 4.

Although various therapeutic manipulations have proven successful in preventing hyperacute rejection, acute vascular rejection poses a more difficult problem, in part, because therapies are needed on an ongoing basis. For this reason, genetic modification of the donor may prove more important for dealing with acute vascular rejection than with hyperacute rejection. The various possible approaches for combating acute vascular rejection are listed in

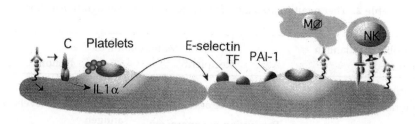

Fig. 4. Pathogenesis of acute vascular rejection. Activation of endothelium by xenoreactive antibodies (Ab), complement (C), platelets and perhaps by inflammatory cells (natural killer [NK] cells and macrophages) leads to the expression of new pathophysiologic properties. These new properties, such as the synthesis of tissue factor and plasminogen activator inhibitor type 1, promote coagulation; the synthesis of E-selectin and cytokines promotes inflammation. These changes in turn cause thrombosis, ischemia, and endothelial injury, the hallmarks of acute vascular rejection. (Reprinted and adapted from *Nature* 1998: 392(Suppl) 11–17, with permission.)

Table 2. Among these approaches, the reduction of Galα1-3Gal in xenotransplant donors may be an important part of the overall strategy, to the extent that Galα1-3Gal proves to be an important antigenic target in acute vascular xenograft rejection. Preliminary studies from our laboratory suggest that the level of antibody binding needed to initiate acute vascular rejection is considerably lower than the level needed to initiate hyperacute rejection.[23] Accordingly, the antigen expression would have to be reduced very significantly to achieve therapeutic benefit for acute vascular rejection. To decrease antigen expression to a sufficient extent would thus probably require knocking out α1,3-galactosyltransferase as discussed above. In addition to lowering antigen expression, it is likely that expression of human complement regulatory proteins will be helpful in preventing acute vascular rejection. Preliminary studies from our laboratory suggest that interfering with the antigen-antibody reaction and controlling the complement cascade may be sufficient to prevent acute vascular rejection for at least some period of time.[40] These

Table 2. Therapeutic strategies for acute vascular xenograft rejection.

Possible Mechanism Targeted	Manipulation of:	
	Recipient	Donor
Antibody-antigen interaction	Specific depletion of xenoreactive antibodies Prevention of xenoreactive antibody synthesis (e.g. cyclophosphamide, leflunomide)	Breeding pigs with inherently low levels of antigen Generating transgenic pigs with low levels of antigen
Complement activation	Systemic anti-complement therapy (e.g. CVF, sCR1, gamma globulin)	Generation of donor pigs transgenic for human complement regulatory proteins
Endothelial cell activation	Administration of anti-inflammatory agents	Inhibition of NFkB function Introduction of protective genes
Molecular incompatibilities	Administration of inhibitors (e.g. inhibitors of complement or coagulation)	Introduction of compatible molecules

goals were accomplished by using animals transgenic for human decay-accelerating factor and CD59 as a source of organs and baboons depleted of immunoglobulin as recipients. Cozzi and associates[33] achieved prolonged survival of xenografts, presumably preventing acute vascular rejection, by using transgenic pigs expressing high levels of decay-accelerating factor and cynomolgus monkeys treated with very high doses of cyclophosphamide. The immunosuppression perhaps prevented the synthesis of antidonor antibodies.

Work in rodents points to the potential involvement of natural killer (NK) cells and macrophages in mediating acute vascular rejection. Although it is conceivable that our own methods and the methods used by White and Wallwork[28] also inhibited the activity of NK cells and macrophages directly, the ability of immunoglobulin manipulation to prevent acute vascular rejection suggests that the involvement of NK cells and macrophages might be less important than *in vitro* studies and studies in rodents have suggested.[45] On the other hand, NK cells might exacerbate the injury triggered by xenoreactive antibodies, as human NK cells have been shown to activate porcine endothelial cells *in vitro*.[46-48]

ACCOMMODATION

Fortunately, the presence of antidonor antibodies in the circulation of a graft recipient does not inevitably trigger acute vascular rejection. Some years ago, we discovered that if antidonor antibodies were temporarily depleted from a recipient, an organ transplant could be established so that rejection did not ensue when the antidonor antibodies were returned to the circulation.[49] This phenomenon was later referred to as "accommodation".[25] Accommodation may reflect a change in the antibodies, in the antigen, or in the susceptibility of the organ to rejection. If accommodation can be established, it may be especially important in xenotransplantation because it would obviate the need for ongoing interventions to inhibit antibody binding to the graft. One potential approach to accommodation may be the use of genetic engineering to reduce the susceptibility of an organ transplant to acute vascular rejection and the endothelial-cell activation associated with it.[45] Unfortunately, successful intervention at the level of such effector mechanisms has yet to be achieved. We have focused instead on disrupting the antibody-antigen interaction, as this approach has brought about accommodation in human subjects.[44,49]

CELLULAR MEDIATED IMMUNE RESPONSES

Organ transplants and cellular and free tissue transplants are subject to cellular rejection. In allotransplantation, cellular rejection is controlled by conventional immunosuppressive therapy, but there is concern that, for several reasons, cellular rejection may be especially severe in xenotransplants. First, the great variety of antigenic proteins in a xenograft may lead to recruitment of a diverse set of "xenoreactive" T cells. Second, the binding of xenoreactive antibodies and activation of the complement system may lead to amplification of elicited immune responses.[50] For example, activation of complement in a graft may cause activation of antigen-presenting cells, in turn stimulating T-cell responses. Still another factor that might amplify the elicited immune response to a xenotransplant involves "immunoregulation", which ordinarily would circumscribe cellular immune responses, but may fail or be deficient across species. Such failure could reflect limitations in the recognition of xenogeneic cells or incompatibility of relevant growth factors, as but two examples.

Induction of immunologic tolerance has been an erstwhile goal of transplant surgeons and physicians. Especially in the case of xenotransplantation, if the current immunosuppressive regimens are not sufficient, induction of immunologic tolerance may be needed. At least three approaches are being pursued: i) the generation of mixed haematopoietic chimerism, ii) the establishment of micro-chimerism by various means, and iii) thymic transplantation.[51-53] The development of mixed haematopoietic chimerism through the introduction of donor bone marrow[54] has worked very well across rodent species,[55,56] although success may be limited by xenoreactive antibodies and the engraftment impaired by incompatibility of host growth factors or micro-environment.[57] Fortunately, there is evidence that these problems can be overcome.[52] Various approaches to peripheral tolerance, such as the blockade of co-stimulation by administration of a fusion protein consisting of a soluble form of CTLA-4 T cell molecule and immunoglobulin (CTLA-4-Ig), are being pursued.

Still another factor in the cellular response to a xenotransplant involves the action of NK cells. Natural killer cell functions can be amplified by cell-surface receptors that recognize Galα1-3Gal.[58] Natural killer cell functions are down-regulated by receptors that recognize homologous major histocompatibility complex (MHC) class I.[59,60] Human NK cells may be especially active against xenogeneic cells because of stimulation on the one hand and failure of down-regulation on the other. The possible involvement of NK cells in xenograft rejection might be addressed by generation of transgenic pigs expressing on the cell surface MHC-like molecules that will more effectively recognize corresponding receptors on NK cells and that will down-regulate the function of NK cells.

How a xenogeneic donor could be modified genetically to enhance the development of tolerance or to limit elicited immune responses is still uncertain. Clearly, efforts to control the natural immune barriers to xenotransplantation may contribute to limiting the elicited immune response. To the extent that recipient T cells recognize donor cells directly, that is, the T-cell receptors of the recipient recognizing native MHC antigens on donor cells, a xenogeneic donor might be engineered in such a way as to reduce co-recognition (through CD4 and CD8) or co-stimulation (through CD28 or other T-cell surface molecules) or to express inhibitory molecules such as CD59 or Fas ligand. These approaches and the expression of inhibitory molecules, which are being considered as approaches to gene therapy in allotransplantation, may well prove more effective in xenotransplantation because inhibitory genes can be introduced as transgenes and thereby expressed in all relevant cells in the graft. Another useful and perhaps necessary approach will involve genetic modifications to allow the survival and functioning of donor bone-marrow cells.

PHYSIOLOGIC HURDLES
TO XENOTRANSPLANTATION

Progress in addressing some of the immunological obstacles to xenotransplantation has brought into focus the question of the extent to which a xenotransplant would function optimally in a foreign host. We have found that the porcine kidney and the porcine lung can replace the most important functions of the primate kidney and primate lung, respectively.[61,62] Subtle defects in physiology across species may nevertheless exist. Organs such as the liver, which secrete a variety of proteins and which depend on complex enzymatic cascades, may prove incompatible with a primate host. Accordingly, one important application of genetic engineering in xenotransplantation may be the amplification or modulation of xenograft function to allow for more complete establishment of physiologic function or to overcome critical defects. For example, recent studies by Akhter and associates[63] and Kypson and co-authors[64] aimed at improving the function of cardiac allografts by manipulation of β-adrenergic signalling, and this technique might be adapted to the xenotransplant to improve cardiac function. On the other hand, most cellular processes and biochemical cascades are intrinsically regulated to meet the overall physiologic needs of the whole individual. The key question, then, is which of the many potential defects actually need to be repaired.

Another potential hurdle to the clinical application of xenotransplantation is the possibility that the xenograft may disturb normal metabolic and physiologic functions in the recipient. For example, Lawson and co-workers[65,66] have shown that porcine thrombomodulin fails to interact adequately with human thrombin and protein C to generate activated protein C. This defect could lead to a prothrombotic diathesis because of failure of generation of activated protein C. Of even greater concern is the possibility that the transplantation of an organ, such as the liver, could add prothrombotic or pro-inflammatory products into the blood of the recipient. Although perhaps a great many physiologic defects can be detected at the molecular level, the critical question will be which of these defects is important at the whole-organ level or with respect to the well-being of the recipient, and which must be repaired by pharmaceutical or genetic means.

ZOONOSIS

The increasing success of experimental xenotransplants and therapeutic trials bring to the fore the question of zoonosis, that is, infectious disease introduced from the graft into the recipient. The transfer of infectious agents from the graft to the recipient is a well-known complication of allotransplantation. To the extent that infection of the recipient in this way increases the risks of transplantation, that risk can generally be estimated and a decision made based on the risk versus the potential benefits conferred by the transplant. The concern about zoonosis in xenotransplantation is not so much the risk to the recipient of the transplant, but the risk that an infectious agent will be transferred from the recipient to the population at large. Fortunately, all of the microbial agents known to infect the pig can be detected by screening and potentially eliminated from a population of xenotransplant donors. There is concern, however, that the pig may harbor endogenous retroviruses, which are inherited with genomic DNA and which might become activated and transferred to the cells of the recipient. For example, Patience and co-authors[67] recently reported that a C-type retrovirus endogenous to the pig could be activated in pig cells, leading to the release of particles that can infect human cell lines. Whether this virus or other endogenous viruses can actually infect across species and whether such infection would lead to disease are unknown, but remain a subject of current epidemiologic investigation. If cross-species infection does prove to be an important issue, genetic therapies might also be used to address this problem. The simplest genetic therapy would involve breeding out the organism, but this approach might fail if the organism were widespread or integrated at multiple loci. Some genetic therapies have been developed to potentially control human immunodeficiency viruses.[68] Although these therapies have generally failed because it has been difficult or impossible to gain expression of the transferred genes in stem cells and at levels sufficient to deal with high viral loads, the application of such therapies might be much easier in xenotransplantation because the therapeutic genes could be delivered through the germline. Ultimately, if elimination of endogenous retroviruses were necessary, it could potentially be accomplished by gene targeting and cloning as discussed above.

A SCENARIO FOR THE CLINICAL APPLICATION OF XENOTRANSPLANTATION

Successful application of xenotransplantation in the clinical arena requires insights not only into immunology, but also physiology and infectious disease, all of which have been discussed briefly here in the context of genetic therapy. In recent years, important advances have been made in elucidating the immunologic hurdles of pig-to-primate transplantation. Although this scientific progress is important and exciting, we believe that xenotransplantation will enter the clinical arena through a step-by-step process. Free tissue xenografting and extracorporeal perfusion of xenogeneic organs in limited clinical trials are already in progress,[69-71] and preliminary evidence is encouraging as porcine free tissue xenografts appear to endure in a human recipient.[71] Temporary or "bridge" organ transplantation will probably follow. Bridge transplants will not address the problem of the shortage of human organs, but incisive analysis of the outcomes of these transplants will provide important information about the remaining immunologic hurdles and the potential physiologic and infectious considerations. With this information, further therapies including genetic engineering may allow the use of porcine organs as permanent replacements. Even then, one can envision ongoing efforts to apply genetic therapies that will optimize graft function and limit the complications of transplantation.

ACKNOWLEDGEMENTS

Supported by grants from the Heart, Lung, and Blood Institute of the National Institutes of Health.

REFERENCES

1. Evans RW. Coming to terms with reality: Why xeno-transplantation is a necessity. In *Xenotransplantation*, ed. JL Platt, ASM Press, Washington D. C., 2001, pp. 29–51.
2. Starzl TE, Fung J, Tzakis A, *et al.* Baboon-to-human liver transplantation. *Lancet* 1993; **341**:65–71.
3. Onishi A, Iwamoto M, Akita T, *et al.* Pig cloning by microinjection of fetal fibroblast nuclei. *Science* 2000; **289**:1188–1190.
4. Polejaeva IA, Chen S, Vaught TD, *et al.* Cloned piglets produced by nuclear transfer from adult somatic cells. *Nature* 2000; **407**: 86–90.
5. Betthauser J, Forsberg E, Augenstein M, *et al.* Production of cloned pigs from *in vitro* systems. *Nature Biotechnology* 2000; **18**:1055–1059.
6. Platt JL. Hyperacute xenograft rejection. In *Medical Intelligence Unit*, ed. RG Austin, Landes, 1995.
7. Platt JL. Immunology of xenotransplantation. In *Samter's Immunologic Diseases*. 6th ed, eds. K. F. Austen, M. M. Frank, J. P. Atkinson, H. Cantor, Lippincott Williams & Wilkins, Philadelphia, 2001, pp. 1132–1146.
8. Platt JL. Hyperacute rejection: Fact or fancy. *Transplantation* 2000; **69**:1034–1035.

9. Cooper DKC, Human PA, Lexer G, *et al*. Effects of cyclosporine and antibody adsorption on pig cardiac xenograft survival in the baboon. *J Heart Transplantation* 1988; 7:238–246.

10. Platt JL, Fischel RJ, Matas AJ, Reif SA, Bolman RM, Bach FH. Immunopathology of hyperacute xenograft rejection in a swine-to-primate model. *Transplantation* 1991; 52:214–220.

11. Platt JL, Turman MA, Noreen HJ, Fischel RJ, Bolman RM, Bach FH. An ELISA assay for xenoreactive natural antibodies. *Transplantation* 1990; 49:1000–1001.

12. Calne RY. Organ transplantation between widely disparate species. *Transplantation Proceedings* 1970; 2:550–556.

13. Sandrin MS, Vaughan HA, Dabkowski PL, McKenzie IFC. Anti-pig IgM antibodies in human serum react predominantly with Galα(1,3)Gal epitopes. *Proc Nat Acad Sci, USA* 1993; 90:11391–11395.

14. Good AH, Cooper DKC, Malcolm AJ, *et al*. Identification of carbohydrate structures that bind human antiporcine antibodies: Implications for discordant xenografting in humans. *Transplantation Proceedings* 1992; 24:559–562.

15. Collins BH, Parker W, Platt JL. Characterization of porcine endothelial cell determinants recognized by human natural antibodies. *Xenotransplantation* 1994; 1:36–46.

16. Parker W, Bruno D, Holzknecht ZE, Platt JL. Characterization and affinity isolation of xenoreactive human natural antibodies. *Journal of Immunology* 1994; 153:3791–3803.

17. Lin SS, Kooyman DL, Daniels LJ, *et al*. The role of natural anti-Gala1-3Gal antibodies in hyperacute rejection of pig-to-baboon cardiac xenotransplants. *Transplant Immunology* 1997; 5:212–218.

18. Alvarado CG, Cotterell AH, McCurry KR, *et al*. Variation in the level of xenoantigen expression in porcine organs. *Transplantation* 1995; 59:1589–1596.

19. Thall AD, Malý P, Lowe JB. Oocyte Galα1,3Gal epitopes implicated in sperm adhesion to the zona pellucida glycoprotein ZP3 are not required for fertilization in the mouse. *The J Biological Chemistry* 1995; 270:21437–21440.

20. Sandrin MS, Fodor WL, Mouhtouris E, *et al.* Enzymatic remodelling of the carbohydrate surface of a xenogeneic cell substantially reduces human antibody binding and complement-mediated cytolysis. *Nature Medicine* 1995; **1**:1261–1267.

21. Sharma A, Okabe JF, Birch P, Platt JL, Logan JS. Reduction in the level of Gal (α1,3) Gal in transgenic mice and pigs by the expression of an α(1,2) fucosyltransferase. *Proc Nat Acad Sci, USA* 1996; **93**:7190–7195.

22. Osman N, McKenzie IF, Ostenried K, Ioannou YA, Desnick RJ, Sandrin MS. Combined transgenic expression of alpha-galactosidase and alpha1,2-fucosyltransferase leads to optimal reduction in the major xenoepitope Galalpha(1,3)Gal. *Proc Nat Acad Sci, USA* 1997; **94**:14677–14682.

23. Parker W, Lin SS, Platt JL. Antigen expression in xeno-transplantation: How low must it go? *Transplantation* 2001; **71**: 313–319.

24. Miyagawa S, Hirose H, Shirakura R, *et al.* The mechanism of discordant xenograft rejection. *Transplantation* 1988; **46**:825–830.

25. Platt JL, Vercellotti GM, Dalmasso AP, *et al.* Transplantation of discordant xenografts: A review of progress. *Immunology Today* 1990; **11**:450–456.

26. Dalmasso AP. The complement system in xenotransplantation. *Immunopharmacology* 1992; **24**:149–160.

27. Dalmasso AP, Vercellotti GM, Platt JL, Bach FH. Inhibition of complement-mediated endothelial cell cytotoxicity by decay accelerating factor: Potential for prevention of xenograft hyperacute rejection. *Transplantation* 1991; **52**:530–533.

28. White D, Wallwork J. Xenografting: Probability, possibility, or pipe dream? *Lancet* 1993; **342**:879–880.

29. Platt JL, Logan JS. Use of transgenic animals in xeno-transplantation. *Transplantation Reviews* 1996; **10**:69–77.

30. Cozzi E, White DJG. The generation of transgenic pigs as potential organ donors for humans. *Nature Medicine* 1995; **1**: 964–966.

31. McCurry KR, Kooyman DL, Alvarado CG, *et al*. Human complement regulatory proteins protect swine-to-primate cardiac xenografts from humoral injury. *Nature Medicine* 1995; **1**: 423–427.

32. Byrne GW, McCurry KR, Martin MJ, McClellan SM, Platt JL, Logan JS. Transgenic pigs expressing human CD59 and decay-accelerating factor produce an intrinsic barrier to complement-mediated damage. *Transplantation* 1997; **63**:149–155.

33. Cozzi E, Yannoutsos N, Langford GA, Pino-Chavez G, Wallwork J, White DJG. Effect of transgenic expression of human decay-accelerating factor on the inhibition of hyperacute rejection of pig organs. In *Xenotransplantation: The Transplantation of Organs and Tissues Between Species*. 2nd ed, eds. DKC Cooper, E Kemp, JL Platt, DJG White, Springer, Berlin, 1997, pp. 665–682.

34. Leventhal JR, Matas AJ, Sun LH, *et al*. The immunopathology of cardiac xenograft rejection in the guinea pig-to-rat model. *Transplantation* 1993; **56**:1–8.

35. Porter KA. Renal transplantation. In *Pathology of the Kidney*. 4 ed, ed. RH Heptinstall, Little Brown and Company, Boston, 1992, pp. 1799–1933.

36. Magee JC, Collins BH, Harland RC, *et al*. Immunoglobulin prevents complement mediated hyperacute rejection in swine-to-primate xenotransplantation. *J Clinical Investigation* 1995; **96**:2404–2412.

37. Lin SS, Platt JL. Immunologic barriers to xenotransplantation. *J Heart and Lung Transplantation* 1996; **15**:547–555.

38. McPaul JJ, Stastny P, Freeman RB. Specificities of antibodies eluted from human cadaveric renal allografts. *J Clinical Investigation* 1981; **67**:1405–1414.

39. Paul LC, Claas FHJ, van Es LA, Kalff MW, de Graeff J. Accelerated rejection of a renal allograft associated with pretransplantation antibodies directed against donor antigens on endothelium and monocytes. *New England Journal of Medicine* 1979; **300**:1258–1260.

40. Lin SS, Weidner BC, Byrne GW, *et al.* The role of antibodies in acute vascular rejection of pig-to-baboon cardiac transplants. *J Clinical Investigation* 1998; **101**:1745–1756.
41. Perper RJ, Najarian JS. Experimental renal heterotransplantation. III. Passive transfer of transplantation immunity. *Transplantation* 1967; **5**:514–533.
42. Cotterell AH, Collins BH, Parker W, Harland RC, Platt JL. The humoral immune response in humans following cross-perfusion of porcine organs. *Transplantation* 1995; **60**:861–868.
43. McCurry KR, Parker W, Cotterell AH, *et al.* Humoral responses in pig-to-baboon cardiac transplantation: Implications for the pathogenesis and treatment of acute vascular rejection and for accommodation. *Human Immunology* 1997; **58**:91–105.
44. Parker W, Saadi S, Lin SS, Holzknecht ZE, Bustos M, Platt JL. Transplantation of discordant xenografts: A challenge revisited. *Immunology Today* 1996; **17**:373–378.
45. Bach FH, Winkler H, Ferran C, Hancock WW, Robson SC. Delayed xenograft rejection. *Immunology Today* 1996; **17**:379–384.
46. Goodman DJ, von Albertini M, Willson A, Millan MT, Bach FH. Direct activation of porcine endothelial cells by human natural killer cells. *Transplantation* 1996; **61**:763–771.
47. Malyguine AM, Saadi S, Platt JL, Dawson JR. Human natural killer cells induce morphologic changes in porcine endothelial cell monolayers. *Transplantation* 1996; **61**:161–164.
48. Malyguine AM, Saadi S, Holzknecht RA, *et al.* Induction of procoagulant function in porcine endothelial cells by human NK cells. *J Immunology* 1997; **159**:4659–4664.
49. Chopek MW, Simmons RL, Platt JL. ABO-incompatible renal transplantation: Initial immunopathologic evaluation. *Transplantation Proceedings* 1987; **19**:4553–4557.
50. Ihrcke NS, Wrenshall LE, Lindman BJ, Platt JL. Role of heparan sulfate in immune system-blood vessel interactions. *Immunology Today* 1993; **14**:500–505.

51. Starzl TE, Demetris AJ, Murase N, Thomson AW, Trucco M, Ricordi C. Donor cell chimerism permitted by immuno-suppressive drugs: A new view of organ transplantation. *Immunology Today* 1993; **14**:326–332.
52. Sachs DH, Sablinski T. Tolerance across discordant xenogeneic barriers. *Xenotransplantation* 1995; **2**:234–239.
53. Zhao Y, Swenson K, Sergio JJ, Arn JS, Sachs DH, Sykes M. Skin graft tolerance across a discordant xenogeneic barrier. *Nature Medicine* 1996; **2**:1211–1216.
54. Sykes M, Lee LA, Sachs DH. Xenograft tolerance. *Immunological Review* 1994; **141**:245–276.
55. Li H, Ricordi C, Demetris AJ, *et al*. Mixed xenogeneic chimerism (mouse + rat —>mouse) to induce donor-specific tolerance to sequential or simultaneous islet xenografts. *Transplantation* 1994; **57**:592–598.
56. Aksentijevich I, Sachs DH, Sykes M. Humoral tolerance in xenogeneic BMT recipients conditioned by a nonmyeloablative regimen. *Transplantation* 1992; **53**:1108–1114.
57. Gritsch HA, Glaser RM, Emery DW, *et al*. The importance of nonimmune factors in reconstitution by discordant xenogeneic hematopoietic cells. *Transplantation* 1994; **57**:906–917.
58. Inverardi L, Clissi B, Stolzer AL, Bender JR, Pardi R. Overlapping recognition of xenogeneic carbohydrate ligands by human natural killer lymphocytes and natural antibodies. *Transplantation Proceedings* 1996; **28**:552.
59. Ljunggren H-G, Karre K. In search of the "missing self": MHC molecules and NK cell recognition. *Immunology Today* 1990; **11**:237–244.
60. Lanier LL, Phillips JH. Inhibitory MHC class I receptors on NK cells and T cells. *Immunology Today* 1996; **17**:86–91.
61. Lawson JH, Platt JL. Xenotransplantation: Prospects for clinical application. In *Dialysis and Transplantation: A Companion to Brenner and Rector's the Kidney*, eds. W Owen, B Pereira, W. B. Saunders Co. Philadelphia, 2000, pp. 653–660.

62. Daggett CW, Yeatman M, Lodge AJ, *et al.* Total respiratory support from swine lungs in primate recipients. *J Thoracic and Cardiovascular Surgery* 1998; **115**:19–27.

63. Akhter SA, Skaer CA, Kypson AP, *et al.* Restoration of β-adrenergic signaling in failing cardiac ventricular myocytes via adenoviral-mediated gene transfer. *Proc Nat Acad Sci, USA* 1997; **94**:12100–12105.

64. Kypson AP, Peppel K, Akhter SA, *et al.* *Ex vivo* adenoviral-mediated gene transfer to the adult rat heart. *J Thoracic and Cardiovascular Surgery* 1998; **115**:623–630.

65. Lawson JH, Platt JL. Molecular barriers to xenotransplantation. *Transplantation* 1996; **62**:303–310.

66. Lawson JH, Sorrell RD, Platt JL. Thrombomodulin activity in porcine-to-human xenotransplantation (abstract). *Circulation* 1997; **96**:565.

67. Patience C, Takeuchi Y, Weiss RA. Infection of human cells by an endogenous retrovirus of pigs. *Nature Medicine* 1997; **3**: 282–286.

68. Sullenger BA, Gallardo HF, Ungers GE, Gilboa E. Overexpression of TAR sequences renders cells resistant to human immunodeficiency virus replication. *Cell* 1990; **63**:601–608.

69. Groth CG, Korsgren O, Tibell A, *et al.* Transplantation of porcine fetal pancreas to diabetic patients. *Lancet* 1994; **344**:1402–1404.

70. Chari RS, Collins BH, Magee JC, *et al.* Treatment of hepatic failure with *ex vivo* pig-liver perfusion followed by liver transplantation. *New England Journal of Medicine* 1994; **331**: 234–237.

71. Deacon T, Schumacher J, Dinsmore J, *et al.* Histological evidence of fetal pig neural cell survival after transplantation into a patient with Parkinson's disease. *Nature Medicine* 1997; **3**:350–353.

Section 5

Developmental Biology and Organ Replacement

Chapter 13

Stem Cells:
Sources and Applications

Lee DK Buttery & Julia M Polak

INTRODUCTION

The general principle of tissue engineering involves combining living cells with a natural or synthetic support or scaffold to produce a three-dimensional living tissue construct that is functionally, structurally and mechanically equal (if not better) to the tissue it has been designed to replace.[1] A major advantage of this approach is that tissues can be designed to grow in such a way that they more precisely match the requirements of the individual in terms of size, shape and immunological compatibility, minimizing the need for further treatment.

In order to achieve effective, long lasting and stable repair of damaged or diseased tissues, there are a number of "basic" criteria that must be satisfied. These include:

i) Generating adequate numbers of cells/size of tissue to be able to fill the defect/complete the repair.
ii) Differentiating the cells towards and maintaining the correct phenotype.

iii) Ensuring the cells/tissues adopt the appropriate three-dimensional organization and produce extracelluar matrix (with provision of structural support if required-resorbable scaffold)

iv) Producing cells/tissues that are structurally and mechanically compliant with the normal demands of native tissue.

v) Achieving full integration with native tissue such as vascularization (if required) and overcoming the risk of immunological rejection.

To a large degree, the ability to satisfy these criteria is dependent on the quality of the starting materials and probably of key importance is a suitable supply of cells. There are a number of different sources of cells that could be used for tissue repair and regeneration. These include mature (non-stem) cells from the patient, "adult" stem cells from the patient such as bone marrow stromal (or mesenchymal) stem cells and embryonic stem (ES) cells/embryonic germ (EG) cells.

Whilst mature cells isolated from tissue biopsies can potentially be used for re-implantation into the same donor thus overcoming the need for immunosuppression, they are probably not the best source of cells for tissue repair. These are generally differentiated or differentiating cells that have low proliferative potential. This makes generating sufficient cells to promote tissue repair potentially difficult. Moreover, these cells are usually committed to a particular cell lineage/phenotype that is restricted to the type of tissue from where they were harvested. This therefore raises issues on accessibility of tissue sites from which cells can be harvested.

Stem cells can overcome many of the limitations of mature cells and consequently, there is currently much interest in the use of stem cells for tissue repair. In this short chapter, we will review the basic characteristics of stem cells — their biology, sources and applications in tissue repair and regeneration.

STEM CELLS

A stem cell can be described as an "immature" or undifferentiated cell that is capable of producing an identical daughter cell.[2,3] Stem cell self-renewal may be perpetuated over many generations resulting in considerable amplification of stem cell numbers (stem cells may actually remain dormant or quiescent for prolonged periods until exposed to particular stimulus). A stem cell is able to produce at least one type (often many types) of highly differentiated or specialized descendent and usually involves formation of a "committed" progenitor cell or transit amplifying cell. A simplified overview of stem cell fate is depicted in Fig. 1. This scheme also helps to illustrate the concept of stem cell potency or range of cell phenotypes to which it can give rise. There are three basic measures of stem cell potency:

i) totipotent — can form all cells/tissues that contribute to the formation of an organism (e.g. the fertilized egg or zygote).
ii) pluripotent — can form most (but not all) cells/tissues of an organism (e.g. ES cells and EG cells).
iii) multipotent — can form a small number of cells/tissues that are usually restricted to a particular germ layer origin (e.g. bone marrow stromal or mesenchymal stem cells).

As will be discussed in the succeeding sections of this chapter, this classification of stem cell potency is not rigid. Indeed, it is evident that the distinction between pluripotent and multipotent is becoming increasingly more blurred with some cells having greater plasticity than previously realized.

What determines stem cell potency is dependent to a large extent on the genetic make-up of the cell and whether it contains the appropriate genetic programme to make a particular cell type. However, this is somewhat "offset" by the influence of the environment in which the stem cell is placed. For example changes in cytokine gradients, cell–cell and cell–matrix contact are important

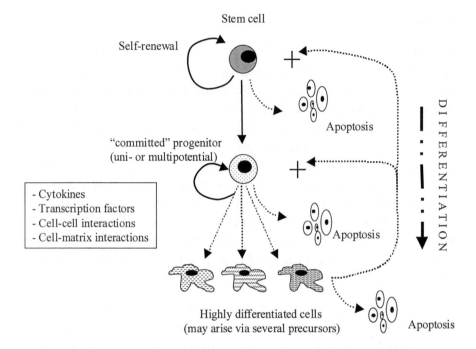

Fig. 1. A simplified overview of stem cell fate. Stem cells are capable of potentially unlimited self-renewal and can also give rise to successively more differentiated progeny that usually involves the generation of various progenitors and precursor cells. The type and number of lineages to which a stem cell can give rise is determined by the genetic characteristics of the stem cell (e.g. pluripotent) and by the environment in which it is placed including cytokine gradients and cell–cell interactions.

in the switching "on" and "off" of genes and gene pathways thereby controlling the type of cell(s) that are generated.

Finally, in this section and as a prelude to the next section it is important to consider location, isolation and derivation of stem cells. So far we have outlined some of the characteristics that make stem cells an "attractive proposition" for tissue repair and regeneration strategies[1] but the concept is based on reliable sources and means of obtaining sufficient numbers of stem cells. If we consider Fig. 1 once again, it is evident that there are mechanisms to activate stem cells or progenitor cells to replace damaged cells. Indeed, in tissues with high attrition rates like bone, blood and skin, the very fact that we are able to maintain these tissues throughout most of our life is indicative of the presence of stem cells or progenitor cells that are able to constantly replace damaged cells and repair tissues.[3] In terms of tissue repair and regeneration this provides us with potential sites from which stem cells might be harvested.

SOURCES OF STEM CELLS

i) "Adult" Stem Cells

There are a number of sources of stem cells that are potentially amenable for use in tissue repair and regeneration. One source of stem cells is from our own bodies. So-called "adult" stem cells[2,3] are resident stem cells that are found in specific niches or tissue compartments and as mentioned above are important cells in maintaining the integrity of tissues like skin, bone and blood. "Adult" stem cells include haematopoietic stem cells,[4] bone marrow stromal (mesenchymal) stem cells,[5] neural stem cells,[6] dermal (keratinocyte) stem cells,[7] fetal cord blood stem cells[8] and several others. Of these, perhaps the best characterized are stem cells derived from the bone marrow. The bone marrow yields two types of stem cell — the haematopoietic stem cell that classically gives rise to the entire blood cell lineage and the marrow stromal (mesenchymal) stem cell that classically gives rise to various connective tissues notably bone and adipose tissue. Here, we will focus primarily on the marrow stromal stem cell.

The existence of marrow stromal stem cells has been known for more than 30 years and initially through the work of Friedenstein[9] these cells were characterized as an adherent sub-population that could be isolated from marrow aspirates. Over the years, these cells have been intensively investigated (for a more detailed review, see Ref. 5) and have consistently been shown to have at least multipotent potential. Under *in vitro* conditions, marrow stromal stem cells are capable of self-renewal for many generations without significant loss of their stem cell characteristics. They are also able to generate several distinct phenotypes including osteoblasts, chondrocytes and adipocytes and this lineage selection can be "controlled" by relatively simple manipulation of the culture conditions and biochemical supplements to which the cells are exposed and maintained. These qualities have enabled marrow stromal stem cells to be used to effect repair of large osseous defects (that are too large to be repaired by endogenous repair mechanisms) after implantation (usually with some sort of scaffold) into animal models.[10] In another example, bone marrow stromal cells have been used as an experimental treatment in children suffering from the debilitating genetic bone disease osteogenesis imperfecta.[11] The characteristic manifestation of the disease is extremely brittle bones due largely to the synthesis, by the osteoblast, of a defective form of collagen-I, which is the major protein present in bone. In the study, subjects received allogeneic bone marrow transplants, after ablation of their own marrow, and after just a few weeks this resulted in a significant improvement in the amount and quality of bone formation. This provides evidence for the engraftment of functional marrow stromal stem cells that in this example were able to generate osteoblasts capable of synthesizing normal bone matrix.

There is now some evidence to indicate that "adult" stem cells isolated from various tissues/niches have greater plasticity and are capable of forming a more diverse range of cell types than previously suspected. Marrow stromal stem cells transplanted into rat brain have been shown to acquire a neural phenotype.[12] Marrow stem cells (possibly both stromal and haematopoietic components) have been shown to contribute to the hepatocyte phenotype in humans[13] and myocardium in rats.[14] Interestingly in the latter example, there was apparently "active" targeting of these cells to the heart after they were infused into the blood stream of animals whose hearts had been experimentally damaged. An even more impressive demonstration of the potency of marrow stem cells has come from a study in mice where the progeny of one single marrow stem cell injected into a mouse were found to contribute to lung, liver, skin, intestine as well as bone and blood.[15] There is also data to show that adult neural stem cells can contribute to the haematopoietic lineage[16] and stem cells isolated from adipose tissue generating bone, cartilage and muscle phenotypes.[17]

These observations raise several interesting issues on the biology of "adult" stem cells and their potential applications for tissue repair. The data provide strong evidence to indicate that "adult" stem cells (at least those discussed here) are pluripotent. The pluripotency exhibited by these cells also occurs in an environmental context such that marrow stem cells within the marrow space do not make neural cells or hepatocytes but once transplanted to the brain or liver, they seem to respond to that environment and acquire a neural or hepatocyte fate. There is also evidence of targeting of adult stem cells such that they are able to migrate towards and take up residence in a tissue that has been damaged in some way and contribute to its repair and regeneration. These are obviously desirable attributes and have fuelled the debate over whether "can adult stem cells suffice"[18] as a cell source for promoting tissue repair and negate the need to explore alternatives such as ES cells (discussed below and in subsequent sections).

ii) ES Cells

Embryonic stem cells are isolated from the inner cell mass of the pre-implantation blastocyst[19,20] and have been derived from mice, non-human primates and humans. They are pluripotent cells, retaining the capacity to generate any and all fetal and adult cell types *in vivo* and *in vitro*.[21-23] In the presence of leukaemia inhibitory factor (LIF) or embryonic fibroblasts, these cells can be maintained and expanded *in vitro* in an undifferentiated, pluripotent state almost indefinitely.[21-23] This provides a potentially unlimited source of stem cells. Moreover, upon withdrawal of LIF, ES cells spontaneously differentiate to form distinct cellular aggregates or embryoid bodies, which contain differentiating cells of ectodermal, endodermal and mesodermal lineage.[21-23] By manipulating the culture conditions under which ES cells differentiate, it has been possible to control and restrict the differentiation pathways and thereby generate cultures enriched for lineage-specific precursors. Utilizing this approach, mouse ES cells have been used to generate a range of distinct phenotypes including haematopoietic precursors,[24] neural cells,[25] adipocytes,[26] muscle cells,[27] myocytes,[28] chondrocytes,[29] pancreatic islets[30] and in our own studies osteoblasts.[31]

For the most part, developments in manipulating ES cells have been based on the use of mouse ES cells. However, the recent isolation of human ES cells[20] has created the realistic opportunity that human ES cells can be developed for tissue repair and regeneration. Although there are some differences in the morphology, propagation and culture requirements for maintaining mouse ES cells and human ES cells[20] are broadly similar in their potential to differentiate into specific cell types.[20,32] Thus, differentiation protocols developed for mouse ES cells are likely to be relevant to human ES cells (see Ref. 33 for a more detailed discussion of human ES cells).

If we look at the ways in which we might control ES cell differentiation and preferentially select one particular phenotype, then there are a number of approaches (see Fig. 2).

A relatively simple approach is to culture in media supplemented with cytokines, hormones or other factors that might be expected to stimulate the cell type of interest. This obviously requires detailed knowledge of the biology of the cell type of interest but as illustrated in our own studies, this biochemical selection approach is relatively effective. Our objective was to generate osteoblasts from mouse ES

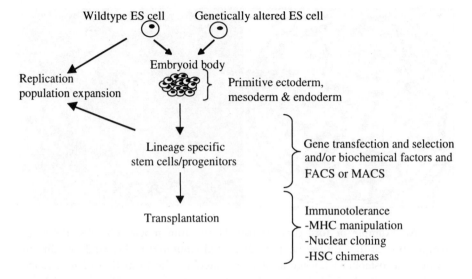

Fig. 2. A schematic representation of the application of ES cells for derivation of specific cell lineages. MHC — major histocompatibility complex; HSC — haematopoietic stem cell.

cells and in our very first series of experiments, we were able to demonstrate derivation of osteoblasts by simply maintaining dispersed embryoid bodies in a culture medium that was/is routinely used to culture explanted osteoblasts (Fig. 3).

Subsequently, we went on to refine the culture conditions and found that whilst the biochemical composition of the medium in which the differentiating ES cells are maintained is an important stimulus for selecting the osteoblast phenotype, the timing of stimulation is crucial.[31] Thus, by delaying the addition of one particular factor, in this example dexamethasone, for up to 14 days after dispersal and culture of the embryoid bodies, we were able to attain a seven-fold enrichment of cells capable of expressing the osteoblast phenotype.

We also showed that co-culture with mouse osteoblasts was an effective inducer of osteogenic differentiation in ES cells. There are

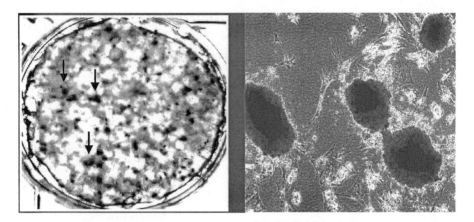

Fig. 3. An example of formation of osteoblasts from mouse ES cells. By simple manipulation of the culture medium, disrupted embryoid bodies could be directed towards the osteoblast lineage as demonstrated by the formation of mineralized bone-like colonies. In the left panel, a lower power image is seen showing enrichment with darkly stained mineralized bone-like colonies. These are seen at higher power in right panel.

unpublished reports from other workers to show that implantation of human ES cells into mouse organs and tissues including brain and liver is an effective method for inducing neural or hepatic phenotypes. Such observations are consistent with those on adult stem cells and further emphasise the influence of environment on stem cells and the acquisition of a particular phenotype.

Biochemical selection, although a useful method for generating *in vitro* cultures enriched with a particular phenotype, is not generally sufficiently effective to exclude other phenotypes. In our experience, we found approximately 60–70% enrichment with osteoblast lineage with the remaining cells comprising a mixed population of other phenotypes like neural cells. If we are to develop ES cells (and adult stem cells) for transplantation therapy, we have to consider methods to further purify or select cells of the desired lineage/phenotype. Methods such as FACS (fluorescence-activated cell sorter) or MACS (magnetic-activated cell sorter) are extremely useful methods to sort and purify a particular cell type or population. It is largely dependent on the cell type of interest expressing a cell surface marker that can be recognised by a fluorescence or magnetic microbead tagged antibody that can then be used to preferentially sort that cell. To be effective it obviously requires the cell surface marker to be fairly unique to the desired cell lineage and ideally the surface marker should not be sensitive to activation upon binding of the tagged antibody. Both FACS and MACS are widely used in clinical haematology to sort sub-populations of heamatopoietic lineage and in terms of our own research interests have been used to purify osteoprogenitors from marrow stromal stem cells.

As an alternative, stem cells (both ES cells and adult stem cells) can be transduced with a lineage specific gene. In this knock-in approach, the gene insert can also carry one of several resistance genes which allow for preferential selection of cell sub-populations that should be restricted to lineage of interest. This method has been used to select neural[25] and cardiomyocytes phenotypes.[28] In the latter example, the selected cardiomyocytes were subsequently implanted into the damaged hearts of mice and shown to form stable grafts.[38]

The potential applications of ES cells for tissue repair and regeneration is further bolstered by the possibilities of using nuclear transfer to render these cells autologous and thus remove risks of tissue rejection, or through the use of human ES cell lines with modification of the major histocompatibility complex. For a more detailed review of transplant therapies using ES cells, see Ref. 33.

There are other sources of pluripotent stem cells that have not been discussed such as EG cells[34] derived from the gonadal ridge of the early fetus or embryonic carcinoma (EC) cells derived from teratocarcinomas.[35] These are important cell sources that could be used therapeutically or as models to further investigate the mechanisms of pluripotency and lineage-restricted differentiation. Again, a more detailed discussion of these cells can be found in Ref. 33.

iii) Cell Reprogramming

In the light of the success of cloning Dolly the sheep and the various other animals that have followed, much interest has been generated in understanding the mechanisms of nuclear cloning/ reprogramming and potentially harnessing them for tissue repair strategies. In nuclear cloning, an enucleated oocyte is fused with the nucleus of a somatic cell which can go on to generate an intact embryo and produce a viable organism that is genetically identical to the donor of the somatic nucleus.[36] The factors and mechanisms that induce this remarkable transformation are not known but it seems likely that a component of the enucleated oocyte can initiate rewinding of the genetic programme of the somatic nucleus to generate a "totipotent cell" (if implanted into a surrogate — see Fig. 4). In terms of tissue repair and regeneration, this has obvious applications and could potentially overcome the limitations and

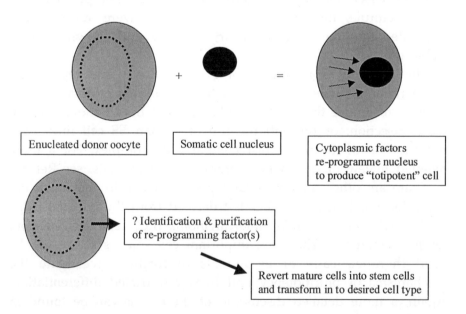

Fig. 4. Simplified representation of the concept of nuclear programming.

ethical issues (discussed below) associated with adult stem cells and ES cells. There are as yet unpublished reports that PPL Therapeutics (part of the group that cloned Dolly) have developed a method to reprogram mature bovine skin cells to become pluripotent stem cells (not necessarily based on the approach presented in Fig. 4 — see commentary in Ref. 18). For more detailed discussion of mechanisms of nuclear reprogramming, see Refs. 37 and 38.

iv) Concluding Remarks

In this short review, we have discussed some of the most recent developments in the use of stem cells for tissue repair and regeneration. There is no doubt that stem cells derived from adult and embryonic sources hold great therapeutic potential but it is clear that there is still much research required before their use in the clinic is commonplace. As mentioned above, there is much debate over adult stem cells versus ES cells and whether we need both. The opinion of these authors and many others working in this field is that it is too early to disregard one or other of these cell sources. There is certainly a need for rationalization but this can only be exercised once we have carefully compared and contrasted these various cells under the appropriate experimental conditions. Some characteristics that might help resolve the issue of cell source can already be applied to the debate. Accessibility of cells is obviously important. In terms of adult stem cells it is already clear that some cells like neural stem cells pose significant difficulties in harvesting (at least in living donors). Even cells that are somewhat more accessible like marrow stem cells involve an invasive aspiration procedure. There are also concerns over the incidence of adult stem cells with for example marrow stromal cells occurring at a frequency of approximately 1 cell in 100,000 cells and this number and their potency might decline with increasing age. With regard to ES cells, the major concern relates to the ethics of their creation and the proposed practice of therapeutic cloning. However, this can be offset by the fact that there are potentially "large" numbers of embryos created by *in vitro* fertilization programmes that are surplus to requirements (normally destroyed) and could potentially be used for derivation of ES cells. Finally, with regard to both adult stem cells and ES cells the stability and risk of forming unwanted tissues or even teratocarcinomas has yet to be fully evaluated.

The following web sites provide useful information on the principles and ethics of stem cells:

http://www.roslin.ac.uk
http://www.doh.gov.uk/cegc/index.htm
http://www.royalsoc.ac.uk/policy/
http://www.nih.gov/news/stemcell/primer.htm
http://www.med.ic.ac.uk/divisions/8/index.html

ACKNOWLEDGEMENTS

Imperial College Tissue Engineering Centre, Chelsea & Westminster Hospital Campus is supported by funding obtained from Medical Research Council, Engineering & Physical Sciences Research Council, Dunhill Medical Trust, Golden Charitable Trust, Chelsea & Westminster Healthcare NHS Trust Charitable Funds and Chelsea & Westminster Healthcare NHS Trust.

REFERENCES

1. Stock UA, Vacanti JP. Tissue engineering: Current state and prospects. *Annu Rev Med* 2001; **52**:443–451.
2. Gehron RP. Stem cells near the century mark. *J Clin Invest* 2000; **105**:1489–1491.
3. Watt FM, Hogan BLM. Out of eden: Stem cells and their niches. *Science* 2000; **287**:1427–1430.
4. Morrison SJ, Uchida N, Weissman IL. The biology of hematopoietic stem cells. *Annu Rev Cell Dev Biol* 1995; **11**: 35–71.

5. Biancho P, Gehron Robey P. Marrow stromal stem cells. *J Clin Ivest* 2000; **105**:1663–1668.

6. Alvarez-Buylla A, Garcia-Verdugo JM, Tramontin AD. A unified hypothesis of the lineage of neural stem cells. *Nat Re Neurosci* 2001; **2**:287–293.

7. Watt FM. Epidermal stem cells as targets for gene therapy. *Hum Gene Therap* 2000; **11**:2261–2266.

8. Gallacher L, Murdoch B, Wu D, Karanu F, Fellows F, Bhatia M. Identification of novel circulating human embryonic blood stem cells. *Blood* 2000; **96**:1740–1747.

9. Friedenstein AJ, Petrakova KV, Kurolesova AI, Folova GP. Heterotropic transplants of bone marrow. Analysis of precursor cells for osteogenic and hematopoietic tissues. *Transplantation* 1968; **6**:230–247.

10. Ohgushi H, Goldberg VM, Caplan AI. Heterotopic osteogenesis in porous ceramics induced by marrow cells. *J Orthop Res* 1989; **7**:566–578.

11. Horwitz EM, Prockop DJ, Fitzpatrick LA, Koo WW, Gordon PL, Neel M, Sussman M, Orchard P, Marx JC, Pyeritz RE, Brenner MK. Transplantability and therapeutic effects of bone marrow-derived mesenchymal cells in children with osteogenesis imperfecta. *Nat Med* 1999; **5**:309–313.

12. Azizi SA, Strokes D, Augelli BJ, DiGirolamo C, Prockop DJ. Engraftment and migration of human bone marrow stromal cells implanted in the brains of albino rats: Similarities to astrocyte grafts. *Prc Natl Acad Sci, USA* 1998; **95**:3908–3913.

13. Alison MR, Poulson R, Jeffery R, *et al*. Hepatocytes from non-hepatic stem cells. *Nature* 2000; **406**:257.

14. Orlic D, Kajstura J, Chimenti S, Jakoniuk I, Anderson SM, Li B. Bone marrow cells regenerate infacrted myocardium. *Nature* 2001; **410**:701–704.

15. Krause DS, Theise ND, Collector MI, *et al*. Multi-organ, multi-lineage engraftment by a single bone marrow-derived stem cell. *Cell* 2001; **105**:369–377.

16. Bjornson CR, Rietze RL, Reynolds BA, Magli MC, Vescovi AC. Turning brain into blood: A hematopoietic fate adopted by adult neural stem cells *in vivo. Science* 1999; **283**:534–537.

17. Zuk PA, Zhu M, Mizuno H, Huang J, Futrell JW, Katz AJ, Benhaim P, Lorenz HP, Hedrick MH. Multilineage cells from human adipose tissue: Implications for cell-based therapies. *Tissue Eng* 2001; **7**:211–228.

18. Vogel G. Can adult stem cells suffice. *Science* **292**:1820–1822.

19. Martin GR. Isolation of a pluripotent cell line from early mouse embryos cultured in medium conditioned by teratocarcinoma stem cells. *Proc Natl Acad Sci, USA* 1981; **78**:7634–7638.

20. Thomson JA, Itskovitz-Eldor J, Shapiro SS, Waknitz MA, Swiergiel JJ, Marshall VS, Jones JM. Embryonic stem cell lines derived from human blastocysts. *Science* 1998; **282**:1145–1147.

21. Smith AG. Culture and differentiation of embryonic stem cells. *J Tiss Cult Methods* 1991; **13**:89–94.

22. Keller GM. *In vitro* differentiation of embryonic stem cells. *Curr Opinion Cell Biol* 1995; **7**:862–869.

23. Wiles MV, Johansson BM. Embryonic stem cell development in chemically defined medium. *Exp Cell Res* 1999; **247**:241–248.

24. Wiles M, Keller GM. Multiple hematopoietic lineages develop from embryonic stem cells in culture. *Development* 1991; **111**: 259–267.

25. Li M, Pevny L, Lovell-Bridge R, Smith A. Generation of purified neural precursors from embryonic stem cells by lineage selection. *Curr Biol* 1998; **8**:971–974.

26. Dani C, Smith AG, Dessolin S, Leroy P, Staccini L, Villageois P, Darimont C, Aihaud G. Differentiation of embryonic stem cells into adipocytes *in vitro. J Cell Sci* 1997; **110**:1279–1285.

27. Rohwedel J, Maltsev V, Bober E, Arnold HH, Hescheler J, Wobus A. Muscle cell differentiation of embryonic stem cells reflects myogenesis *in vivo*: Developmentally regulated expression of myogenic determination genes and functional expression of ionic currents. *Dev Biol* 1994; **164**:87–101.

28. Klug MG, Soonpaa MH, Koh GY, *et al*. Genetically selected cardiomyocytes from differentiating embryonic stem cells form stable intracardiac grafts. *J Clin Invest* 1996; **98**:216–224.

29. Kramer J, Hegert C, Guan K, Wobus AM, Muller PK, Rohwedel J. Embryonic stem cell-derived chondrogenic differentiation *in vitro*: Activation by BMP-2 and BMP-4. *Mech Devl* 2000; **92**: 193–205.

30. Lumelsky N, Blondel O, Laeng P, Velasco I, Ravin R, McKay R. Differentiation of embryonic stem cells to insulin secreting structures similar to pancreatic islets. *Science* 2001; **292**: 1389–1393.

31. Buttery LD, Bourne S, Xynos JD, Wood H, Hughes FJ, Hughes SP, Episkopou V, Polak JM. Differentiation of osteoblasts and *in vitro* bone formation from murine embryonic stem cells. *Tissue Eng* 2001; **7**:89–99.

32. Itskovitz-Eldor J, Schuldiner M, Karsenti D, Eden A, Yanuka O, Amit M, Soreq H, Benvenisty N. Differentiation of human embryonic stem cells into embryoid bodies compromising the three embryonic germ layers. *Mol Med* 2000; **6**:88–95.

33. Odorico JS, Kaufman DS, Thomson JA. Multilineage differentiation from human embryonic stem cell lines. *Stem Cells* 2001; **19**:193–204.

34. Shamblott MJ, Axelman J, Littlefield JW, Blumenthal PD, Huggins GR, Cui Y, Cheng L, Gearhart JD. Human embryonic germ cell derivatives express a broad range of developmentally distinct markers and proliferate extensively *in vitro*. *Proc Natl Acad Sci, USA* 2001; **98**:113–118.

35. Andrews PW. Teratocarcinomas and human embryology: Pluripotent human EC cell lines. Review article. *APMIS* 1998; **106**(1):158–167.

36. Lanza RP, Cibelli JB, West MD. Prospects for the use of nuclear transfer in human transplantation. *Nat Biotechnol* 1999; **17**: 1171–1174.

37. Colman A, Kind A. Therapeutic cloning: Concepts and practicalities. *TIBTECH* 2000; **18**:192–196.
38. Kikyo N, Wolfee AP. Reprogramming nuclei: Insights from cloning, nuclear transfer and heterokaryons. *J Cell Sci* 2000; **113**:11–20.

Chapter 14

Possible Production of Spare Parts Using Methods of Developmental Biology

JMW Slack

The last 15 years have seen a revolution in the understanding of developmental mechanisms at the molecular level.[19] This has raised hopes that it may soon be possible to apply this knowledge to various practical goals, among which might be the production of human organs for transplantation. There is a particular appeal to methods that might start from the patient's own cells and thus show full immunological compatibility. Unfortunately the most obvious and acceptable routes involve processes that remain poorly understood, in particular growth control in postnatal development, the mechanism of cell differentiation and the organisation of tissues into organs. On the other hand, routes for which the biology is reasonably well understood may not be possible for legal and ethical reasons.

I shall briefly consider three possible routes: stimulation of tissue regeneration *in vivo*, growth of spare parts *in vitro*, and growth of organ cultures from eggs.

STIMULATION OF TISSUE REGENERATION
IN SITU

This is likely to be possible if there is a single target cell population and it does not matter much about the precise spatial structure of the regenerate. For example, it is already possible to stimulate new blood vessel formation by implantation of slow release pellets containing angiogenic factors such as fibroblast growth factors or vascular endothelial growth factor.[10] There has been considerable interest in the application of growth factors to promote healing of surface wounds,[8] and of bone morphogenetic proteins to promote healing of recalcitrant fractures.[20] But at present, it is not possible to conceive of a method for regenerating a solid organ because of basic ignorance about the normal mechanisms of growth control. Recent work has underlined the importance of the insulin signalling pathway in regulation of size, but we still have no idea about what systemic controls there are on the relative proportions of body parts.[2]

To illustrate the deficiencies in our knowledge, consider the difference between the regeneration behaviour of the liver and pancreas.[15] The liver regenerates rapidly following surgical reduction and as a result of this, the volume is restored but the shape is not. The volume is controlled by a circulating hormone, of which hepatocyte growth factor (HGF) is an important component. However, the main source of HGF is endothelial cells and macrophages of the kidney and lung, and the mechanism by which the liver senses its size and by which it transmits this information to the other organs to provoke HGF release remains unknown.

The pancreas regenerates slowly and incompletely following surgical reduction. There is no good evidence for systemic controls. However ablation of exocrine cells alone, for example by protein synthesis inhibitors, is followed by rapid regeneration, while ablation of beta cells alone by streptozotocin is followed by very little regeneration.[16] This suggests that there are short range inhibitory effects which could potentially be overcome, but we currently do not know what they are.

GROWTH OF BODY PARTS *IN VITRO*

It is often thought that tissue culture technology should allow the production of spare parts by culture *in vitro*.[5] But again, there are severe difficulties based on lack of knowledge. First, there is the problem of controlling the pathway of cell differentiation in the desired way. For example, pancreatic beta cells for treatment of diabetes, or dopaminergic neurons for treatment of Parkinson's disease, are not the sort of cell that can easily be grown in culture. They arise as terminal cell types from their respective stem cells and methods for steering them down the correct pathway of differentiation are still rudimentary.[14,21]

The growth of a solid organ *in vitro* involves even more problems than the production of a single cell type. Firstly, organs are always composed of multiple cell types. In some cases, these are multiple differentiation products of the organ-specific stem cells, for example the exocrine and the endocrine cells of the pancreas.[9] But in other cases, they are cells of different embryological lineages. For example, a lung would need a smooth muscle layer in each bronchiole, and a supporting layer of connective tissue around the alveoli. This means that epithelial and mesenchymal cells would need to be grown together in a harmonious way that mimicked their growth *in vivo*. Furthermore, organ cultures grown in the laboratory from embryonic buds are always very small. The main reason for this is that they depend on diffusion of nutrients from a bath of tissue culture medium. To grow a solid organ *in vitro* would require the provision of a nutrient supply through a vascular system.

It is possible to imagine circumventing these difficulties through advances in tissue engineering. For example, admixtures of cells could initially be grown on a three dimensional mesh-like biomaterial, and the capillary endothelial cells might be persuaded to colonise the matrix by stimulation with suitable growth factors. Some possibilities are discussed in chapter 1 by Hench *et al*. But again, it is still a long way to go and this technology is straying somewhat from the potential applications of developmental biology.

GROWTH OF BODY PARTS *IN VIVO*, FROM EGGS

The production of body parts from eggs probably is possible, at least in principle. But there are serious ethical problems. It is likely to involve human cloning from somatic cells, germline genetic modification, and surrogate mothers to incubate the organ cultures (for details, see chapter 13 by Buttery & Polak). Public distaste for all of these procedures may make it impossible to proceed and would certainly necessitate a strict regulatory framework. Whether it would be legal is currently unclear because of uncertainty about whether an organ culture derived from an egg is or is not a human embryo. If it were judged to be a human embryo, then the hypothetical procedure as outlined here would be illegal under current UK law for no less than three reasons: genetic modification of human embryos, cloning of human embryos and implantation of a human embryo into an animal.

It will certainly never be acceptable to create an intact human being for the purposes of organ donation, as this would be equivalent to murder. But it is possible to imagine ways in which we might use our understanding of early development to reprogram a human egg so that it developed not into a human foetus, but into an "organ culture" consisting just of the thoracico-abdominal part of the body. The putative procedure is shown in Fig. 1. It is based on the presumption that cloning from human tissue culture cells is possible. Based on existing experience with sheep and cattle,[1,13] it seems likely that cloning from foetal fibroblasts would work with acceptable

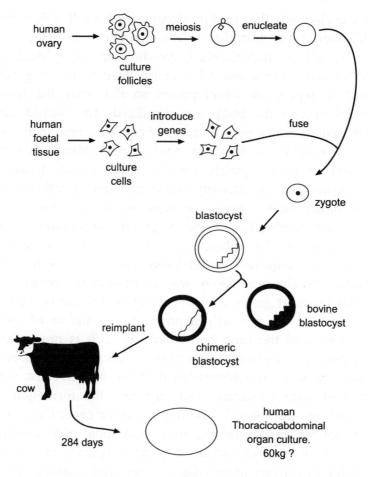

Fig. 1. Possible procedure for the production of human thoracicoabdominal organ cultures. The human genetic material consists of nuclei from tissue culture cells into which have been inserted genes to suppress development of head and limbs.

effciency. So an "organ farm" might start from a number of human foetal fibroblast lines which were of known HLA type and had been screened for potential infectious agents. In principle, it would be attractive to clone from cells of the individual patient but it remains unclear whether routine cloning from adult cells is possible and this would also introduce a considerable time delay.

As it would be unthinkable to produce an intact human being simply for spare parts, the cells would need to be be genetically modified to reprogram development so that only the thoracico-abdominal part of the body was produced. This would have no nervous system, hence no possibility of consciousness, pain or other feelings. It would also have no external resemblance to a human being, and so might qualify for the same ethical status as the human tissues that are currently in use for transplantation purposes, such as blood banks, frozen corneas, frozen gametes, or donor organs in the ice box. The necessary genetic reprogramming would be easier to achieve by adding components than by removing them, since the latter requires the relatively rare event of homologous recombination. The most likely way to achieve the required result would be to insert genes encoding dominant negative versions of transcription factors normally required for formation of the head, such as Otx, and the limb buds, like Lhx-2. It is now routine to make dominant negative transcription factors by swapping the activation domain for a repression domain, or *vice versa*, and it is known that such constructs can suppress the formation of the corresponding body parts.[6,11] So the source of genetic material is a foetal fibroblast line containing these additional genes, which will be under the control of suitable regionally specific promoters.

In order to convert these cells into an organ culture, a nucleus needs to be introduced into a human egg (technically a metaphase II oocyte). But even using foetal cells as nuclear donors, the process is likely to be relatively inefficient. Recent experiments on farm animals report about 5–20% of nuclear transplants forming blastocysts and about 20% of advanced pregnancies resulting from blastocyst reimplantation.[17,18] This suggests a requirement for 100–200 nuclear transplantations to be sure of getting a pregnancy. It is very unlikely that enough donors will be found to supply the necessary eggs, suggesting that a means of culturing preantral oocytes will need to be developed. There are 10^5 immature oocytes in a human ovary, so the few ovaries that might perhaps be recovered from young accident victims could potentially produce a very large number of eggs. In mice oocytes have been cultured, ovulated, fertilized and developed normally if taken from animals >8 days after birth.[3] Another source that has been considered are the oogonia present in the products of termination of pregnancy. This is another ethical hot potato but the extended culture period and support for terminal differentiation required is well beyond what is currently possible and is probably unrealistic for the forseeable future.

Although success cannot be guaranteed without trying it, the method for enucleation and cell fusion that has been used to clone farm animals may work for human material and would result in the formation of a zygote whose genetic constitution was that of the genetically modified tissue culture cells. *In vitro* culture methods for human preimplantation embryos already enable culture to blastocyst stage. Further growth of the resulting organ culture would require reimplantation into a human surrogate mother but this route is most unlikely to be followed for both aesthetic and safety reasons. As a more acceptable alternative, it would be worth investigating the possibility of making chimeric embryos in which the inner cell mass came from the human organ culture and the outer trophectoderm was of bovine origin. Such a reconstituted embryo might grow in cows and do so at a rate exceeding the growth in human hosts. It would have the advantages that the reimplantation technique is already routine, and nutrition of the culture would be cheap and simple as it would require merely feeding the cow. Whether it would work is uncertain. The only precedents are the sheep-goat chimeras and the blastocysts reconstituted from different mouse species, both experiments carried out in the 1980s.[4,12] In both these examples, the species gap is much less than the gap between human and bovine.

If it were feasible, the organ culture might grow to a size of a newborn calf, about 60 kg, over the 284 day gestation period, and its constituent organs would then be of a suitable size for transplantation into adults. It is interesting to speculate whether such a size would in fact be reached by a human organ culture fed by a bovine placenta. An indication that it might is that young organs grafted to mature hosts show a fairly rapid catch-up growth which greatly exceeds the growth rate that they would have experienced in situ.[7] An organ farm is likely to have cohorts of cows incubating organ cultures at different stages. They would probably be induced on a particular day of the week so that availability of organs of particular HLA types was known in advance.

Development of a procedure of this type may well not be possible. It would certainly require a substantial investment in the necessary research and development. Quite apart from the substantial legal and ethical difficulties that have been mentioned, there could be further problems obtaining regulatory approval for the therapeutic use of such a complex living material. So the costs of development would be substantial. But it is equally clear that the potential profits would be astronomical, so this area might prove attractive to a biotechnology company willing to take a big risk.

REFERENCES

1. Cibelli JB, Stice SL, et al. Cloned transgenic calves produced from nonquiescent fetal fibroblasts. Science 1998; 280:1256–1258.
2. Conlon I, Raff M. Size control in animal development. Cell 1999; 96:235–244.
3. Eppig JJ, O'Brien M, et al. Mammalian oocyte growth and development in vitro. Mol Reprod Dev 1996; 44:260–273.
4. Fehilly CB, Willadsen SM, et al. Interspecific chimerism between sheep and goat. Nature 1984; 307:634–636.
5. Gage FH. Cell therapy. Nature 1998; 392 Suppl:18–24.
6. Isaacs HV, Andreazzoli M, Slack JMW (1999). Anteroposterior patterning by mutual repression of orthodenticle and caudal

type transcription factors. *Evolution and Development* 1999; **1**: 143–152.

7. Kam I, *et al.* Evidence that host size determines liver size: Studies in dogs receiving orthotopic liver transplants. *Hepatology* 1987; **7**:362–366.

8. Mellin TN, Cashen DE, *et al.* Acidic fibroblast growth factor accelerates dermal wound healing in diabetic mice. *J Invest Dermatol* 1995; **104**:850–855.

9. Percival AC, Slack JMW. Analysis of pancreatic development using a cell lineage label. *Exp Cell Res* 1999; **247**:123–132.

10. Risau W, Flamme I. Vasculogenesis. *Ann Rev Cell Dev Biol* 1995; **11**:73–91.

11. Rodriguez-Esteban C, Schwabe JWR, De la Pena J, Ricon-Limas, DE, Magallon J, Botas J, Izpisua-Belmonte JC. Lhx2, a vertebrate homologue of apterous, regulates vertebrate limb outgrowth. *Development* 1998; **125**:3925–3934.

12. Rossant J, Chapman VM. Somatic and germline mosaicism in interspecific chimeras between Mus musculus and Mus caroli. *J Embryol Exp Morph* 1983; **73**:193–205.

13. Schnieke AE, Kind AJ, *et al.* Human factor IX transgenic sheep produced by transfer of nuclei from transfected fetal fibroblasts. *Science* 1987; **278**:2130–2133.

14. Schuldiner M, Yanuka O, Itskovitz-Eldor J, Melton DA, Benvenisty N. Effects of eight growth factors on the differentiation of cells derived from human embryonic stem cells. *Proc Natl Acad Sci, USA* 2000; **97**:11307–11312.

15. Sidorova VF. (1978). *The Postnatal Growth and Restoration of Internal Organs in Vertebrates.* PSG Publishing Co Inc, Calcutta.

16. Slack JMW. Developmental biology of the pancreas. *Development* 1995; **121**:1569–1580.

17. Stice SL, Robl JM, Ponce de Leon FA, Jerry J, Golueke PG, Cibelli JB, Kane JJ (1998). Cloning: New breakthroughs leading to commercial opportunities. *Theriogenology* 1998; **49**:129–138.

18. Wolf E, Zakhartchenko V, Brem G. Nuclear transfer in mammals: Recent developments and future perspectives. *J Biotechnol* 1998; **65**:99–110.
19. Slack JMW. *Essential Developmental Biology.* Blackwell Science, Oxford.
20. Wozney JM, Rosen V. Bone morphogenetic protein and bone morphogenetic protein gene family in bone formation and repair. *Clin Orthopaed Rel Res* 1998; **346**:26–37.
21. Yan J, Studer L, McKay RDG. Ascorbic acid increases the yield of dopaminergic neurons derived from basic fibroblast growth factor expanded mesencephalic precursors. *J Neurochem* 2001; **76**:307–311.

Section 6

Safety of Human Tissues and Cells

Chapter 15

Safety of Human Tissues and Cells for Transplantation

Ruth M Warwick & John N Kearney

INTRODUCTION

Assuring the safety and effectiveness of tissue grafts and cell transplantation becomes increasingly important as their use increases. The risks associated with these grafts include the potential for transmission of infections, malignancies, auto-immune diseases and theoretical risks associated with some diseases of unknown aetiology. This chapter relates to the known and potential risks of disease transmission by tissue grafts and cell transplantation and the methods that can be employed to minimise those risks.

Most of this chapter is written from the perspective of tissue banking. In terms of cell transplantation, it includes bone marrow and cord blood transplantation with only minor reference to procedures involving *in vitro* cell proliferation e.g. tissue engineering. *In vitro* cell proliferation carries with it a whole range of safety considerations beyond the scope of this chapter, but it should be borne in mind that the principles described in this chapter apply. If allogeneic cells are enormously expanded in number and if it is the intention that the cells are used in many recipients, then the microbiological testing should reflect the increased size of the population at risk. Whilst cells are in culture, there is the opportunity for microbiological contamination and also for expansion of microbiological agents. When cells are cultured *in vitro*, the relevant processes must be monitored. This should include the demonstration of lack of malignant transformation and that the relevant biological properties of the cells are maintained. In the case of tissue engineering, these principles also apply including in any commercial environment.

KNOWN RISKS FROM THE DONOR AND THE MANUFACTURING PROCESS

(a) Infectious Diseases

There have been many reported transmissions of viral, bacterial and fungal agents by most types of tissue or organs commonly transplanted.[1,2] For example, bone[3,4] and skin[5] have transmitted HIV. If cadaver tissue donor selection was random without reviewing the medical and behavioural history, the number of infected patients in the potential donor pool would be much higher. It has been estimated that the prevalence of HIV infection in cadaveric donors, as detected by HIV antibody screening alone, could be as high as 0.6%.[6] Frozen bone[7] and bone marrow, respectively,[8] have transmitted hepatitis C and B. Tuberculosis has been transmitted by bone and possibly by heart valves.[9] Dura mater[10] and cornea have transmitted Creutzfeldt-Jakob disease and the latter has also transmitted rabies.[11] Organ grafting has been associated with HHV8 transmission.[12] Numerous incidents of fungal and bacterial infections have also been reported by tissue grafting, sometimes of donor origin but more often acquired during tissue procurement, processing and storage.

HTLV is also potentially transmissible by grafting or transplantation because it has been shown to be transmissible by sexual intercourse, breast-feeding, intravenous drug use with shared needles and by blood transfusion. The sero-conversion rate following transfusion with infected blood is between 12.8 and 63.4%.[13,14] Late sequelae of HTLV infection include T-cell leukaemia or lymphoma, and tropical spastic paraparesis.[15] The prevalence of HTLV in antenatal populations has been reported as 0.14% in the West Midlands[16] and 0.2–0.26% in London.[17,18] This is especially relevant when multi-racial targeting of umbilical cord blood donors is undertaken because the prevalence of HTLV is up to 1% in some ethnic groups. Any effect of recipient immunosuppression at the time of infection, as would occur in bone marrow or cord blood transplantation, on the rate of development of HTLV related disease is unknown. However, profound immunosuppression, and a relatively large infective dose, may accelerate disease onset.

Malaria can be transmitted by blood transfusion and theoretically by any blood-containing tissue graft or by cryopreserved cells such as cord blood or by fresh cells such as bone marrow. The incidence of congenital malaria in areas in which the disease is endemic is estimated to be <1%[19] and parasites have been detected in the cord blood of 35% of infants whose mothers have infected placentas.[20] Miller and Telford have pointed out that congenital malaria does occur but its frequency depends on various factors including co-existing infections and maternal immunity.[20] In the context of cord blood donation, the possibility of malarial transmission from an infected mother to the donor of cord blood and thereby to a transplant recipient does exist.

Other protozoa have been transmitted; *Histoplasma capsulatum* by organ transplantation,[21] as has *Toxoplasma gondii.*[22]

(b) The Potential for Other Types of Disease Transmission

Organ donations have transmitted malignancy and this is at least in part due to the transplantation of large numbers of viable cells.[23-25] These direct transmissions from donor to recipient have occurred against an HLA barrier. Transmission of malignancy can occur from mother to infant,[26,27] and hence, theoretically, to the immuno-suppressed recipient of an umbilical cord blood transplant. It is not surprising that this can occur considering the well documented trafficking of cells between the mother and foetus. In fact, it is surprising that vertical transmission has been described so rarely, for example in melanoma and aggressive lymphomas.[26,27]

Of relevance to cord blood, foetal cells in the maternal circulation may be very long lived and male foetal progenitor cells have been shown to persist in maternal blood for decades after delivery.[28] It seems that pregnancy may establish a long-term low-grade chimaeric state in the mother.[29] Foetal DNA and cells have been identified in skin lesions from women with systemic sclerosis.[30] Foetal DNA has also been demonstrated in the maternal skin in the context of polymorphic skin rashes of pregnancy, demonstrating the potential capacity of foetal cells to "engraft" via the maternal circulation, circumventing the HLA barrier.[31] The degree of chimerism may depend on the dose of foetal cells crossing the placental barrier or on the HLA compatibility between the foetus and the mother.[32] However, whether maternal tolerance of foetal cells is mirrored by the tolerance of maternal cells by the infant's immune system is less clear. There is some evidence that fetal cells are destroyed in the maternal circulation.[33] However, it is not clear how well maternal cells survive in the foetal circulation and what relevance this may have for vertical transmission of not only malignant diseases but also of other disease categories such as auto-immune diseases. Even theoretical persistence of maternal malignant cells in the fetal circulation is reason to reject a maternal donor of cord blood who has a history of malignancy. These observations also have direct relevance where fetal stem cells are used for "tissue" or "organ" culture if they are contaminated by the low-level presence of maternal cells.

In addition, a genetic predisposition to disease in the mother may be inherited by an infant cord blood donor with a genetic predisposition to disease in the mother. This is an alternative means of potential transmission as against by cellular trafficking. For example, a maternal carcinoma might be due to an underlying genetic predisposition to malignancy, especially if there is an associated family history of malignancy at an early age. This predisposition may be inherited by a cord blood donor and then transplanted to the recipient by cord blood transplantation. An example is maternal hepatic carcinoma. This might be associated with either environmental exposure to a carcinogen such as aflatoxin but could also be associated with an autosomal recessive condition such as Wilson's disease which might not be recognised at the time of the cord blood donor's birth or even at six months or one year.

(c) Diseases of Unknown Aetiology

Crohn's Disease and ulcerative colitis, and some neurological degenerative disorders such as Parkinson's disease are examples of diseases where the aetiology is not clear although in some cases, an infectious aetiology is suspected. In these cases, it is difficult to accept donors not only because the risk of disease transmission cannot be accurately assessed but also because the sufferers may have undergone many therapeutic procedures or received medication and have been chronically sick. Long term immobilisation or poor nutrition may affect the quality of some tissues, particularly bone. Other tissues appear not to be affected by such diseases e.g. cornea banks. Hence, it is for this reason that donations from individuals considered unsuitable by some multi-tissue banks are often accepted by other banks.

(d) Unknown/Theoretical Risks

Some donors can be demonstrated to carry viruses that can be transmitted to others but the viruses are not proven to be pathogenic e.g. hepatitis G.[34,35] Others carry viruses which are not routinely tested for but which may have a pathogenic role in some circumstances. An example is HHV8, which has been shown to have associations with sarcoidosis,[36,37] Kaposi's Sarcoma,[38-40] myeloma[41] and in lymphoma and Castleman's disease.[42,43] However, it has been argued that detection of the virus does not necessarily prove causality.[44]

(e) Donor Xenogenic Exposure

There are many potential sources of xenogenic exposure for patients which may lead to third party exposure if such patients become donors of blood or tissues. In many cases, the patients may not be aware that they have been exposed and are therefore unable to declare the exposure when they or their families are questioned as part of the selection of donors to ensure the safety of donated material for transplantation. These major theoretical concerns relate to exposure to novel viruses (e.g. animal retroviruses) and prions. It should be noted that none of the sterilisation techniques routinely applied to xenogenic tissues (e.g. glutaraldehyde, gamma irradiation and ethylene oxide) would inactivate prions. In fact, glutaraldehyde probably stabilises prions. In addition, it should not be assumed that industry standard dosages of sterilising agents e.g. 2.5 MRad (25 KGy) of gamma irradiation will inactivate all types of viruses at physiological titres. Nevertheless, because cross-infection with viruses has not been reported for xenogenic tissues that have been "sterilised" using these methods, the UK Department of Health currently considers them to be acceptable for transplantation from the point of view of non-viral infective agents. Xenogenic materials used in tissue grafts or implants can be categorised as follows: –

Extracted components from xenogenic tissues, e.g. bovine collagen used in a variety of synthetic products including some dermis replacement material used in burns, plastic surgery and other surgical indications. It is crosslinked and terminally "sterilised" and is used to coat nylon mesh in another skin substitute.

Modified animal tissues, e.g. porcine heart valves that have been crosslinked with glutaraldehyde and "sterilised" by gamma irradiation. In the past, other animal tissues have been used to manufacture heart valves including bovine pericardium, and animal tissues have also been used to make haemostatic felt and catgut sutures.

Animal serum and other extracts as cell growth supplements. If cells for tissue engineering are to be cultured *in vitro* prior to implantation, the commonest supplement added to the culture medium is foetal calf serum (or simply calf serum or serum from other animals). The source of the serum is now said to be from controlled herds and from countries without BSE. It is not clear if this was always the case, or when change to superior practice for the selection of animals for the serum supply was implemented.

In the UK, this cell growth activity largely relates to cultured skin for burns and leg ulcers, and cultured chondrocytes for articular cartilage regeneration. Some hospital laboratories have carried out these types of culture since the early 1980s. In the USA, commercial companies do it on a larger scale. One development in the 1980s was to substitute bovine pituitary extract for foetal calf serum for culture of skin cells, in "serum free medium". It is not clear how the suppliers of growth medium sourced the bovine pituitaries at the time, or how many patients received grafts cultured by this method either in the UK or world-wide. At the end of the culture period, most laboratories culture the skin grafts in medium without serum, in an attempt to remove xenogenic proteins but the efficacy of this step is not clear.

Animal Cells: Culturing of human keratinocytes usually requires a layer of feeder cells. The most commonly used feeder cells are mouse 3T3 fibroblasts (J2 strain). The feeder cells are usually treated with either Mitomycin C or irradiation to prevent them from proliferating but allowing them to continue to metabolise. Although attempts are made to remove these cells prior to transplantation, this is not guaranteed.

It may be that additional consideration is needed to consider whether it is sensible to only exclude potential donors on the basis of having received an obvious xenograft when a much larger population will have received indirect xenogenic exposure risk. Unfortunately, many of these patients will not be aware of their exposure. Also, if for example catgut suture exposure were included in the cohort of donors for exclusion, the impact would presumably be huge. Furthermore, there is no hard evidence that components extracted from tissues (e.g. collagen/gelatin) are any less risky than the whole processed tissues. These components are found in a huge diversity of surgical products. Presumably, the risk from such products would be commensurable with exposure to pharmaceutical gelatin in medication. However, on the favourable side, "viable" xenogenic exposure to 3T3 fibroblasts does not appear to be problematic as these cells have been well characterised and hopefully do not present a significant threat. Nonetheless, these aspects require consideration for the future of cell culture and tissue engineering in general.

(f) Variant Creutzfeldt Jacob Disease (vCJD)

vCJD occurs predominantly in the young. Cadaver tissue donors come from a wide age spectrum, the average being 51 years in the authors' tissue bank. Paul Brown[45] has suggested that there is uncertainty whether latent cases of vCJD may be capable of producing transmission by cross contamination in surgery or be transmitted by blood, tissues or organs. It has been suggested that between 63 and 136,000 cases of vCJD may occur in the Great Britain population.[46] On this basis, the number of vCJD infected individuals is not likely to be large, but still remains unknown. The likelihood of transmission of vCJD through tissue donation and transplantation therefore also remains unknown. The number of UK cadaver tissue donors is relatively small (about 400 for the English National Blood Service per annum) but each donor may supply grafts for up to 40 recipients. The potential for amplification of disease transmission is therefore high. Review of recent consecutive Tissue Services donors illustrates current sources of cadaver tissue donors. The cause of death in these donors was 32% cardiovascular accidents, 28% cardiac deaths, 20% road traffic accidents, 4% suicide and 16% "other" causes.

There is no firm evidence that CJD can be transmitted in humans by transfusion and although animal experiments have provided evidence of the transmissibility of BSE by blood transfusion,[47-49] there have been no equivalent experiments using human tissue transplants. However, human TSEs, e.g. classical CJD, have been transmitted parentally by dura mater, and pericardium.[1] vCJD first described in the 1990s differs in many clinical respects from classical CJD but may share its transmissibility.

REDUCING THE RISK OF DISEASE TRANSMISSION

Ensuring the microbiological safety of individuals, donations of tissues and cells for transplantation depends on the exclusion of high risk donors by a process of selection, and then testing for a range of microbiological infections such as HIV and Hepatitis B and C and Syphilis. HTLV testing is commonly undertaken in many European countries and in the US and will soon commence in the UK. It is also expected that in due course, a test for variant Creutzfeldt Jacob Disease (vCJD) will be developed although currently no risk factors have been identified that will reduce the risk of transmitting this disease with any certainty. Surveillance of known cases may provide guidance in this field in due course. Because of constant improvement in epidemiological information, it is recommended that an advisory structure for all tissue donor selection, testing and infections aspects should be employed to constantly review new information as it becomes available.

(a) Microbiological Risks and the Selection and Testing of Donors

The principles for selection of living donors of tissues and cells, e.g. bone marrow donors, are very similar to those for allogeneic blood. The process for cord blood and cadaveric donors is different. Whilst the medical history can be elicited directly from the blood or bone marrow donor, this is not the case for some other donations. In the case of a cadaver donor, the history is taken from a relative, and in the case of a cord blood donor, the history is taken from the mother as a surrogate for the donor infant.

There must be strict documented protocols for the selection of donors. There should be standard questionnaires which are administered by trained staff to either the donor, and donor surrogate, such as the mother in the case of cord blood donors, or in the case of cadaver donors, to an appropriate family member. The questionnaire should address the recent and past medical history, the circumstances of death, relevant family history, and behavioural risk factors. Additional information should be sought from the primary healthcare professional, from the medical notes and any other relevant sources. In the case of a cadaver donor, the post mortem report is a relevant source. The donor or the donor family must provide consent for this information to be sought and the information must be held in a confidential and secure manner.

At present, screening for HTLV is not currently mandatory for blood or tissue donors in the UK, but for tissue or cord, blood donors can be undertaken on archived samples stored at the time of donation. The UK is imminently moving to HTLV screening for blood donors and this will impact on the safety of donations. Such screening is undertaken in many European countries.

A validated malarial antibody test has been used in blood donor screening[50] and has also been applied to the testing of tissue donors and of cord blood donors who come from malaria endemic areas. Potential exposure to the various malarial parasites must have occurred long ego, enough for malarial antibody to be detectable. Overt malarial infection in the mother of a cord blood donor at the time of delivery would be reason for exclusion although vertical transmission of malaria is unlikely in a mother who is not acutely ill at the time of cord blood donation. If the malaria exposure is recent and therefore detection of seropositivity cannot be relied upon or, if the mother is confirmed to be seropositive, transplant clinicians can be informed of the potential risk and be alerted to consider malaria as the possible cause of a pyrexia of unknown origin in the recipient. This may not be appropriate in the case of a tissue donation.

Malaria testing is not a mandatory test, and the risk of malaria and its treatment, is substantially less than not having a matched donor available if a bone marrow transplant is indicated. These arguments need consideration in deciding whether to reject a cord blood or bone marrow donation at a small risk of transmitting malaria.

The arguments for taking small risks in the context of finding a rare matched cord blood or bone marrow donation cannot be easily extended to the banking of routine tissue donations. In these circumstances, strict donor selection criteria can be applied together with a validated malarial antibody screening test if available. There is also evidence[51,52] that the use of 25 KGy gamma irradiation can be considered for non-viable tissues to irradicate any residual malaria parasites (as well as any contaminating bacteria).

The risk from surgical or cadaver tissue donors or from cord blood donors may be greater than with regular volunteer blood donors. Although they are also volunteers, they or their relatives are approached and asked if they wish to participate in the donation programme. In some circumstances, ethnic minorities are specifically recruited and may come from areas where specific viral infections are endemic. For example, potential donors from Afro-Caribbean countries are more likely than indigenous UK donors to be infected with HTLV. A structured interview with a standard questionnaire and full documentation of responses between the candidate donor and trained staff will reduce this risk. The interview forms the foundation of donor selection. After this, the mandatory micro-biological tests are applied.

There are particular issues relating to testing. Firstly, the risk of a window period infection (i.e. an infectious but undetectable donation due to recent infection in the donor before antibodies to the infection can be recognised) may be significantly addressed by identifying individuals with behavioural risk.

Secondly, the validity of the blood sample used for testing is a critical factor. Microbiological test kits have, in general, been validated for use employing blood samples from blood donors or patients. They have not been validated, in general, for use with blood from cadavers and a recent survey has shown a high incidence of false reactive results with cadaver samples, especially where the samples are obtained more than 24 hours after the cessation of the circulation.[53]

Thirdly, cadaver donors have not infrequently suffered trauma and have been transfused and, especially where there is both blood loss and the use of infusions, the donor may be plasma diluted resulting in invalidity of the sample for testing. In these cases, a plasma dilution algorithm must be employed.[54]

The sample may also have been taken from the site of an infusion and could have been diluted by the infusate. All samples must be taken distant to the site of infusions to avoid this complication.

Strict protocols for the taking of samples and their inspection should be documented and employed.

(b) Genetic and Auto-Immune Diseases in the Donor

Transplantation particularly of viable cells especially in the context of profound immuno-suppression has the potential to transmit diseases including non-microbiological ones. Bone marrow transplantation (BMT) or equivalent cord blood transplantation (CBT) may provide a treatment for a variety of genetic disorders. Sickle cell disease, other homozygous or double heterozygote haemo-globinopathies, thalassaemia, the immune deficiencies, Fanconi's anaemia and the metabolic diseases all fall into this category. Allogeneic and autologous transplantation have also been used for treatment of solid malignant tumours and immune disorders.[55] However, the greatest indication for BMT or CBT is for malignant disease, mostly of the bone marrow e.g. for leukaemia. There are also many autologous transplants undertaken for auto-immune disorders.[55] It is not surprising that diseases have been transmitted by BMT including the auto-immune mediated disorders such as myasthenia gravis[56] and thrombocytopenia[57] and diabetes.[58] Any of these groups of diseases, which could theoretically be treated by BMT or CBT, could also be transmitted by that therapeutic procedure.

Careful donor selection using a standardised questionnaire with a view to eliciting family history of malignancy, leukaemia or an inherited metabolic disease will reduce this risk. A careful family medical history and appropriate investigation may help inform the decision whether or not such a donor should be accepted. These factors may also contribute to the appropriate evaluation of second malignancies in the transplant recipient population.[59] Haemoglobinopathy screening should be undertaken routinely for all infant donors of cord blood and adult donors of bone marrow if the individual is from an ethnic background known to be at risk. In multi-racial societies, there should be blanket screening.

There are potentially some other genetic diseases in the donor, which could be identified by programmes such as DNA screening. For example, acute lymphoblastic leukaemia (ALL) has been shown to be associated with an abnormal fusion protein that may be present before birth and may precede the development of ALL by several years.[60,61] Screening for these fusion proteins or for their corresponding gene could be considered a useful safety criterion for cord blood banking but only if their presence was clearly predictive of disease. This is not currently the case and these abnormal proteins may be only one of several interacting factors involved in ALL aetiology. At the present time, such donation screening and the resulting requirement for counselling of donors with a positive screen would be fraught with uncertainty. However, the principle does apply and other candidate disease markers may be identified in the future.

In the case of tissue donation or donation of non-haemopoietic cells, the risks of transmitting a genetic disease are different. Tissue donation often involves non-viable frozen material. These donations are unlikely to contain viable lymphocytes or proliferative cells and this means that the risk of transmission of auto-immune disorders or malignant disease would be remote. However in the case of, for example, viable pancreatic islet cells for the treatment of an allogeneic recipient,[62] it might be considered advisable to select a donor at low risk of diabetes by virtue of their family history or other markers of susceptibility.

There have been transmissions of malignancy by viable organs and this is also theoretically possible for bone marrow. Donor selection guidelines have been produced by the Council of Europe regarding the risk from donors with a malignancy.[63] These restrictions are also often applied to some tissue donations although the risk may be lower from non-viable tissue with regard to this type of disease transmission. However, disease transmission risks may remain if a donor malignancy has a viral origin e.g. T cell leukaemia.[64]

(c) Issues Relating to Prions

1. A number of actions may improve the safety of the tissue supply with regard to vCJD transmission, including leucodepletion of bone and the use of disposable instruments where feasible. The shelf life of some tissues is very long e.g. three years for bone and heart valves and possibly decades for cord blood. Archived samples could be tested for vCJD in the future, long after the tissues were donated, when blood tests for vCJD have been developed. With cadaver tissue donors, there is the potential for acquiring additional analytes, for example, peripheral reticulo-endothelial tissue, such as appendix, tonsil or a brain biopsy for testing purposes.

2. The testing of cadaver tissue donors for vCJD is fraught with ethical issues, albeit probably less than with living blood donors. Although there may be no consequences of a positive test for a cadaveric donor, specific family consent for testing for vCJD may be required in the current climate. Medico-legal and counselling advice is needed on this issue particularly because consideration is needed for the donor family if in the future a genetic predisposition to vCJD were to be identified. Recipient issues also exist. If a donor of organs and tissues were found to have a positive test result and the organs had already been transplanted before the test results were available, there would be recipient counselling issues. If retrieval staff have a needlestick injury with an implicated donor, there will also be counselling and clinical management issues for the staff member.

3. There are also implications for the supply of tissue. Analyte acquisition may require special procedures e.g. brain tissue might only be obtained from donors undergoing an autopsy. Only a proportion of the current cadaver donors are also autopsy cases. The remaining donors would either be excluded or complex procedures for obtaining analyte introduced. Clearly, current living donors of surgical by-product tissues could not be tested

in this fashion. Peripheral reticular tissue such as tonsil may have significant abnormal PrP^{SC} present before the brain is affected by vCJD so that a tonsil biopsy test might be more sensitive than using a brain biopsy. The concentration of PrP^{SC} in tonsil is 10% of that seen in brain in the active clinical phase.[65]

4. Tests for prion exist but sensitivity is the major issue. Only one report claims to have detected PrP^{SC} in the blood of sheep with scrapie.[66] Separating normal prion (PrP^{C}) from PrP^{SC} has been done using protocadherin 2, described as a high affinity cell surface receptor for PrP and by plasminogen.[67]

(d) Counselling Issues

Any donor selection or testing policy will result in some donors being rejected from the programme. It will, from time to time, identify individuals who carry infections or diseases of importance for the donor or the mother (in the case of cord blood), and other family members or contacts. Tissue banks have a duty of care to the donors. In the consent procedure, the bank should undertake to confidentially inform donors or relevant family members in the event of any positive test results if they are of relevance to their health. With the donor or family's permission, other professionals such as the general practitioner or specialist referral units will be informed. Where there are possible infectious sequelae for the donor, there should be the facility for immediate referral to a pertinent specialist centre. Where universal or selective screening programmes from cord banks identify mothers with HIV or hepatitis B, optimal care can be provided for the mother/infant pairs. Not all obstetric units have these HIV and hepatitis screening programmes and selective screening programmes may fail to identify all infected mothers. There may be liability issues following a cord blood bank's late identification of risk families and the lost opportunity for preventative action to have been taken. There are additional counselling issues when additional test results at the time of selection of a cord blood for transplantation become available or when a new test is introduced for tissues that may be banked for long periods before issue. In such cases, it may be difficult or impossible to contact the donors. Other related issues arise when a genetic disease or predisposition to disease is identified. The counselling issue may be difficult and will vary from case to case. The issues may increase as more DNA based tests for genetic diseases or genetic susceptibilities (e.g. for vCJD) become available. There should be systems in place for donors to give consent for such tests before they are undertaken.

(e) Inactivation/Sterilisation

Very high densities of micro-organisms are present in close proximity to epithelial surfaces of the human body including the lumen of the gut and the epithelial appendages (e.g. hair follicles) of the skin and the birth canal in the case of umbilical cord blood. Within the body, in addition to transient bacteraemia caused by breakdown of the epithelial barrier (e.g. brushing teeth, abrasions, etc.), there are a range of intracellular viruses and bacteria that can persist for varying periods, in some cases permanently. Although great efforts are made to screen donors for the most significant viral diseases and to avoid contaminating retrieved tissues by post-mortem translocation of bacteria from the gut or from the skin during the retrieval process, there remains a finite risk. With respect to viruses, the main risks come from viruses which are not routinely screened for or not currently known. The use of screening tests, developed for fresh blood which are not as reliable when applied to post-mortem blood, needs consideration. Finally, recent infections may not be detectable by current screening tests. Regarding bacteria and fungi the major risks are the possibility of gross contamination prior to or during the retrieval process or minor contamination during retrieval or subsequent tissue processing steps after which the microbes are allowed to proliferate to infectious-dose levels.

These microbial risk factors have led to the establishment of a set of principles applied to tissue banking to reduce microbiological infection, which are listed below in an escalating sequence of risk reduction.

1. Take steps to ensure the donor tissues are not grossly contaminated prior to retrieval.
2. Minimise contamination during retrieval.
3. Following retrieval:

   - avoid adding further microbiological contaminants;
   - take steps to prevent microbiological proliferation during processing and long term storage;
   - apply one or more steps to reduce the microbiological load (preferably below the infectious dose) (disinfection);
   - or eliminate all micro-organisms (sterilisation).

This section will consider the latter three principles.

(f) Prevention of Microbiological Proliferation

Regardless of whether disinfection or sterilisation procedures are to be applied to the tissue, it is imperative that any micro-organisms present are not allowed to proliferate during the various transportation and processing steps. Refrigeration is the commonest technique used to inhibit microbiological growth. At temperatures between 0–4°C, very few micro-organisms exhibit significant growth. Those that can grow will do so very slowly so that if the total refrigeration time allowable is short (circa 48 hours) and providing the specified temperatures are maintained, then little or no microbiological growth is expected. A second benefit of refrigeration is a reduction in the metabolism of the tissue cells, and the rate of enzymatic and other detrimental reactions.

For tissues to be used as non-living implants, a further reduction in temperature will progressively inhibit both microbiological growth and other detrimental spoilage processes. The lower the temperature, the longer the tissue may be stored without risk of deterioration. Low temperature storage may also be applied to viable tissues if appropriate validated cryopreservation procedures are used during the cooling phase.

Antibiotics that are bacteriostatic, or used at bacteriostatic concentrations, may be used to inhibit microbiological growth. However, it is important that they are used under conditions that facilitate their bacteriostatic effects e.g. it is pointless adding an antibiotic to a low temperature transport solution if it is ineffective at low temperatures.

Other methods that inhibit microbiological growth are either too damaging to cells or tissue matrices, or are only suitable for application to the final long-term banking method as opposed to the transportation and processing phases. This includes lyophilisation techniques. Most bacteria and fungi can only grow within narrow limits of water activity. If the osmotic pressure of the environment is increased by the addition of salts, sugars or other solutes; or by the physical removal of water (e.g. by freeze drying), eventually growth will be inhibited. Generally, fungi are more tolerant of low water activity than bacteria, although there are halophilic bacteria that can live in very high osmotic environments. However, most pathogens grow optimally in body fluids, which have an osmolarity of circa 280–300m Osm/Kg H_2O. Tissues that are freeze dried to less than 5% (w/w) residual water can be banked long term at room temperature without any prospect of microbiological proliferation. An alternative to freeze-drying is chemical drying using very high concentrations of solutes that effectively dehydrate the tissue. One example is high concentration (98%) glycerol.[68]

(g) Disinfection

In addition to ensuring that any micro-organisms present are unable to proliferate, a second approach is to reduce the bioburden using a disinfection step. This may be applied prior to extensive processing in order to decontaminate the tissue prior to its introduction into the aseptic processing laboratory, particularly where multiple processing steps are required (e.g. for bone tissue). Alternatively, it may be utilised as a terminal disinfection step just prior to long term storage (e.g. antibiotic disinfection of cryopreserved tissues).

A number of disinfection procedures have been applied to tissue grafts and implants. The following encompasses the most commonly used methods.

(i) Heat

Heat is particularly useful for disinfecting bone, where the dense mineral matrix would pose significant diffusional barriers to e.g. fluid disinfectants. Pasteurisation temperatures of circa 60°C have a long history of use for biological products, for example, albumin treated at 60°C for 10 hours.[69] This treatment completely inactivates many groups of vegetative bacteria and viruses. It has been shown to at least reduce the titre of those viruses and bacteria not completely eliminated. Although application of higher temperatures may be more effective, adverse effects on the structure and biological function of bone begin to appear. In his early studies on bone induction by demineralised bone matrix, Urist warned that temperatures above 70°C produced heat shrinkage of bone collagen. This indicates denaturation.[70] In a later study, osteoinductive capacity of bone declined markedly when the bone had been subjected to temperatures above 60°C[71] Burwell (1966) also demonstrated that boiled iliac bone induced little new bone formation.[72]

(ii) Alcohols

Several alcohols have been shown to possess antimicrobiological properties. Generally, the alcohols have rapid bactericidal activity but are not sporicidal and have poor activity against many viruses.[73]

(iii) Hypochlorite

Hypochlorite is effective against bacteria viruses and spores providing a satisfactory concentration of free chlorine is achieved. It has been advocated for disinfection of amniotic membrane.[74] Its disadvantages include inactivation by organic material and instability at low concentration.[75]

(iv) Glutaraldehyde

At a sufficiently high concentration (2% alkaline), glutaraldehyde is considered a sterilising agent providing it can fully permeate the material. Potential disadvantages are that it crosslinks collagen molecules[76] and may form cyclic polymers within the matrix. The latter may de-polymerise and release toxic monomers. Although still used for processing porcine heart valve xenografts (primarily for its crosslinking ability), it is little used for treating allografts.

(v) Mercurials

Various mercurials e.g. "Cialit" have been advocated for the disinfection and long term preservation of e.g. cartilage, middle-ear ossicles and dura mater.[77] However, the antimicrobiological activity has been found to be inadequate.

(vi) Glycerol

High concentration glycerol (98%) has been used to preserve allogeneic skin,[68] presumably by chemical dehydration. Although it is sometimes claimed to disinfect or even sterilise the skin, it is not considered as a disinfectant in any of the microbiological texts and its sporicidal activity is poor.

(vii) Antibiotics/Antimycotics

For the preservation of viable tissues (e.g. eye corneas, skin, heart valves, etc.), the only disinfection step compatible with continued cell viability is the use of cocktails of antibiotics/antimycotics. These must be carefully chosen to ensure that they cover the whole spectrum of possible contaminants and that the mixtures are not antagonistic. They must be effective at the chosen incubation temperature and consideration is given to the shelf life of the mixture at the chosen long-term storage temperature. Obviously, antibiotic cocktails will not disinfect any viruses within the tissue.

In addition to being effective against the contaminating bacteria and fungi, the antibiotic mixtures should not be cytotoxic towards the tissue cells. Although there are plenty of targets on prokaryotic bacteria not shared by the eukaryotic human cells for antibiotics to attack, this is not always the case for antimycotics directed against the eukaryotic fungi. Type of antifungal, dosage and interaction with the other antibiotics must be assessed for cytotoxicity.[77]

(viii) Cryopreservation

For certain other viable grafts e.g. bone marrow and cord blood, the imperative is to preserve as many stem cells as possible. This is optimised by immediate cryopreservation after collection. Nevertheless, samples are cultured for microbiological contamination. A positive culture would not necessarily result in discard of the donation. The donated material is potentially so precious that the clinical user and recipient may decide to use the donation but also to treat the recipient with appropriate antibiotics which are demonstrated to be effective against the identified contaminant.

(h) Terminal Sterilisation

Terminal sterilisation is the sterilisation of the tissue sealed within its final packaging such that following the sterilisation procedure, the package is not re-opened until used surgically.

In order to be considered a true sterilisation procedure, it must be effective against all groups of micro-organisms that may be present in or on the tissue including bacteria, fungi and viruses (prions are excluded from this discussion). In addition, in the UK tissue banking standards, there is a requirement that the procedure must be able to inactivate the numbers of microbes normally present on the tissue (bioburden) plus an additional 6 $\log^{10}$ of the most resistant microbe. This overkill gives a sterility assurance that less than one in a million tissue grafts will be positive. Obviously if one or more disinfection steps have been applied prior to terminal sterilisation, the average bioburden might be very low. However, there is still a requirement to demonstrate the additional 6 $\log^{10}$ inactivation of the most resistant microbiological type.

The two commonest terminal sterilisation procedures for tissues have been the use of either ionising irradiation or ethylene oxide gas. For the latter, the most resistant microbiological forms appear, from the literature, to be bacterial spores. By measuring inactivation rates for spores placed within tissue matrices, Kearney *et al.* were able to calculate the decimal reduction rate (D-value) for a commercial ethylene oxide steriliser.[78,79] The results showed that in theory, up to 10^{30} spores could be inactivated in a normal sterilisation cycle. Therefore, this worst-case scenario demonstrated complete compliance with the definition for sterilisation defined above, by a number of orders of magnitude.

For irradiation sterilisation, the literature demonstrates a much wider range of D-values for the various microbiological groups, with a strong correlation to the size of the micro-organism, the smaller being the more resistant. The most resistant microbiological group is therefore the viruses. For example, the D10 values reported for bacteria and bacterial spores ranged from 0.03 KGy to 10 KGy, whereas for viruses values ranged between 1.7 to 13 KGy.[80] The latter (foot and mouth virus) would thus require a radiation dosage of 78 KGy (7.8 MRad) just to provide the 6 $\log^{10}$ of overkill. This dosage would cause severe damage to tissue matrices. It is therefore doubtful whether the commonly used dose of 25 KGy (2.5 MRad) can be considered a sterilising procedure with respect to all possible pathogenic viruses that could be present. However, this dose would represent a significant effect on protozoan parasitic diseases such as Malaria and Chagas' Disease.

Other agents e.g. peracetic acid, that are capable of inactivating all groups of micro-organisms including spores, have been used to decontaminate tissue implants, but may have a detrimental affect on the integrity of some of the constituents of the tissue such as collagen.

Aqueous solutions are more difficult to use as terminal sterilants owing to the difficulty in removing the agent at the end of the exposure period without compromising the tissue. This would not be impossible particularly for peracetic acid, as the breakdown products are natural non-toxic compounds i.e. acetic acid, water and oxygen. At the end of the sterilisation process, rapid breakdown of the peracetic acid can be achieved by the addition of an inactivator.

(i) Risk Assessments

Whenever a new risk arises or becomes evident, it is important to re-evaluate all products, processes and procedures in a formal way. This usually entails the application of a risk assessment. There are many different models for performing risk assessments. However, most will address the following:

1. The first stage is to identify all of the risks associated with the use of a tissue or cell graft. For example, this may include risks of transmission of known viruses and unknown viruses, bacteria,

parasites and malignancies, transfer of toxic components or generation of toxic components during processing etc. In addition, there may be a risk that the tissue graft may not function appropriately in the body.

Identifying all of the risks is important so that ultimately an assessment of the risks and benefits can be objectively assessed. For example, it is pointless applying a sterilisation technique to e.g. an eye cornea to reduce risks of disease transmission, if it kills the corneal cells and renders the cornea non-functional.

2. Each risk factor is assigned a weighting (numerical or otherwise) indicating the severity of consequences i.e. impact score for each of the risks. Then a probability weighting or score is assigned to each risk indicating how likely it is to occur for the given tissue graft or procedure. In this way, low impact risks with low probability of occurrence will have little influence on the score whereas high impact, high probability risks will contribute greatly to the overall score. The total risk for a particular graft or procedure can then be derived by combining impact and probability scores (e.g. multiplications) for each risk factor. The total score gives the overall risk to the recipient for that graft/procedure.

3. If the overall risk is considered high, changes can be considered and similarly analysed. Although there are subjective elements in this procedure, it does at least provide a model for considering balance of risk and for the ranking of magnitude of risk.

(j) Regulation and Guidance for Tissue and Cell Banking

Many countries have recognised the need for regulation of tissue banking to ensure safety of human grafts or transplants for patients and for the ethical management of donors and their relatives. In 1993, following the discovery of commercially imported tissue allografts infected with hepatitis B virus, the US Interim Rule[81] was implemented by the Food and Drug Administration (FDA) which allowed FDA inspection of tissue banks, as well as the recall and destruction of tissue. This Interim Rule focused on assuring the safety of tissue for transplantation with emphasis on the micro-biological aspects. The Final Rule was implemented in 1997. Good Manufacturing Practice (GMP) requirements were not promulgated for tissue banks in the US where broader legislation is now under consideration. FDA regulation was recently reviewed following a US scandal involving the commercial tissue banking sector. This showed inadequacies in the FDA inspection system and professional organisation accreditation. A number of recommendations were made including that all tissue banks should be registered, and that the FDA should complete an initial inspection of all tissue banks. It was also recommended that there should be an appropriate minimum cycle for tissue bank inspections and that the FDA should work with State and professional associations for inspection and accreditation.

In individual European countries, there is wide variation in many aspects of the regulation of tissue banking but only one Europe-wide Recommendation. This was issued by the Council of Europe and pertains to ethical and safety issues.[82] Practices are also not uniform in Europe. For example, some countries follow the practice of "opting in" for organ donation which requires next of kin consent in contrast to the adoption of "presumed consent" or "opting out" in accordance with the recommendation of the Council of Europe in 1978. A summary of relevant laws and guidance from the UK, and Europe is given in Table 15.1.

In addition to the US's FDA regulation and the guidance in Tables 15.1, there are tissue banking regulation or transplant laws in Belgium, France, Spain and Greece. In Germany and Australia, a more pharmaceutical approach to tissues has been taken. Also, there are Council of Europe tissue banking standards covering safety and quality of organs, tissues and cells which were published in 2001. There are also relevant medical devices directives. The impact of the signing of the Amsterdam Treaty means that regulation of

Table 15.1. UK laws and guidance for tissue banking.

1. Guidelines for the blood transfusion services in the United Kingdom — 5th Edition 2001 available from the UK Stationery Office ISBN-0-11-702555-0 2001
2. A Code of Practice for Tissue Banks. UK Department of Health 2001
3. Decontamination of medical devices HSC 2000/032 18th October 2000
4. Committee on Microbiological Safety of Blood and Tissues for Transplantation, Department of Health: Guidance on the microbiological safety of human tissues and organs used in transplantation. *NHS Executive*, August 2000
5. Variant Creutzfeldt-Jakob Disease (vCJD): Minimising the risk of Transmission HSC 1999/178 13th August 1999
6. Controls Assurance in infection control: Decontamination of Medical Devices HSC 1999/179 13th August 1999
7. Report of the Medical Research Council Working Group: Operational and Ethical guidelines for collections of human tissue and biological samples for use in research Nov 1999
8. Consensus statement of Recommended Policies for uses of Human Tissue in Research, Education and Quality Control with notes reflecting UK law and practices prepared by a Working Party of the Royal College of Pathologists and the Institute of Biomedical Science 1999
9. The Retention and Storage of Pathological Records and Archives. Report of the Working Party of the Royal College of Pathologists and the Institute of Biomedical Science. Second Edition 1999
10. Brithtish Association for Tissue Banking 1999. Standards www.batb.org.uk
11. Duties of a doctor comprising the four GMC booklets: Confidentiality October 1995, Serious Communicable Diseases October 1997, Good Medical Practice July 1998, Seeking Patient's Consent: The Ethical Considerations November 1998
12. A code of practive for the diagnosis of brain stem death including guidelines for the idetification and manegement of potential organ and tissue donors. March 1998 Department of Health

Table 15.1 Continued

13. Transmissible spongiform encephalopthy agents: Safe working and the prevention of infection. Advisory Committee on Dangerous Pathogens Spongiform Encephalopathy Advisory Committee. March 1998 The Stationery Office. ISBN 0-11-322166-5
14. Date Protection Act 1984
15. Anatomy Act 1984
16. The Department of Health Caldicott Committee Report on the Review of Patient-Identifiable Information December 1997
17. Rules and Guidance for Pharmaceutical Manufacturers and Distributors (EEC Orange Guide). The Stationery Office 1997
18. Human Tissue Ethical and Legal Issues Published by the Nuffield Council on Bioethics 1995
19. Human Organ Transplants Act 1989
20. Coroners Act 1988
21. The Human Tissue Act 1961
22. Protecting health care workers and patients from hepatitis B. NHS Management Executive. HSG(93)40 and Addendum to HSG(93)40
Relevant Council of Europe Documents and European Community Guidance
1. Convention for the Protection of Human Rights and Dignity of the Human Being with regard to the Application of Biology and Medicine: Convention on Human Rights and Biomedicine — Oviedo, 4IV.1997 — European Treaty Services/164
2. Standardisation of Organ Donor Screening to Prevent Transmission of Neoplastic Diseases — ISBN 92-871-3485-5 — Council of Europe — December 1997
3. State of the Art Report on Serological screening methods for the most relevant microbiological diseases of organ and tissue donors — SP-PB(96)21-E Council of Europe — Strasbourg 1997

Table 15.1 Continued

4. European Pharmacopeia monograph on Solutions for organ preservation
5. Recommendation No. R(94) 1 on Human Tissue Bank
Relevant Commission of the European Communities Document
1. The rules governing medicinal products in the E.C. Vol. IV Good Manufacturing Practice for medicinal Products. Luxemboung: Office for Official Publication of the EC, 1992. ISBN 92-826-3180-X

tissues by the European Commission is likely in the next few years and is very likely to cover both tissues and cells. This will improve the diverse national European approaches, which result in a difficult environment for the safe development and issue of human tissues, and tissue engineered products.

A recent international cord blood banking survey[83] revealed marked variations in non-commercial CB banking practice, even within the same country and this may also apply to the commercial cord blood banking sector as well. In the UK, this is also true of many other activities relating to procurement, processing, testing[53] and storage of tissue products of human origin. Registration, agreeing standards, inspection and the accreditation of tissue and cell banks could achieve standardisation. There have been many international efforts to agree relevant standards and to move to a system of peer review of umbilical cord blood banks, and related activities dealing with haemopoietic stem cells.

In the UK, a regulatory framework for non-commercial therapeutic tissue and cell banking commenced in 2001. This involves inspection of tissue & cell banks and facilities by the Medicines Control Agency according to the Department of Health's Code of Practice for Tissue Banks[84] and the UK MSBT recommendations.[85] These two documents include all aspects of microbiological safety and of good manufacturing practice. These UK measures will help to ensure the safety and quality of tissues and cells for all patients. It will also ensure documented high standards allowing confident movement of tissues and cells across national boundaries.

There is considerable European debate regarding whether non-viable human tissues should be classified as medical devices, as discussed by the European Confederation of Medical Suppliers Association Industry Task Force Meeting, in June 1995. Whilst such a move would ease the movement of tissues within Europe, it could promote the evolution of tissue banking towards commercialisation. Treating donated tissue as a product for sale would diverge from the direction taken for blood and could seriously hamper the recruitment of volunteer donors. Previously, human tissues have been excluded from the Medical Devices Directive 93/42 EEC, due to concerns regarding safety, ethics and sourcing; these arguments still apply.

CONCLUSION

There are numerous potential infectious and disease transmission risks along the path from the recruitment of a cell or tissue donor to the final selection of material for transplantation. Several strategies are needed to reduce the risk of transmission of disease. The interview of the donor or of the donor family, in the case of cadaver donors, forms the basis for the selection of donors at low risk of blood borne infections and other diseases. Subsequent testing of the donor for the mandatory markers enables further reduction of the risk. Additional information to elicit risk factors for infectious and genetic diseases may also be sought from a variety of other agencies.

Any cost-effective strategy to maximise the safety of tissues and cells for transplantation must take heed of the conflicts that arise to ensure safety for transplantation. There needs to be an ethical and cost efficient balance between donor considerations and the safety of the recipient.

REFERENCES

1. Eastlund T. Infectious disease transmission through cell, tissue and organ transplantation: Reducing the risk through donor selection. *Cell Transplantation* 1995; **4**:455–477.
2. Tomford WW. Transmission of disease through transplantation of musculoskeletal allografts. *J Bone Joint Surg* 1995; **77a**: 1742–1754.
3. Simonds RJ, Holmberg SD, Hurwitz RL, Coleman TR, Bottenfield S, Conley LJ, Kohlenberg SH, Castro KG, Dahan BA, Schable CA, Rayfield MA, Rogers MF. Transmission of human immunodeficiency virus type 1 from a seronegative organ and tissue donor. *N Engl J Med* 1992; **326**:726–732.
4. Karcher HL. HIV transmitted by bone graft. *BMJ* 1997; **314**:1300.
5. Clarke JA. HIV transmission and skin grafts. *Lancet* 1987; **1**:983.
6. Buck BE, Malinin TI, Brown MD. Bone transplantation and human immunodeficiency virus. An estimate of risk of acquired immunodeficiency syndrome (AIDS). *Clin Orth Rel Res* 1989; **240**:129–136.
7. Conrad EU, Gretch D, Obermeyer KR, Moogk M, Sayers M, Wilson J, Strong DM. The transmission of hepatitis C virus by tissue transplanatation. *J Bone Joint Surg* 1995; **77a**:214–224.
8. Tedder RS, Zuckerman MA, Goldstone AH, Hawkins AE, Fielding A, Briggs EM, Irwin D, Blair S, Gorman AM, Patterson KG, Linch DC, Heptonstall J, Brink NS. Hepatitis B transmission from contaminated cryopreservation tank. *Lancet* 1995; **346**: 137–140.
9. Anyanwu CH, Nassau E, Yacoub M. Miliary tuberculosis following homograft valve replacement. *Thorax* 1976; **31**: 101–106.
10. Centres for disease control: Update: Creutzfeldt-Jakob disease in a second patient who received a cadaveric dura mater graft. *MMWR* 1989; **38**:37–43.

11. Centres for disease control: Human-to-human transmission of rabies via corneal transplant — France. *MMWR* 1980; **29**:25–26.

12. Regamey N, Tamm M, Wernli M, Witschi A, Thiel G, Cathomas G, Erb P. Transmission of herpesvirus 8 infection from renal transplant donors to recipients. *N Engl J Med* 1998; **339**:1358–1363.

13. Sullivan MT, Williams AE, Fang C, Grandinetti T, Poieszbj, Ehrlich GD. The American Red Cross HTLV-I/II collaborative study. Transmission of human T-lymphotropic virus types I and II by blood transfusion. A retrospective study of recipients of blood components (1993 through 1988). *Arch Intern Med* 1991; **151**:2043–2048.

14. Okochi K, Sato H, Hinuma Y. A retrospective study on transmission of adult T cell leukaemia virus by blood transfusion: Seroconversion in recipients. *Vox sang* 1984; **46**:245–253.

15. Pagliuca A, Pawson R, Mufti GJ. HTLV-1 Screening in Britain. *BMJ* 1995; **311**:1313–1314.

16. Nightingale S, Orton D, Ratcliffe D, Skidmore S, Tosswill J, Desselberger U. Antenatal survey of seroprevalence of HTLV-I infections in the West Midlands, England. *Epidemiol Infect* 1993; **110**:379–387.

17. Tosswill JHC, Ades AE, Peckham C, Mortimer PP, Weber JN. Infection with human T cell leukaemia/lymphoma virus type I in patients attending an antenatal clinic in London. *BMJ* 1990; **301**:95–96.

18. Banatvala JE, Chrystie IL, Palmer SJ, Kenney A. A retrospective study of HIV, hepatitis B, and HTLV-1 infection at a London antenatal clinic. *Lancet* 1990; **335**:859–860.

19. Redd SC, Wiriona JJ, Steketee RW, Breman JG, Heymann DL. Transplacental transmission of plasmodium falciparum in rural Malawi. *Am J Trop Med Hyg* 1996, **55**:Supp:57–60.

20. Miller IJ, Telford S. Letter to the editor, Congenital Malaria. *N Engl J Med* 1997; **336**:72.

21. Limaye AP, Connolly PA, Sagar M, Fritsche TR, Cookson BT, Wheat LJ, Stamm WE. Transmission of *Histoplasma Capsulatum* by organ transplantation. *NEJMed* 2000; **343**:1163–1166.

22. Speirs GE, Hakim M, Calne RY, Wreggit TG. Relative risk of donor-transmitted Toxoplasma gondii infection in heart, liver, and kidney transplant recipients. *Clin Transplantation* 1988; **2**: 257–260.

23. Healy PJ, Davis CL. Transmission of tumours by transplantation. *Lancet* 1998; **352**:2–3.

24. Frank S, *et al*. Transmission of glioblastoma multiforme through liver transplantation. *Lancet* 1998; **352**:31.

25. Penn I. Transmission of cancer from organ donors malignancy in transplanted organs. *Nephrologia* 1995; **XV**:205–213.

26. Catlin EA, Roberts JD, Erana R, Preffer FI, Ferry JA, Kelliher AS, Atkins L, Weinstein HJ. Transplacental transmission of natural killer cell lymphoma. *N Eng J Med* 1999; **341**(2):85–91.

27. Resnik R. Cancer during pregnancy. *N Engl J Med* 1999; **341**(2): 120–121.

28. Bianchi DW, Zickwolf GK, Weil GJ, Sylvester S, DeMaria MA. Male progenitor cells persist in maternal blood for as long as 27 years postpartum. *Proc Natl Acad Sci, USA* 1996; **93**:705–708.

29. Lo YMD, Lo ESF, Watson N, Noakes L, Sargent IL, Thilaganathan B, Wainscoat JS. Two-way cell traffic between mother and foetus: Biologic and clinical implications. *Blood* 1996; **88**:4390–4395.

30. Artlett CM, Smith JB, Jimenez SA. Identificaton of foetal DNA and cells in skin lesions from women with systemic sclerosis. *N Engl J Med* 1998; **338**:1186–1191.

31. Aractingi S, Berkane N, Bertheau P, Le Goue, Dausset J, Uzan S, Carosella ED. Foetal DNA in skin of polymorphic eruptions of pregnancy. *Lancet* 1998; **352**:1898–1901.

32. Nelson JL, Furst DE, Maloney, Gooley T, Evans PC, Smith A, Bean MA, Ober C, Bianchi DW. Microchimerism and HLA compatible relationships of pregnancy in scleroderma. *Lancet* 1998; **351**:559–562.

33. Bonney EA, Matzinger P. The maternal immune system's interaction with circulating foetal cells. *Journal of Immunology* 1997; **158**:40–47.

34. Alter MJ, Gallagher M, Morris TT, Moyer LA, Meeks AL, Krawczynski K, Kim JP, Margolis HS. Acute non-A-E hepatitis in the US and the role of hepatitis G virus infection. *NEJ Med* 1997; **336**:741–746.

35. Alter HJ, Nakasuji Y, Melpolder J, Wages J, Wesley R, Shih JW, Kim JP. The incidence of transfusion-associated hepatitis G virus infection and its relation to liver disease. *NEJMed* 1997; **336**: 747–754.

36. Alberti LD, Piatelli A, Artese L, Favia G, Patel S, Saunders N, Porter SR, Scully CM, Ngui S-L, Teo C-G. Human herpesvirus 8 variants in sarcoid tissues. *Lancet* 1997; **350**:1655–1661.

37. Alberti LD, Porter SR, Piatelli A, Scully CM, Teo C-G. Human herpesvirus 8 and sarcoidosis. *Lancet* 1998 (Letter); **351**:1589–1590.

38. Monto H. Human herpesvirus 8 — Let the Transplantation physician beware. *N Engl J Med* 1998; **339**:391–392.

39. Moore PS. The Emergence of Kaposi's Sarcoma — Associated Herpesvirus (Human Herpesvirus 8). *N Engl J Med* 2000; **343** (19):1411–1413.

40. Sitas F, Phil D, Carrara H, Beral V, Newton R, Phil D, Reeves G, Bull D, Jentsch U, Pacella-Norman R, Bourboulia D, Whitby D, Boshoff C, Weiss R. Antibodies against human herpesvirus 8 in black south African patients with cancer. *N Engl J Med* 1999; **340**:1863–1871.

41. Berenson JR, Vescio RA. Controversies in Hematology. Herpesvirus and Multiple Myeloma. HHV-8 is present in multiple myeloma patients. *Blood* 1999; **93**(10):3157–3159.

42. Dupin N, Diss TL, Kellam P, Tulliez M, Du M-Q, Sicard D, Weiss RA, Isaacson PG, Boshoff C. HHV-8 is associated with a plasmablastic variant of Castleman disease that is linked to

HHV-8-positive plasmablastic lymphoma. *Blood* 2000; **95**(4): 1406–1412.

43. Dupin N, Diss TL, Kellam P, Tulliez M, Du M, Sicard D, Weiss RA, Isaacson PG, Boshoff vCID. HHV-8-positive plasmablastic lymphoma blood. *Blood* 2000; **95**:1406–1312.

44. Tarte K, Chang Y, Klein B. Kaposi's Sarcoma-Associated Herpesvirus and Multiple Myeloma: Lack of criteria for causality. *Blood* 1999; 3159–3166.

45. Brown P. Bovine Spongiform Encephalopathy and variant Creutzfeldt-Jakob Disease. *BMJ* 2001; **322**:841–844.

46. Ghani AC, Ferguson NM, Donnelly CA, Anderson RM. Predicted vCJD mortality in Great Britain. *Nature* 2000; **406**:583–584.

47. Brown P, Rohwer RG, Dunstan BC, MacAuley C, Gajdusek DC, Drohan WN. The distribution of infectivity in blood components and plasma derivatives in experimental models of transmissible spongiform encephalopathy. *Transfusion* 1998; **38**:810–816.

48. Brown P, Cervenakova L, McShane LM, Barber P, Rubenstein R, Drohan WN. Further studies of blood infectivity in an experimental model of transmissible spongiform encephalopathy with an explanation of why blood components do not transmit Creutzfeldt-Jakob Disease in humans. *Transfusion* 1999; **39**:169.

49. Houston F, Foster JD, Chong A, Hunter N, Bostock CJ. Transmission of BSE by Blood transfusion in Sheep. *Lancet* 2000; **356**:999.

50. Chiodini PL, Hartley S, Hewitt PE, Barbara JA, Lalloo K, Bligh J, Voller A. Evaluation of a malaria antibody ELISA and its value in reducing potential wastage of red cell donations from blood donors exposed to malaria with a note on a case of transfusion-transmitted malaria. *Vox Sang* 1997; **73**(3):143–148.

51. Waki S, Yonome I, Suzuki M. Plasmodium yoelii: Induction of attenuated mutants by irradiation. *Experimental Parasitology* 1986; **62**:316–321.

52. de Fátima M, da Cruz F, Teva A, da Cruz E, Espindola-Mendes, dos Santos LG, Daniel-Ribeiro CT. Inactivation of Plasmodium falciparum parasites using γ-irradiation. *Mem Inst Oswaldo Cruz, Rio de Janeiro* 1997; **92**(1):137–138, 1996; **92**(1):139, 1983; **56**: 339–345.

53. Stanworth SJ, Warwick RM, Ferguson M, Barbara JA. A UK survey of virological testing of cadaver tissue donors. *Vox Sang* 2000; **79**:227–230.

54. Eastlund T. Hemodilution due to blood loss and transfusion and relability of cadaver tissue donor infectious disease testing. *Cell and Tissue Banking* 2000; **1**:121–127.

55. Shmitz N, Gratwohl A, Goldman JM. Allogeneic and autologous transplantation for haematological disease, solid tumours and immune disorders. Current practice in Europe in 1996 and proposals for an operational classification. *BMT* 1996; **17**: 471–477.

56. Smith CIE, Aarli JA, Biberfeld P, Bolme P, Christensson B, Gahrton G, Hammerstrom L, Lefvert AK, Lonnqvist B, Mattell G, Pirskannen R, Ringden O, Svanborg E. Myasthenia Gravis after Bone Marrow Transplantation. Evidence for a donor origin. *N Engl J Med* 1983; **309**:1565–1568.

57. Waters AH, Metcalfe P, Minchinton RM, Barratt AJ, James DCO. Autoimmune thrombocytopenia acquired from allogeneic bone-marrow graft: Compensated thrombocytopenia in bone marrow donor and recipient. *Letter to the Lancet* 1983; 1430.

58. Lampeter EF, McCann SR, Kolb H. Transfer of diabetes type 1 by bone marrow transplantation. *Lancet* 1998; **351**:568–570.

59. Curtis RE, Rowlings PA, Deeg HJ, Shriner DA, Socie G, Travis LB, Horowitz MM, Witherspoon RP, Hoover RN, Soboncinski KA, Fraumeni JF, Boice JD. Solid Tumours after Bone Marrow Transplantation. *N Engl J Med* 1997; **336**:897–904.

60. Greaves MF. Aetiology of acute leukaemia. *Lancet* 1997; **349**: 344–349.

61. Ford MA, Bennett CA, Price CM, Bruin MCA, Van Wering ER. Foetal origins of the tel-aml1 fusion gene in identical twins with leukaemia. *Proc Natl Acad Sci, USA* 1995; 4584–4588.
62. Serup P, Madsen OD, Mandrup-Poulsen T. Islet and stem cell transplantation for treating diabetes. *BMJ* 2001; **322**:29–32.
63. Standardisation of Organ Donor Screening to Prevent Transmission of Neoplastic Diseases — ISBN 92-871-3485-5 - Council of Europe - December 1997.
64. Chen Y-C, Wang C-H, Su I-J, Hu C-Y, Chou M-J, Lee T-H, Lin D-T, Chung T-Y, Liu C-H, Yang C-S. Infection of human T-Cell Leukaemia Virus Type I and development of human T-Cell Leukaemia/Lymphoma in patients with hematologic neoplasms: A possible linkage to blood transfusion. *Blood* 1989; **74**(1):388–394.
65. Hill AF, Butterworth RJ, Joiner S, *et al*. Investigation of variant Creutfeldt-Jacob disease and other human prion diseases with tonsil biopsy samples. *Lancet* 1999; **353**:183–189.
66. Schmerr MJ, Jenny AL, Bulgin MS, Miller JM, Hamir AN, Cutlip RC, Goodwin KR. Use of capillary electrophoresis and fluorescent labelled peptides to detect the abnormal prion protein in the blood of animals that are infected with a transmissible spongiform encephalopathy. *J Chromatography* 1999; **853**:207–214.
67. Fischer MB, *et al*. Binding of disease-associated prion to plasminogen. *Nature* 2000; **408**:479–483.
68. De Backere ACJ. Euro skin bank: Large scale skin-banking in Europe based on glycerol-preservation of donor skin. *Burns* 1994; **20**:(suppl 1) S4–S9.
69. Cuthbertson B, Reid KG, Foster PR. Viral contamination of human plasma and procedures for preventing virus transmission by plasma products. In *Blood separation and plasma fractionation*, ed. JR Harris, Wiley-Liss Inc, New York, 1991.
70. Urist MR. Bone: Formation by autoinduction. *Science* 1965; **150**: 893–899.

71. Urist MR, Silverman BF, Buring K, Dubuc FL, Rosenberg JM. The bone induction principle. *Clin Orthop Rel Res* 1967; **53**: 243–283.

72. Burwell RG. Studies in the transplantation of bone VIII. Treated composite homograft-autografts of cancellous bone: An analysis of inductive mechanisms in bone transplantation. *J Bone Joint Surg* 1966; **48B**:532–566.

73. Hugo WB, Russell AD. Types of antimicrobial agents. In *Principles and Practice of Disinfection, Preservation and Sterilisation*, Blackwell Scientific publications, Oxford, 1982.

74. Robson MC, Krizek TJ. Clinical experiences with amniotic membranes as a temporary biologic dressing. *Connecticut Medicine* 1974; **38**(9):449–451.

75. Babb JR, Bradley CR, Ayliffe GAF. Sporicidal activity of glutaraldehyde and hypochlorites and other factors influencing their selection for the treatment of medical equipment. *J Hosp Infect* 1980; **1**:63–75.

76. Cookson BD, Hoffman PN, Price T, Webster M, Fenton O. "Cialit" as a tissue preservative: A microbiological assessment. *J Hosp Infect* 1988; **11**:263–270.

77. Aguirregoicoa V, Kearney JN, Davies GA, Gowland G. Effects of antifungals on the viability of heart valve cusp derived fibroblasts. *Cardiovascular Research* 1990; **23**(12):1058–1061.

78. Kearney JN, Franklin UC, Aguirregoicoa V, Holland KT. Evaluation of ethylene oxide sterilisation of tissue implants. *J Hosp Infection* 1989; **13**:71–80.

79. Kearney JN, Bojar R, Holland KT. Ethylene oxide sterilisation of bone alloimplants. *Clinical Materials* 1993; **12**:129–135.

80. Silver GJ. Sterilization by ionizing radiation. In *Disinfection, Sterilization and Preservation*, ed. Block SS. Lea and Febiger, Philadelphia, 1983.

81. American Association of Tissue Banks Information Alert. McLean VA, American Association of Tissue Banks. March 7, 1995; **1**.

82. Recommendations of the Committee of Ministers to Member States on Human Tissue Banks 1994; R94(1).
83. Stanworth SJ, Warwick RM, Fehily D, Persaud C, Armitage S, Navarrete C, Contreras M. An international survey of unrelated umbilical cord blood banking *Vox Sanguinis* 2001; **80**:236–243.
84. Department of Health. A Code of Practice for Human Tissue Banking for Therapeutic Purposes. MCA Information Centre, 10th Floor Market Towers, London SW8 5NQ, United Kingdom, April 2000.
85. MSBT — Guidance on the Microbiological Safety of Human Organs, Tissues and Cells used in Transplantation. www.doh.gov.uk/msbt/index.htm.

Index